BASIC PHYSICS
OF THE
SOLAR SYSTEM

This book is in the
ADDISON-WESLEY SERIES IN
THE ENGINEERING SCIENCES

Space Science and Technology

Consulting Editors

HOWARD W. EMMONS   JAMES FLETCHER
S. S. PENNER

BASIC PHYSICS
OF THE
SOLAR SYSTEM

by

V. M. BLANCO

and

S. W. McCUSKEY

Department of Astronomy
Case Institute of Technology

ADDISON-WESLEY PUBLISHING COMPANY, INC.

READING, MASSACHUSETTS, U.S.A.

LONDON, ENGLAND

To Our Wives

PREFACE

The aim of this book is to present in a concise way some of the basic physical and dynamical aspects of the solar system. We have attempted to strike a level of presentation intelligible to those engineers and scientists, not specialists in astronomy, who are interested in the rapidly expanding area of space technology. Thus we have endeavored to form a link between the purely descriptive elementary books on astronomy and the current specialized literature with which such readers may have to deal. To attain this goal, we have discussed the basic principles of celestial mechanics to a much greater depth than would be the case in a standard course in astronomy.

The book has evolved out of a series of lectures given by the staff of the Department of Astronomy, Case Institute of Technology. These lectures were presented to groups consisting of research scientists, engineers, and supervisory personnel in the Cleveland area. The men and women in these groups were all engaged in research programs in which a basic understanding of the astronomy of the solar system is essential. They all had an educational background in science and technology, but one which included only a limited acquaintance with astronomy.

In selecting the material to be presented in this book, we have been guided by our experience with these lectures. In no sense do we claim completeness; the field is vast. The elementary concepts which we found necessary for clarity of exposition have been built up from basic definitions. Emphasis throughout the book is placed on the methods used by astronomers and on the degree of precision with which various astronomical data have been determined. References to published source material are included for those readers who may wish to study the original research papers or the more detailed treatises on the subjects covered.

To provide the reader with a working knowledge of the astronomical coordinate systems and of elementary celestial mechanics, examples of computations are presented as well as practice problems. We have assumed that the reader possesses a knowledge of the mathematics and physics usually covered in the curricula of the first two years of engineering or physical science.

In the descriptive portions of the text, the amount of detail varies inversely as the distance from the earth to the object described. Hence, more attention has been given to the earth and its environs than to the distant planets. This reflects the major interest of persons engaged in space research and also, in an approximate manner, the degree to which the various bodies of the solar system, with the exception of the sun, have been studied. Out of the vast body of knowledge about the sun, we have emphasized those aspects that are more directly related to the planets and interplanetary space.

We wish to acknowledge the help of our colleagues, Dr. J. J. Nassau and Dr. J. Stock, who took part in the lecture series mentioned above, and whose lecture notes were used in the planning of this book. The authors, however, assume full responsibility for the selection and presentation of the material included here. We also record our thanks to Dr. C. B. Stephenson and Mr. Robert Hobbs for reading and criticizing various portions of the book.

V.M.B.

S.W.M.

CONTENTS

LIST OF PLATES

CHAPTER 1

ASTRONOMICAL COORDINATE SYSTEMS

1–1 The horizon system of coordinates. Imagine a northern observer on the surface of the earth and preferably on the open ocean. The sky overhead appears to be a spherical dome centered on the observer. This dome on which the celestial objects seem to be located is part of the *celestial sphere* which surrounds the earth and which for practical purposes may be considered so large that the earth's dimensions are negligible in comparison. The celestial sphere provides a convenient framework for fixing the relative positions of the heavenly bodies. Where a plumb line at the observer's position, extended upward to the sky, would pierce the dome is the observer's *zenith*. The point on the celestial sphere directly opposite the zenith is the *nadir*. Because of the oblateness of the earth, the line joining zenith and nadir does not necessarily pass through the center of the earth.

The plane perpendicular to the zenith-nadir line at the observer's position intersects the celestial sphere in the *horizon*. This is a great circle whose points are everywhere 90° from the zenith. Figure 1–1 shows the horizon circle NESW, together with other essential reference points. In the figure, Z is the zenith, R represents a star or other celestial object, O represents the observer, and P is the point where a northward prolongation of the earth's axis of rotation appears to pierce the sky. This point is called the *north celestial pole*. It is marked very closely by a moderately bright star, Polaris. In the southern hemisphere there is no comparable "pole star."

A great circle through the zenith and the north celestial pole intersects the horizon in the north point, N, and the south point, S. This is the observer's *meridian*.

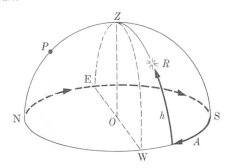

Fig. 1–1. The horizon system of coordinates.

1

The position of R at any instant is given by its *altitude, h,* and its *azimuth, A.* The altitude is measured along a vertical circle, perpendicular to the horizon, that passes through the object R and through the zenith. It is the *angle in degrees measured from the horizon along the vertical circle to the celestial object.* The so-called *zenith distance* is $90° - h$. The azimuth is the *angle in degrees measured from the north point toward the east along the horizon to the foot of the vertical circle passing through the celestial object.* Sometimes the azimuth is measured from the south point of the horizon either to the east or to the west. We adopt here the first of the two definitions.

These coordinates are purely local. As the observer moves on the earth's surface, his zenith and horizon change. Thus, two observers separated from each other will not agree about the altitude and azimuth of an object in the sky. Furthermore, the celestial sphere and the stars on it appear to rotate for any observer, an effect caused by the earth's rotation. Hence the altitude and azimuth of any celestial object change with time.

1–2 Geographic and geocentric coordinates. The angle between the direction of the plumb line and the plane of the earth's equator is called the *astronomical latitude,* φ. As shown in Fig. 1–2, this is equal to the altitude of the pole. The line $O\Sigma'$ is both in the observer's horizon plane and in the plane defining his meridian. The point C is the earth's center, and the line $C\Sigma\Sigma'$ lies in the plane of the earth's equator. The point P' is the

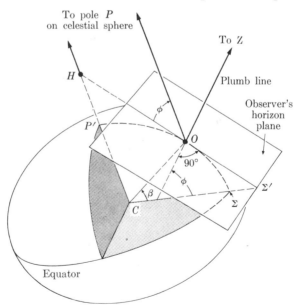

Fɪɢ. 1–2. Geographic and geocentric latitudes.

north terrestrial pole, and H is the north point of the observer's horizon.

Geodetic measurements on the surface of the earth indicate irregular variations in the direction of gravity, depending upon locality. These *station errors* are due to variations of density and form in the earth's crust. Since the direction of gravity is coincident with the direction of a plumb line, such station errors displace the astronomical zenith of the observer from that which would be observed if the earth were a perfectly symmetrical distribution of mass. Errors of 5 seconds of arc are not uncommon, and errors up to 50 seconds of arc are found occasionally.

The *geographical latitude* of the observer is the astronomical latitude corrected for station error. The *geographical longitude* is the *angular distance in degrees measured along the terrestrial equator from the intersection of a fixed meridian to the intersection of a meridian through the position of the observer.* The fixed meridian is usually taken to be that through Greenwich, England. Longitude is taken to be positive if measured westward from Greenwich, and negative if measured eastward. We shall denote longitude by L.

In the treatment of satellite problems, geocentric coordinates are frequently necessary. *Geocentric latitude* is the angle at the center of the earth between the plane of the equator and a line joining the center of the earth to the observer's position. In Fig. 1–2 this is denoted by β. *Geocentric longitude* is the same as geographical longitude. The relationship between geographical and geocentric latitudes will be studied in more detail in Chapter 3.

1–3 Motions of the earth. The principal motions of the earth are its axial rotation and its revolution around the sun. As viewed from a point far above the north pole, the earth would appear to rotate counterclockwise, that is, from west to east. As seen from a point on the earth's surface, all objects on the celestial sphere rise in the east, traverse the dome of the sky on *diurnal circles* parallel to the earth's equator, and set in the west. An object such as the moon, which moves with respect to the stars because of its own orbital motion, describes a path in the sky which is not strictly a diurnal circle.

The earth makes one complete circuit of its orbit in space around the sun in about 365.25 days. This orbit is an ellipse with the sun at one focus. Since we observe from the earth, the orbital motion becomes apparent as a motion of the sun among the stars. In Fig. 1–3, let E_1, E_2, and E_3 be successive positions of the earth in its orbit around the sun, S. From these positions the sun appears to be projected on the celestial sphere at S_1, S_2, and S_3. Once in approximately 365.25 days the sun (observed from the earth) appears to have made a complete circuit with respect to the stars.

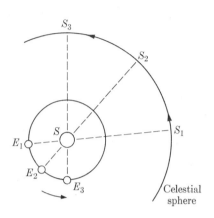

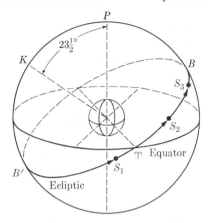

Fig. 1–3. Apparent motion of the sun   Fig. 1–4. Apparent path of the sun
due to the orbital motion of the earth.   along the ecliptic.

The apparent great-circle path of the sun on the celestial sphere during
the course of the year is called the *ecliptic*. The plane of the earth's motion
is the *plane of the ecliptic*. As seen from a point far to the north of the
ecliptic plane, the sun appears to move counterclockwise with respect to
the earth.

If we project the plane of the earth's equator until it intersects the
celestial sphere, the great circle so obtained is the *celestial equator*. Figure
1–4 shows the earth at the center of the celestial sphere. The point P
is the north celestial pole; K is the pole of the ecliptic. The axis of the earth
maintains an angle of about 23°.5 with the perpendicular to the ecliptic.
In other words, the plane of the equator is tilted 23°.5, approximately,
to the plane of the ecliptic. This angle is known as the *obliquity of the
ecliptic*. Its precise value is given in the *American Ephemeris and Nautical
Almanac** for each year.

In Fig. 1–4, successive positions of the sun on the ecliptic are denoted
by S_1, S_2, S_3. The point where the sun crosses the equator from south
to north about March 21 each year is called the *vernal equinox*. We denote
it by ♈†. The opposite point in the sky is the *autumnal equinox*. The
sun reaches this point about September 22 each year. The point B where
the sun attains its greatest northerly departure from the equator is the
summer solstice for a northern observer. By analogy, the point B' is called
the *winter solstice*. The sun reaches these points on or about June 21 and
December 21, respectively.

---

* This publication (hereafter referred to as the *American Ephemeris*), which
is issued yearly, may be obtained from the Superintendent of Documents,
U.S. Government Printing Office, Washington 25, D.C. Price \$4 (cloth).

† The vernal equinox is also referred to as the "First Point of Aries."

Because of the attractions of the sun and moon on the equatorial bulge of the earth, the points K and ♈ are not fixed on the celestial sphere. The point ♈ moves westward along the ecliptic at a rate of 50″.26 per year. This is known as the *precession of the equinox*. As a consequence, the pole P of the equator makes a revolution around the pole K of the ecliptic in about 26,000 years.

Superposed on the steady precession is an approximately periodic motion due to the variable way in which the sun and moon pull on the earth's equatorial bulge during the course of a month. This effect is called *nutation*. As a result of precession and nutation, the point P describes a wavy motion on the celestial sphere. We shall study later the dynamical causes of these effects in more detail. For the present we note them only to discuss their influence on the coordinates of a celestial object.

1–4 The equator system. The local nature and the time dependence of the coordinates of a celestial object in the horizon system are eliminated by choosing the celestial equator as the fundamental circle of reference. In Fig. 1–5, let EΣQW be the celestial equator. The observer is at O. The north celestial pole is at P, and the great circle NPΣS is the observer's meridian. As in Fig. 1–1, the observer's horizon is designated by NESW.

The great circle which passes through P and the celestial object R and is perpendicular to the equator is called an *hour circle*. Let δ denote *the angular distance in degrees from the equator along the hour circle to the object R*. This is the *declination* of R. It is taken as positive when R is north of the equator, and negative when R is south. Declination in this coordinate system is analogous to geocentric latitude on the surface of the earth.

The right ascension (α or R.A.) of R is the angular distance in degrees measured from the vernal equinox along the celestial equator toward the east to the foot of the hour circle through R.

As the sphere of the sky appears to rotate due to the earth's rotation on its axis, α and δ remain fixed because R and ♈ take part in the rota-

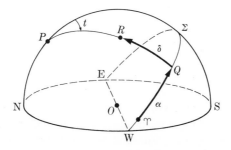

Fig. 1–5. The equator system of coordinates.

tion. Thus the right ascension and declination of a star, R, as given, say in an almanac, remain sensibly constant. They are coordinates *on the sky referred to an origin located on the celestial sphere*, and hence are not influenced by the position of the observer on the earth; together they constitute the *equator system* of coordinates. The principal changes which occur in α and δ over long intervals of time are those due to precession of the equinoxes and to nutation.

Included in the equator system is a semilocal system of coordinates consisting of the declination and the *local hour angle, t.* The latter is the *angular distance in degrees measured along the equator from the observer's meridian toward the west to the foot of the hour circle through the celestial object.* In Fig. 1–5 the hour angle is the arc ΣQ and also the angle t at the pole. As the celestial sphere rotates from east to west, the hour angle of any fixed celestial object increases uniformly with time. The hour angle is useful in discussions of time, and in transforming coordinates from the horizon to the equator system.

1–5 Local time and its variations. The rotation of the earth on its axis provides the basic unit for the measurement of time. We assume that this rotation is uniform. Only over the course of centuries will the basic angular rotation rate change appreciably. The time units in which we shall be interested are *sidereal, apparent solar,* and *mean solar.* As the names imply, these depend upon the object in the sky used as a clock, namely a star or the sun.

(a) *Sidereal time.* The *sidereal day (24 sidereal hours)* is defined as the interval of time between two successive passages of the vernal equinox across a given observer's meridian. Because of the precession of the equinox, this period of time is shorter by about 1/120 of a second than it would be if it were defined in terms of a point fixed with respect to the stars. To an observer on the earth's surface, the entire celestial sphere appears to rotate once in approximately 24 sidereal hours. Because the motion of the vernal equinox with respect to the stars is not uniform, the sidereal day varies in length very slightly. Zero hours local sidereal time is the instant at which the vernal equinox crosses the upper meridian. As the celestial sphere rotates and carries the vernal equinox westward, the hour angle of ♈ increases. Hence we may define *sidereal time* concisely as the *hour angle of the vernal equinox.*

(b) *Apparent solar time.* The *apparent solar day* is the interval of time between two successive transits of the sun's center over the lower meridian of the observer. The local apparent solar day, therefore, begins at midnight. Apparent local noon is the instant at which the center of the sun is on the observer's upper meridian. Succinctly stated, *local apparent solar time is the hour angle of the sun's center plus 12 hours.* If this sum is greater than 24 hours, one subtracts 24 hours from it.

As a result of the earth's orbital motion, the sun appears to move eastward among the stars. It covers 360° in about 365 days, so that the motion is approximately 1° per day. But this motion is not uniform. Owing to the ellipticity of the earth's orbit, the apparent motion of the sun toward the east varies from day to day. Furthermore, the sun moves along the ecliptic, a great circle inclined to the equator. Hence, even if the sun did move uniformly along the ecliptic, the projection of its motion on the equator would not be uniform. These irregularities cause the time intervals between successive passages of the sun over the meridian to vary. In other words, the apparent solar day is not constant.

To alleviate this difficulty, a *fictitious* or *mean sun* is postulated, which is assumed to move at a uniform rate eastward along the equator, making a complete circuit from vernal equinox to vernal equinox in the same time it takes the real sun to make the same circuit along the ecliptic. This interval is called the tropical year (365.2422 mean solar days).

(c) *Mean solar time.* A *mean solar day*, or *civil day*, is the interval of time between two successive lower transits of the mean sun. *Mean solar time is defined to be the hour angle of the mean sun plus 12 hours.*

Since observations are necessarily made on the real sun, we must have a method of converting local apparent solar time into mean solar time, and vice versa. This is accomplished by defining the *equation of time* as the difference at any instant between the two. Let LAT denote the local apparent time and let LCT denote the local mean, or civil, time. Then the equation of time, E, is given by

$$E = \text{LAT} - \text{LCT} \qquad (1\text{–}1)$$

The quantity E is tabulated in the *American Ephemeris* for each day in the year.

For convenience in succeeding discussions we summarize the definitions above in equation form. Let t_θ be the hour angle of the vernal equinox, t_A be the hour angle of the real sun, t_M be the hour angle of the mean sun. Furthermore, let θ_L denote the local sidereal time. Then, by definition,

$$\theta_L = t_\theta \qquad , \qquad (1\text{–}2)$$

$$\text{LAT} = t_A + 12 \qquad , \qquad (1\text{–}3)$$

$$\boxed{\text{LCT} = t_M + 12} \; . \tag{1-4}$$

A few examples will serve to clarify these concepts.

EXAMPLE 1. On March 21 the sun is at the vernal equinox. When the sun is setting, what are the LAT and the θ_L? The sun and vernal equinox are obviously at the west point of the horizon. Then $t_\theta = 6^h = \theta_L$, and $t_A = 6^h$. Therefore LAT $= 18^h$. Strictly speaking, the observed hour angles t_θ and t_A at this moment would be slightly less than 6^h because of refraction by the earth's atmosphere. Refraction, which is neglected here, will be considered in Chapter 3.

EXAMPLE 2. What is the sidereal time at midnight LCT on September 21? At this instant the mean sun is at the autumnal equinox which is on the lower meridian. Hence the vernal equinox is on the upper meridian, and the sidereal clock reads 0^h.

EXAMPLE 3. On October 1 the LCT is 10 a.m. What is the LAT? What is the hour angle of the sun? For this date, $E = +16^m$. Therefore, by Eq. (1-1), LAT $= 10^h16^m$ a.m., and, by Eq. (1-3), $t_A = 22^h16^m$.

(d) *Standard time and longitude.* All the time units defined above are local. An observer moving on the surface of the earth would necessarily require clocks which would have to be adjusted continuously. For convenience, therefore, *standard time zones* have been established on the earth. Within each zone the same time is used. In the United States the time zones are centered approximately on the 75th, 90th, 105th, and 120th meridians of longitude. These are the Eastern, Central, Mountain, and Pacific time zones, respectively.

It should be apparent from Fig. 1-6 that the difference in longitude between any two positions on the surface of the earth is the difference in their local times. Let A and B be two such positions on the earth. Let the meridian planes through A and B be projected onto the celestial sphere in the great circles $P\Sigma_1 P'$ and $P\Sigma_2 P'$. For the observer at A, the sidereal time is given by $\Sigma_1 \Upsilon$. For the observer at B, it is $\Sigma_2 \Upsilon$. The difference is

$$\Delta L = \Sigma_2 \Sigma_1 = \theta_{L2} - \theta_{L1},$$

which is the same as the arc on the terrestrial equator between the meridians of A and B, or the difference in their longitudes.

The reader will note that the sun might have been used in place of the vernal equinox with the same result.

Since the earth rotates $360°$ in 24 hours, the rate is $15°$ per hour. The hour angle of the vernal equinox, or of any other celestial object, increases

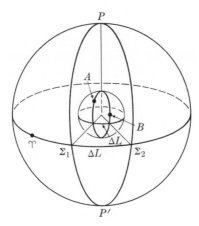

FIG. 1–6. Relationship between longitude and local time.

at this rate. Hence any angular measurements parallel to the celestial equator on the sky can be expressed either in degrees or in hours. These units are used interchangeably. Thus clocks on the 75th meridian west of Greenwich would read five hours less than those at Greenwich. Local time at any place would not differ by more than about 30 minutes from the standard time of the zone in which the place is located if the time zones were defined symmetrically about every 15 degrees in longitude. For political reasons, however, the actual boundaries between the various time zones may differ in local time by more (or less) than 30 minutes. Table I (a, b) of the appendix is convenient for converting time units into arc units, and vice versa.

Time	Arc
1 hr =	$15°$
1 min =	$15'$
1 sec =	$15''$

The meridian of longitude 180° from Greenwich is known as the *international date line*. Again for political reasons, however, the actual date line is somewhat irregular. An observer moving westward from Greenwich would find the time at the date line to be 12 hours *earlier* than the time at Greenwich. An observer moving eastward from Greenwich would find the time at the date line to be 12 hours *later* than the time at Greenwich. Therefore, crossing this line produces a discontinuity of 24 hours, or one day. Ships sailing westward and crossing the date line omit one day from their logs; those crossing the line eastward add an extra day.

1–6 Conversion of time. Consider the instant on September 21 or 22 when the mean sun is exactly at the autumnal equinox and the vernal equinox is crossing the upper meridian of an observer. His sidereal clock reads 0^h and his LCT is also 0^h because the mean sun will be on the lower meridian. This situation is pictured in Fig. 1–7. The mean sun is at M_1, and the diurnal rotation carries both ♈ and M in the direction of the arrows. The observer is at O; the plane NESW is the horizon.

One sidereal day later the vernal equinox again is on the upper meridian. But, because in this interval the mean sun has moved eastward to M_2, the local civil clock does not read 0^h. It reads about 23^h56^m. Thus the sidereal clock appears to gain on the civil clock by approximately 4 minutes per day.

At the end of another sidereal day, when the vernal equinox is on the upper meridian, the mean sun is at M_3. The local civil time then is about 23^h52^m. In the course of a month this loss of the civil in comparison with the sidereal clock amounts to approximately two hours. In the course of one tropical year, the difference will amount to precisely one day. Hence the sidereal clock will be 24 hours ahead of the local civil clock. Thus the tropical year of 365.2422 mean solar days contains 366.2422 sidereal days.

A given time interval in one system, sidereal or mean, can be converted readily into the other. Let I_M denote the number of mean solar units in any given interval of time; let I_θ denote the number of sidereal units in the same interval. Then

$$\frac{I_M}{I_\theta} = \frac{365.2422}{366.2422} = 0.99726957 \qquad (1\text{–}5)$$

or

$$\frac{I_\theta}{I_M} = \frac{366.2422}{365.2422} = 1.00273791. \qquad (1\text{–}6)$$

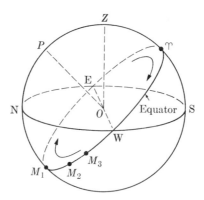

FIG. 1–7. Gain of sidereal time relative to mean solar time.

Either Eqs. (1–5) and (1–6) or Tables II and III of the appendix may be used to convert an interval from sidereal to mean solar units, and vice versa.

EXAMPLE 1. The local civil time at a given place is 3 p.m., and the sidereal clock at the same instant reads $12^h42^m22^s$. (a) Two civil hours later, what does the sidereal clock read? (b) When the sidereal clock reads $0^h00^m00^s$, what does the civil clock read?

(a) By Eq. (1–6), or Table III, $I_\theta = 2^h + 00^m19^s.713$. Hence the sidereal clock reads $14^h42^m41^s.7$ when the civil clock reads 5 p.m.

(b) Here $I_\theta = 24^h - 12^h42^m22^s = 11^h17^m38^s$. By Eq. (1–5) or Table II, $I_M = 11^h15^m47^s.0$. Hence the civil clock reads $2^h15^m47^s.0$ a.m. when the sidereal clock reads $0^h00^m00^s$.

We shall illustrate in Example 2 other representative problems in time conversion. The data required for these are contained in the sample pages of the *American Ephemeris* shown in Table IV of the appendix.

EXAMPLE 2. The Eastern Standard Time on November 16, 1960, in Cleveland (longitude = $5^h26^m16^s$) is $9^h32^m15^s$ p.m. Find: (a) The Local Civil Time in Cleveland; (b) The Local Civil Time in Greenwich; (c) The Apparent Solar Time in Cleveland; (d) The Greenwich Sidereal Time; (e) The hour angle in Cleveland of the star Vega (R.A. = $18^h35^m35^s$).

(a) Since Cleveland is *west* of the 75th meridian, the time locally will be *slower* than Eastern Standard Time by an amount equal to the longitude difference

$$\Delta L = 5^h26^m16^s - 5^h00^m00^s = 26^m16^s.$$

Hence LCT (Cleveland) is $21^h5^m59^s$, or $9^h5^m59^s$ p.m. (November 16).

(b) By similar reasoning the Local Civil Time in Greenwich is 5^h *later* than at the 75th meridian; hence,

$$GCT = 2^h32^m15^s \text{ a.m. (November 17).}$$

(c) At 0^h GCT* the equation of time is $+15^m04^s.5$. (See the last column of Table IV (a) shown in the appendix or on p. 33 of the 1960 *American Ephemeris*.) This quantity is decreasing at the rate of about $0^s.5$ per hour. Hence at 2^h32^m a.m. GCT, it will be $+15^m02^s$ to the nearest second. The Local Apparent Time in Cleveland then is given by

$$LAT \text{ (Cleveland)} = 9^h05^m59^s + 15^m02^s = 9^h21^m01^s \text{ p.m.}$$

---

* The 0^h Ephemeris Time appearing in the *American Ephemeris* for 1960 is 35^s faster than GCT, which is also called Universal Time. For the computation of the equation of time, this factor is neglected here since the error amounts to less than $0^s.1$. Ephemeris and Universal Times are explained in Section 1–7.

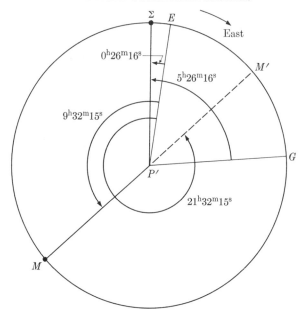

FIG. 1–8. Time diagram illustrating Example 2, parts (a) and (b).

(d) From p. 17 of the *American Ephemeris*,* at 0^h U.T. (Universal Time = Greenwich Civil Time) November 17, the Greenwich Sidereal Time is $3^h44^m12^s.2$. At 0^h on November 18 it is $3^h48^m08^s.7$, the difference being $3^m56^s.5$. Therefore at $2^h32^m15^s$ a.m. GCT, the Greenwich Sidereal Time is given by

$$3^h44^m12^s.2 + \frac{(2^h32^m15^s)}{24}(3^m56^s.5) + 2^h32^m15^s.0$$

or

$$6^h16^m27^s.2 + 25^s.0 = 6^h16^m52^s$$

to the nearest second. Instead of the interpolation carried out here, Table IX of the *American Ephemeris* (Table III of the appendix) may be used.

(e) To find the hour angle of the star Vega in Cleveland at the given instant, we must first determine the Local Sidereal Time in Cleveland. This is obtained from the last result by subtracting $5^h26^m16^s$, the longitude of Cleveland. Thus,

$$\theta_L \text{ (Cleveland)} = 6^h16^m52^s - 5^h26^m16^s$$
$$= 0^h50^m36^s.$$

---

* A copy of this page is given in the appendix, Table IV(b). The *apparent* sidereal time in the *American Ephemeris* is identical with the sidereal time defined previously. Mean sidereal time is defined in Section 1–7.

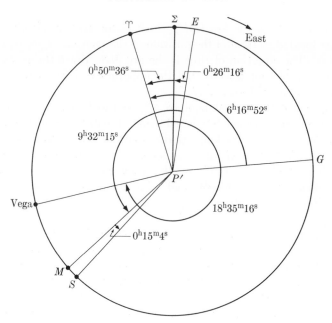

Fɪɢ. 1–9. Time diagram illustrating Example 2, parts (c), (d), and (e).

Since the star Vega has a R.A. of $18^h35^m35^s$, it must have already crossed the Cleveland meridian at the given moment and have an hour angle of

$$24^h50^m36^s - 18^h35^m35^s = 6^h15^m1^s.$$

Further examples of this nature may be found in the explanatory notes published in the *American Ephemeris*.

A *time diagram* is useful in solving problems of this type. Figure 1–8 is a view of the celestial equator as seen from the south pole P'. The line extending upward from P' to Σ represents the observer's meridian; Σ is the intersection of the observer's meridian with the equator. East is to the right of Σ. For Example 2 of this section, the line $P'\Sigma$ represents the Cleveland meridian. Hence we may place the Greenwich meridian, $P'G$, east of Σ at an angle equal to the longitude of Cleveland. Similarly, the eastern standard meridian $P'E$ is placed east of Σ. In its diurnal motion, any celestial object sweeps out counterclockwise angles in the diagram, passing first the Greenwich meridian, then the 75th meridian, and finally the local meridian.

We may now place the mean sun M and its opposite point M' on the diagram. In Example 2, EST is $9^h32^m15^s$ p.m., and the mean sun is $9^h32^m15^s$ west of E. From the data shown, it is possible to compute the arc ΣM (or the arc $\Sigma MGM'$ in the 24^h-clock), the Local Civil Time in

Cleveland, as well as the arc GM', which is the Greenwich Civil Time. The points located so far and the position of the true sun S are reproduced in Fig. 1–9. The equation of time is the arc MS. Since the positive sign of the equation of time tells us that the apparent time is greater than the mean time, S is placed *west* of M.

We may place the vernal equinox ♈ on the figure, the arc $G♈ = 6^h16^m52^s$ being the Greenwich Sidereal Time. Finally the star Vega may be placed in the diagram. We know that it must be $18^h35^m16^s$ east of ♈ according to the definition of right ascension. The data shown permit easy visualization and computation of the Cleveland Local Sidereal Time (arc $\Sigma ♈$ measured westward) and the local hour angle of Vega (arc Σ − Vega) at Cleveland.

1–7 Universal Time, Julian dates, and Ephemeris Time. In many applications, for example in meteorological analysis, data collected simultaneously in different time zones must be intercompared. To avoid the necessity of allowing for time-zone differences, Universal Time is generally used in planning and reporting such observations. Universal Time is the Greenwich Mean Solar Time expressed in the 24-hour system beginning with 0^h at midnight. Thus, at a given instant, Universal Time (U.T.) is the hour angle of the mean sun $+ 12^h$ for an observer on the Greenwich meridian. This time is now used in a great number of scientific experiments.

For astronomical observations extending over many years, it is often necessary to determine how many days have elapsed between any two given moments. The different lengths that the calendar years may have and the past changes in the calendar (for example, Julian to Gregorian) make such computations difficult. In 1582, Joseph Scaliger, a French chronologist, proposed the system of *Julian day numbers*, in which January 1 of the year 4713 B.C. was taken as day number 1. This early date was arbitrarily chosen to ensure that all well established historical events known to Scaliger had positive date numbers. The *American Ephemeris* carries Julian date designations for every day of the year. By an older convention adopted by astronomers, a new Julian day begins at Greenwich mean noon. The Julian day number starting at Greenwich noon, January 1, 1960, is 2,436,935. A Julian century contains exactly 36,525 days.

The problem of how to regulate precisely a clock keeping mean solar time is a major one. The hour angle of the mean sun cannot be observed directly because the mean sun has no physical reality. However, from a great many observations of the true sun's position among the stars, formulas have been developed that yield the position of the mean sun as a function of sidereal time. In determining Universal Time one observes the sidereal time and then converts to U.T. by means of the established formulas.

From an observation of the Greenwich hour angle of a star of known right ascension, the Greenwich hour angle of the vernal equinox can be derived. If the vernal equinox were fixed with respect to the stars or even if it progressed at a uniform rate on the celestial sphere, the definition of time would be a simple matter. As a result of nutation, the position of the vernal equinox does not advance uniformly among the stars. Hence the definition of a sidereal day as the time between successive passages of the vernal equinox over the observer's meridian leads to sidereal days of unequal length. To remedy this a *Greenwich Mean Sidereal Time* has been defined which is the hour angle of a fictitious mean vernal equinox that moves uniformly along the equator at a rate that precession alone would produce if nutation did not exist. The difference in right ascension between the true equinox and the mean equinox is called the *equation of the equinoxes*, and is in practice computed from a formula that takes the nutation effects into account. The equation of the equinoxes, which amounts to only a fraction of a second, is listed for every day of the year in the *American Ephemeris*. Sidereal time measured by the true vernal equinox is called *apparent sidereal time;* if the mean equinox is used, it is called *mean sidereal time.* The period of time referred to as *one sidereal day* in the previous paragraphs is actually reckoned in mean sidereal time.

The use of mean sidereal time obviates any variations that the sidereal day may have as a result of nutation. Variations in the period of rotation of the earth, however, will still cause changes in the length of the day. That such variations occur has been well established. First, there is a secular increase in the period of rotation of about 0.0007 seconds per century caused by tidal friction. (A *secular* change is one which is so slow that for practical purposes it may be considered to be proportional to time over extended periods. Most secular changes in astronomy, however, are actually periodic, but their periods are extremely long.) *Tidal friction* is the name given to the slowing action affecting the earth as it turns against the tides. As can be expected from the conservation of angular momentum, this retardation is accompanied by a reciprocal phenomenon, namely an increase in the moon's mean distance from the earth. In addition to secular changes, there are seasonal changes with a total amplitude of about 7 milliseconds, which apparently are caused by meteorological conditions. These have been well established by direct measurements of the earth's period of rotation with ultra-accurate atomic clocks [1].* There are also irregular fluctuations of unknown origin in the period of rotation, amounting to 5 milliseconds in the last 60 years.

---

* Numbers in brackets refer to bibliographical references listed at the end of each chapter.

The effect of these variations on the uniformity of Universal Time may be visualized as follows. Imagine a plane containing the earth's axis and the Greenwich meridian. As the earth rotates, this plane sweeps the celestial sphere from west to east. Let us call θ the angle measured eastward along the celestial equator from the true vernal equinox to this plane. At a given instant, Universal Time, denoted by T, may be related to θ by the equation

$$\theta = \theta_0 + aT + K, \tag{1-7}$$

where θ_0 is the value of θ at the instant we choose to call 0^{h} U.T., a is the rate of angular rotation of the earth, and K is the correction that takes into account the precession and nutation effects. The observation of θ makes it possible to find T, but there is the complication that K is also a function of T. Fortunately, K is a small quantity, and the equation can be solved by successive approximations. If T is in mean solar seconds, then

$$a = \frac{2\pi}{86400} 1.0027378118868. \tag{1-8}$$

In this relationship, 86,400 is the number of seconds in one sidereal day, and the constant 1.0027378118868 is used to convert sidereal to mean solar time [2]. This relationship constitutes the definition of a. From the foregoing discussion we see that Universal Time, which is derived from observed values of θ, will not be uniform if the earth does not rotate at a constant rate. The known variations in this period are slight, and for most purposes Universal Time is a perfectly suitable standard of time that can be derived rapidly from astronomical observations.

For very precise timekeeping, however, a new definition of time, called *Ephemeris Time*, has been adopted by international agreement. It is based not on the rotation of the earth on its axis, but on the periods of revolution of the planets about the sun and of the moon about the earth. An equation, $\theta = f(T)$, relates the observations to the time, but now θ is the angle described by the line joining, say the earth and the sun when measured from a given reference direction. To obtain the Ephemeris Time at a given instant, the equations representing the motion of the earth, including the effects of the attraction of the other planets, must be solved in terms of T. A large number of such solutions yields θ as a function of T, and observation of a given θ determines the value of T at the moment of the observation. One second in Ephemeris Time has been made equal to the average mean solar second as derived by 300 years of astronomical observations. In the determination of Ephemeris Time, observations of the moon have proved to be the most practical. Since the observations required to determine Ephemeris Time cover a long period of years, the relationship between Universal and Ephemeris Times cannot be determined instantaneously. The practice is to find what difference, ΔT, between these times has existed in the past and to extrapolate the results to the present and future. The *American Ephemeris* lists the value of ΔT, which in 1960 amounts to about 35^s (ΔT = Ephemeris Time − Universal Time).

1–8 Celestial latitude and longitude, and heliocentric coordinates. The equator system of coordinates described in Section 1–4 is that used for all observational work from the earth. For discussions of planetary motions and other phenomena in the solar system, however, it is convenient to use the ecliptic plane as the fundamental plane, the ecliptic circle as the fundamental circle, and to introduce secondary circles through the poles of the ecliptic. These are known as *circles of celestial longitude*. In this system centered on the earth, the *celestial longitude is defined to be the angle in degrees measured eastward along the ecliptic from the vernal equinox to the intersection of the ecliptic and the circle of celestial longitude through the object.*

In contrast to geographical longitude, celestial longitude is reckoned from 0° to 360°. The *celestial latitude* is the distance in degrees *from the ecliptic to the object, measured along the circle of celestial longitude passing through the object.* It is positive when the object is north and negative when the object is south of the ecliptic.

We shall denote celestial longitude by λ and celestial latitude by β.

Observations are necessarily made from the earth, and hence the coordinate systems so far introduced are *geocentric*. For describing planetary positions, however, the sun is a more appropriate origin. Therefore we introduce a heliocentric system, as shown in Fig. 1–10. The *heliocentric longitude, l,* of P is the angle *measured from the vernal equinox eastward*

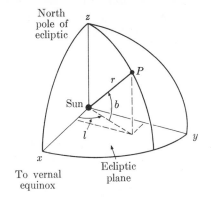

Fig. 1–10. Heliocentric coordinates.

along the ecliptic to the foot of the circle of celestial longitude through P.
This is analogous to the celestial longitude in the geocentric system. The
*heliocentric latitude, b, is the angle north or south of the ecliptic from the
ecliptic to the object P.* This polar coordinate system is superposed on a
cartesian system, as shown in the figure. The radius vector r is the distance from the sun to P. In a later section we shall define appropriate
units for r.

1–9 Coordinate transformations. By spherical trigonometry the coordinates of a celestial object in one system can be transformed into another
system.

EXAMPLE 1. Suppose that the altitude h and azimuth A of a star are
to be determined, given its right ascension α and declination δ, together
with the time and place of observation. In Fig. 1–11 let R be the star,
and let φ be the latitude of the observer at O whose zenith is Z. Then by
the cosine law for the spherical triangle PZR,

$$\sin h = \sin \varphi \sin \delta + \cos \varphi \cos \delta \cos t. \qquad (1\text{–}9)$$

Here t is the hour angle of the star. When the time is known, say LCT,
the local sidereal time θ_L can be found (Section 1–6). Then $t = \theta_L - \alpha$.
All quantities on the right-hand side of Eq. (1–9) are known, and h can
be determined.

Similarly, we have

$$\frac{\sin (360° - A)}{\sin (90° - \delta)} = \frac{\sin t}{\sin (90° - h)},$$

which reduces to

$$\sin A = -\cos \delta \sec h \sin t. \qquad (1\text{–}10)$$

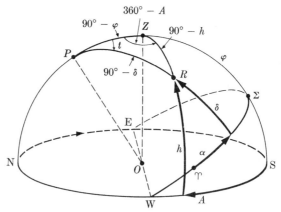

FIG. 1–11. The astronomical triangle for coordinate transformations.

Equations (1–9) and (1–10) yield the required h and A for the star. This problem can, of course, be reversed. The spherical triangle PZR is called the *astronomical triangle*.

EXAMPLE 2. Given α and δ for a celestial object, find its celestial longitude and latitude, λ and β. This situation is pictured in Fig. 1–12. The observer on the earth is at O. The obliquity of the ecliptic is ϵ or arc KP in the figure. The pole of the ecliptic is K; that of the equator is P. In the spherical triangle KPR, the angle $90° + \alpha$ and the sides $90° - \delta$ and ϵ are given. Hence we find

$$\sin \beta = \sin \delta \cos \epsilon - \cos \delta \sin \epsilon \sin \alpha, \tag{1–11}$$

$$\sin \lambda \cos \beta = \sin \delta \sin \epsilon + \cos \delta \cos \epsilon \sin \alpha, \tag{1–12}$$

$$\cos \lambda \cos \beta = \cos \delta \cos \alpha. \tag{1–13}$$

Division of Eq. (1–12) by Eq. (1–13) yields $\tan \lambda$. These equations may be solved for α and δ in terms of λ and β with the results:

$$\sin \delta = \cos \epsilon \sin \beta + \sin \epsilon \cos \beta \sin \lambda, \tag{1–14}$$

$$\sin \alpha \cos \delta = -\sin \epsilon \sin \beta + \cos \epsilon \cos \beta \sin \lambda, \tag{1–15}$$

$$\cos \alpha \cos \delta = \cos \beta \cos \lambda. \tag{1–16}$$

The coordinates for the sun, moon, and planets in the equator system (α, δ) are listed each year in the *American Ephemeris*. Heliocentric coordinates are also given so that it is a relatively straightforward matter to obtain the positions of members of the solar system at any desired time. For very accurate specification of positions, numerous interpolations in the tabulated material are required. Furthermore, corrections due to

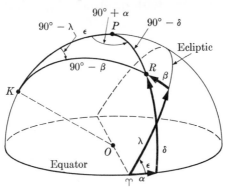

FIG. 1–12. Transformation from (α, δ) to (λ, β).

precession, nutation, and other effects must be made. The reader is referred to more extensive works [3] on practical astronomy for a discussion of these points.

1–10 Fundamentals of celestial navigation on the earth. The determination of an observer's geographical position (L, φ) by observation of celestial objects constitutes navigational astronomy. In this section an outline of three methods will be considered. More detailed discussion will be found elsewhere [4].

The tools required for celestial navigation are a sextant or similar device to measure altitudes, a clock or chronometer which keeps Greenwich Civil Time (or its equivalent), and a reference book such as the *American Ephemeris and Nautical Almanac* or the abridged version, the *American Nautical Almanac*. The latter provides data on the coordinates of celestial objects in a form particularly useful for navigation.

(a) *Method of meridian altitudes.* When a star R (or the sun, moon, or a planet) passes the meridian NPZS of an observer O (Fig. 1–13), its altitude is at a maximum. We observe that the meridian altitude for an object on the upper meridian is

$$h_m = 90° - \varphi + \delta. \tag{1–17}$$

Thus the value of h_m determined by means of a sextant, together with the declination δ of the object taken from the *Nautical Almanac*, yields the latitude φ. The situation portrayed in Fig. 1–13 is true for an observer in the northern hemisphere and for $\delta < \varphi$. The reader may deduce the appropriate relationship for $\delta > \varphi$, bearing in mind that h_m is always less than 90°.

The longitude is found by observation of the sun (or another celestial object) as it crosses the meridian. At the instant of crossing, the local apparent time is $12^h0^m0^s$. With the date known and the equation of time taken from the *Nautical Almanac*, LCT = LAT − E and, since the ob-

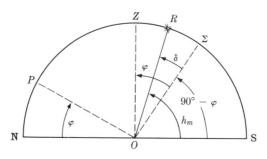

Fig. 1–13. Determination of latitude by a meridian altitude.

server's chronometer or clock is keeping GCT, the difference LCT — GCT is the observer's longitude.

If a star is observed at the instant of crossing the meridian, the local sidereal time $\theta_L = \alpha$, the right ascension of the star. This can be converted into LCT by the method of Example 2, Section 1–6, and the longitude from Greenwich found as above. Observations of the moon and planets are reduced in the same way as those for the stars. The proper right ascension and declination at the time of observation must, of course, be obtained for the moon or planets from the data given in the *Nautical Almanac*.

(b) *Method of single altitude at a given time.* We assume in this method that the latitude, φ, is known. In the northern hemisphere φ can be determined easily by observation of the star Polaris. It has been mentioned earlier and it is apparent from Fig. 1–13 that the altitude of the pole equals the latitude of the observer. The star Polaris marks very closely the pole for an observer in the northern hemisphere. The small correction required to obtain the true altitude of the pole from an observed altitude of Polaris is given in the *Nautical Almanac*. In the southern hemisphere no pole star is available. The latitude might be found by a meridian altitude observation as in (a) above.

A single measurement of altitude, h, and the corresponding GCT, together with the known latitude, φ, and known declination of the object, δ, provide the required data for solution of the astronomical triangle for the local hour angle t. For the hour angle, we have (Fig. 1–14)

$$\cos t = \frac{\sin h - \sin \varphi \sin \delta}{\cos \varphi \cos \delta}. \tag{1–18}$$

Then $\text{LAT} = 12^{\text{h}} + t_A$ if the sun is observed, and from this, one can determine LCT. As in method (a), the longitude is LCT — GCT. If a star is observed, the local sidereal time $\theta_L = \alpha + t_*$ is converted into LCT by the method described in Section 1–6.

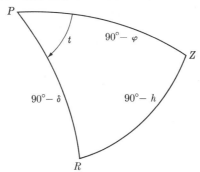

FIG. 1–14. Determination of t from a single altitude h.

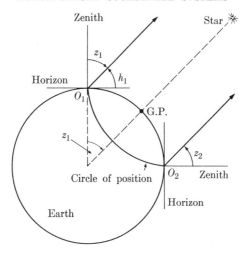

Fig. 1–15. Relation between zenith distance and a position circle.

(c) *Method of position circles.* Consider an observer O_1 on the earth (Fig. 1–15), who measures the altitude h_1 of a star (or the sun) at a given instant. For another observer located at G.P. in the figure, the star will be directly overhead at the same instant. He stands at the *substellar point* or, in the case of the sun, the *subsolar point.* The G.P. denotes "geographical position" of the substellar point.

It is clear from Fig. 1–15 that the *zenith distance* for the star is $z_1 = 90° - h_1$, and that this is the same as the angle at the center of the earth between lines to zenith and star. The earth is here considered to be spherical.

By definition, one nautical mile on the earth's surface (the earth considered as a sphere) equals one minute of arc at the earth's center. Hence the distance in nautical miles on the earth's surface from O_1 to G.P. is z_1, expressed in minutes of arc. One nautical mile equals 6080.27 feet or 1.152 statute miles or 1.853 kilometers.

Now consider an observer at O_2 (O_2 and O_1 are equidistant from G.P.). If he, too, observed the star at the same instant, he would find the zenith distance z_2 to be equal to z_1 because the lines of sight to the star are parallel. In fact, there are infinitely many positions on a circle of radius G.P. to O_1 where simultaneous observations would yield a zenith distance z_1. This circle is called a *position circle.* A single observation of the zenith distance of a star, therefore, locates the observer on a position circle.

The center of the circle can be obtained from the coordinates of the object observed together with the time. The latitude of G.P. equals the declination of the object. The longitude of G.P. can be obtained

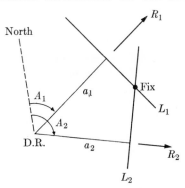

FIG. 1–16. Determination of a fix from two observed altitudes.

from the known GCT in the following ways:

(1) When the sun is observed, we have the known GCT + E = GAT and, for the Greenwich hour angle,

$$t_{AG} = \text{GAT} - 12^{\text{h}}.$$

Let t_{AG} here be either positive or negative, depending on whether GAT is greater or less than 12^{h}. Then the longitude of G.P. is t_{AG}.

(2) When a star or planet is observed, we have the known GCT converted to θ_G (Section 1–6), and the Greenwich hour angle is

$$t_{*G} = \theta_G - \alpha,$$

where α is the right ascension of the star as given in the *Nautical Almanac*. Here t_{*G} is the longitude of G.P.

The observation of a second object from the given position O_1 places the observer on a second circle of position. The observer then is located at one or the other of the intersections of these two circles. We assume that he knows his approximate position so that a choice between the two possible intersections can be made.

In practice it is obviously inaccurate and not feasible to draw position circles on a globe in order to implement the navigational method described above. The navigator works on a chart and knows his approximate position within a few miles by *dead reckoning*. On the scale of the chart, the radius of the position circle is so large that an arc near the observer can be replaced by a straight line segment. This is called a *line of position*. The intersection of two lines of position is called a *fix*, and determines on the chart the position of the observer. We shall illustrate by an example the application of this standard method of navigation.

EXAMPLE. In this example we shall omit many of the numerical details and illustrate only the method for an observer who is fixed in geographical position while he sights on two stars. In Fig. 1–16, let D.R.

denote the dead-reckoning position of the observer. At a certain GCT he observes a star R_1 at an altitude of, say 60°. The corresponding observed zenith distance $z_0 = 30° = 1800'$ of arc. For the same instant and for his D.R. position, he *computes that the zenith distance should be* $z_c = 31° = 1860'$ of arc. *This means that he is actually 60 nautical miles closer to the substellar point than his D.R. position indicates.* Therefore, he lays off on the scale of his chart a distance $a_1 = 60$ nautical miles in the direction of the star R_1 and draws a position line L_1, as shown in Fig. 1–16. The azimuth A_1 from the north defining the direction to the star can be computed from the time and the data for the star given in the *Nautical Almanac.* Tabular and graphical methods are available to expedite these calculations in practice.

Similarly, the navigator measures the altitude h'_0, and hence obtains the zenith distance z'_0 for a second star R_2. He computes z'_c and the azimuth, A_2, for this star, determines the *altitude intercept,* a_2, and lays off a position line appropriate to the star R_2.

The intersection of the lines L_1 and L_2 on his map gives his latitude and longitude, the coordinates of the fix.

Suitable modifications of this method which make it possible to obtain a *running fix* for a ship or plane in motion are described in standard works on celestial navigation.*

1–11 Interplanetary celestial navigation. The methods described in the preceding section for celestial navigation on the earth must be considerably modified for use in interplanetary space. We shall outline here a few of the possibilities [5]. Only determinations of position by observation of celestial objects will be considered. Inertial navigation in which the position and velocity of a space craft are determined by instruments capable of detecting and integrating vehicle accelerations will not be discussed.

Because of their high degree of constancy in position on the celestial sphere, the stars form a convenient reference system for space navigation. Although some of them have inherent motions of their own (*proper motion*), these are small in general. For the brighter stars which would be most suitable for navigation purposes, excessive proper motion can be allowed for.

For simplicity, consider a space vehicle R located in the solar system relative to a planet P and to the sun S, as shown in Fig. 1–17. We choose the sun as the origin of a heliocentric coordinate system. What we wish to find *from observations at R* are the coordinates r, l_R, b_R. While the direction of reference has been indicated as that toward the vernal equinox,

---

* Details and numerical examples may be found in the book by Kells, Kern, and Bland, cited earlier in this section.

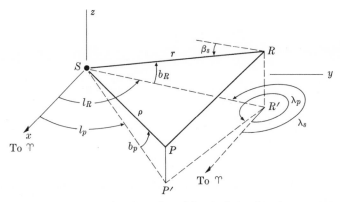

Fɪɢ. 1–17. Determination of position in interplanetary space.

one should understand that a bright star near the ecliptic would serve just as well as the vernal equinox. One other star near the ecliptic would be required to define the xy-plane.

From the vehicle R, let the observer measure

λ_s, the apparent longitude of the sun,

λ_p, the apparent longitude of the planet,

$-\beta_s$, the apparent latitude of the sun.

For navigation in the solar system the last quantity would presumably be small.

We assume that the navigator has an accurate chronometer, so that he can get from the *American Ephemeris* the following data for the instant of observation:

l_p, the heliocentric longitude of the planet,

b_p, the heliocentric latitude of the planet,

ρ, the distance from sun to planet.

If we project the space triangle SPR into $SP'R'$ in the xy-plane, as shown in Fig. 1–18, we observe that

$$\frac{r \cos b_R}{\sin \theta} = \frac{\rho \cos b_p}{\sin \psi}.$$

Furthermore, $b_R = -\beta_s$ is an *observed quantity*. Therefore,

$$r = \frac{\rho \cos b_p \sin \theta}{\cos \beta_s \sin \psi}. \tag{1–19}$$

Now $\psi = \lambda_p - \lambda_s$, an *observed quantity*, and $\varphi = l_R - l_p$.   Hence

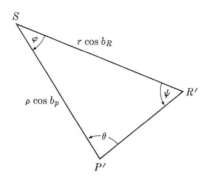

FIG. 1–18. Triangle for determination of interplanetary position

$\theta = 360° - \lambda_p + l_p$. Substituting for ψ and θ in Eq. (1–19), we find

$$r = \frac{\rho \cos b_p \sin (l_p - \lambda_p)}{\cos \beta_s \sin (\lambda_p - \lambda_s)}. \tag{1–20}$$

All quantities in this equation are either known or measured. Therefore r can be determined.

The heliocentric coordinates of R are the radius vector, r, given by Eq. (1–20), together with

$$b_R = -\beta_s, \tag{1–21}$$

$$l_R = \lambda_s - 180°. \tag{1–22}$$

This is obviously an oversimplified version of a very complex problem in instrumentation. We cannot go into further details here. But it may be instructive to discuss briefly a few of the errors that are likely to arise in the application of the trigonometric method just described.

For simplicity, consider the vehicle R to be in the plane of the ecliptic, so that $b_R = -\beta_s = 0$. Then $r = f(\lambda_p, \lambda_s)$, and we find

$$\frac{\partial r}{\partial \lambda_p} = \frac{\rho \cos b_p \sin (\lambda_s - l_p)}{\sin^2 (\lambda_p - \lambda_s)}, \tag{1–23}$$

$$\frac{\partial r}{\partial \lambda_s} = \frac{\rho \cos b_p \sin (l_p - \lambda_p) \cos (\lambda_p - \lambda_s)}{\sin^2 (\lambda_p - \lambda_s)}. \tag{1–24}$$

If ϵ_p and ϵ_s are the probable erros in λ_p and λ_s, respectively, the error in r is given by

$$\epsilon_r^2 = \left(\frac{\partial r}{\partial \lambda_p}\right)^2 \epsilon_p^2 + \left(\frac{\partial r}{\partial \lambda_s}\right)^2 \epsilon_s^2. \tag{1–25}$$

We have assumed here that there are no errors in the tabulated l_p, b_p, and ρ taken from the *American Ephemeris*. It is obvious from Eqs. (1–23) and (1–24) that the navigator would wish to choose $\lambda_p - \lambda_s$ as near 90° or 270° as possible to minimize the inherent error in r due to the measurements.

EXAMPLE. Let us denote the earth by P and assume that on a certain date $\rho = 1$ A.U., $b_p = 0$, and $l_p = 45°$. Furthermore suppose that the navigator observes $\lambda_s = 285°$, $\lambda_p = 300°$, $\beta_s = 0°$. Then Eq. (1–20) yields $r = 3.73$ A.U. If the angular measurements are made with an accuracy of $1'$ of arc, or 2.9×10^{-4} radians, the error in r given by Eqs. (1–23), (1–24), and (1–25) is 8.4×10^5 km. If the angles could be measured to an accuracy of $20''$ of arc, the resulting error in r would be 2.8×10^5 km. This is more than two-thirds the distance from the earth to the moon.

In order to come anywhere near the order of accuracy here assumed for angular measurement, the space vehicle would have to be rotationally stable. Any celestial observations would best be made by photographic means and as nearly instantaneously as possible.

The method sketched above is only an indication of the type of problem involved in space navigation. There are obviously other methods of finding the distance from the sun to the space vehicle. We shall mention a few.

(a) The *sun's angular diameter* varies inversely as the distance from S to R if the ship is not too close to the sun. At a distance of 1 A.U., a measurement of the solar diameter made with an accuracy of $1''$ of arc would locate the space craft within 8×10^4 km along the radius vector. A question unsettled at present is whether the sun's diameter varies. There is some evidence of a periodic variation with the sunspot cycle, with a total amplitude somewhat less than $3''$ of arc [6].

(b) The *intensity of solar radiation* varies inversely as the square of the distance from the sun. In principle, therefore, a measurement of the radiation would yield a distance from the sun. This would be a more sensitive distance measure than that of the solar diameter. With photoelectric techniques and under ideal circumstances, it is not difficult to keep the probable error to ± 0.002 magnitudes. Such an error would result in an error of 1.5×10^5 km in a distance measurement. If the sun's brightness varies, the use of this method would result in corresponding variations in the distance measurement.

(c) The *equilibrium blackbody temperature* of a body in space varies inversely as the square root of the distance from the sun. Theoretical temperatures can be derived easily for a spherical heat-conducting body of known reflecting power, as shown in Section 2–7. At the earth's distance from the sun, an error of $0°.1$ C would result in an error of 10^5 km in a distance measured this way.

Evidence that the sun varies in brightness is indicated by recent photo-electric observations of the brightness of Uranus and Neptune [7]. The variation may have a periodicity related to the well-known sunspot cycle, and its amplitude in terms of the total radiant energy may be about 2%. Abbot [8] has concluded from direct solar radiant energy measurements that during periods of high sunspot activity the observed rate of solar radiant energy may decrease by as much as 2%. His work, however, has been criticized by Sterne and Dieter [8]. If these variations are real, they would have to be taken into account in any application of methods (b) and (c).

These are a few of the ways by which the radius vector from sun to space craft can be determined by physical measurement. We have, of course, neglected here any discussion of tracking by radar or similar devices.

PROBLEMS

Section 1–4:

1–1. State the α and δ of the west point of the horizon when the vernal equinox is at the east point.

1–2. For an observer at $\varphi = +33°$, what is the altitude of the Σ-point? What is the zenith distance of the pole?

1–3. The azimuth of a star is 210° and its altitude is 35° as observed from a point at $\varphi = +10°$. By means of a sketch similar to Fig. 1–5, estimate the star's hour angle and declination.

1–4. What is the value of t_A and the approximate azimuth of the sun on June 21 at sunset for an observer at latitude $\varphi = +30°$?

1–5. The right ascension and declination of the sun on September 1 are 10^h38^m and $+8°.6$, respectively. By means of a sketch, estimate its altitude and azimuth for an observer at $\varphi = -30°$ at 3 p.m. LAT.

Section 1–5:

1–6. The local civil time at a place of longitude L is $10^h16^m12^s$ a.m. at the instant at which the GCT is $11^h26^m45^s$. What is L?

1–7. What is the local apparent time at a place of longitude $+110°40'30''$ when it is noon LCT on December 11, 1960?

1–8. What is the hour angle of the sun on November 25, 1960, at Tokyo $(L = -9^h18^m58^s)$ when the local civil clock reads 9 a.m.?

1–9. What is the Ephemeris Time at which the sun passes the winter solstice in December, 1960?

Section 1–6:

1–10. The LAT at Washington $(L = +5^h08^m16^s)$ is $2^h45^m12^s$ p.m. on November 30, 1960. What is the LCT at Washington for the same instant?

1-11. The sidereal clock at Greenwich reads $18^h46^m10^s$ on November 20, 1960. What is the corresponding reading of the Greenwich civil clock? What is the hour angle of the real sun at the given instant?

1-12. The star Deneb, $\alpha = 20^h39^m34^s.5$, is crossing the meridian of an observer at Denver ($L = +6^h59^m48^s$). What is the local sidereal time? What is the local civil time if the event occurs on December 16, 1960?

1-13. What is the reading of the local sidereal clock in Cleveland ($L = +5^h26^m16^s$) on December 1, 1960, when the Eastern Standard Time is $12^h00^m00^s$?

1-14. On December 30, 1960, what is the reading of a clock set for Eastern Standard Time when the sidereal clock reads 0^h?

1-15. The local civil time at San Francisco ($L = +8^h09^m43^s$) is 9 p.m. on November 15, 1960. What is the reading of the local sidereal clock at the same instant?

Section 1-9:

1-16. By means of Eqs. (1-9) and (1-10), calculate h and A for the sun for the conditions given in Problem 1-5 and compare with the estimated values.

1-17. What is the azimuth of the rising sun for an observer at $\varphi = +60°$ on November 17, 1960? At what altitude will it cross the observer's meridian?

1-18. Find the celestial longitude, λ, of the sun at 0^h Ephemeris Time on December 1, 1960, by using Eqs. (1-11), (1-12), and (1-13). The obliquity of the ecliptic is $23°27'8''.3$.

Section 1-10:

1-19. A navigator notes that the sun crosses his meridian at an altitude of $35°02'$ on November 27, 1960, at $3^h10^m30^s$ p.m. GCT. What is his longitude and latitude?

1-20. The star Altair ($\alpha = 19^h48^m6^s$; $\delta = +8°43'$) was observed at an altitude of $72°14'$ crossing the meridian of a place at $11^h07^m12^s$ p.m. GCT on December 1, 1960. Find the longitude and latitude of the place.

1-21. A navigator in D.R. position ($L = +124°50'$, $\varphi = +33°30'$) observed Regulus and Antares and obtained the following results:

Regulus	*Antares*
$h_0 = 32°10'$,	$h_0 = 20°50'$,
$h_c = 31°50'$,	$h_c = 20°35'$,
$A = 270°$,	$A = 150°$.

Find the altitude intercepts and determine the coordinates of the fix by plotting on rectangular coordinate paper.

REFERENCES

1. W. MARKOWITZ, "Variations in Rotation of the Earth, Results Obtained with the Dual-Rate Moon Camera and Photographic Zenith Tubes," *Astron. J.* **64,** 106, 1959.

2. G. M. CLEMENCE, "Ephemeris Time," *Astron. J.* **64,** 113, 1959.

3. J. J. NASSAU, *Textbook of Practical Astronomy*, 2nd ed., Chapters IV and VI. New York: McGraw-Hill Book Company, Inc., 1948.

4. L. M. KELLS, W. F. KERN, and J. R. BLAND, *Navigation*, Chapters VIII through XII. New York: McGraw-Hill Book Company, Inc., 1943.

5. M. VERTREGT, "Orientation in Space," *J. British Interplanetary Soc.* **15,** 324, 1956.

6. G. P. KUIPER, *The Sun*, p. 18. Chicago: University of Chicago Press, 1953.

7. H. L. JOHNSON and B. IRIARTE, "The Sun as a Variable Star," *Lowell Observatory Bulletin*, No. 96, 1959.

8. T. E. STERNE and N. DIETER, "The Constancy of the Solar Constant," *Smithsonian Contributions to Astrophysics*, Vol. 3, No. 3, p. 9, 1958. The paper by C. G. Abbot is found in the same issue on p. 13.

9. H. N. RUSSELL, R. S. DUGAN, and J. Q. STEWART, *Astronomy*, Vol. I, rev. ed., *The Solar System*. Boston: Ginn and Co., 1945.

10. W. M. SMART, *Textbook on Spherical Astronomy*, 4th ed. Cambridge: Cambridge University Press, 1949.

CHAPTER 2

THE PLANETS AND THEIR SATELLITES

2–1 Orbital motions of the planets. *Orbital elements* are quantities used in describing the size, shape, and orientation in space of the orbit of any object moving under gravitational attraction. The elements also include quantities that permit the computation of the position of the object in its orbit at a given time. The orbit is assumed to lie in a plane and to be a conic section. Strictly speaking, this is justified only in cases where the orbit represents a solution to the classical problem of two point masses whose accelerations are the result of their mutual gravitational attraction. These motions are approximated very well by the members of the solar system. Each planet moves about the sun in an orbit that agrees closely with the theoretical path the planet would follow if it and the sun were alone in space, and if their masses were concentrated in two points. The last requirement is satisfied by bodies whose mass is distributed with spherical symmetry.* The same is true of each satellite and its parent planet. The motions of all bodies of the solar system resemble closely the theoretical motions that are expected for bodies taken two at a time in the sense of sun-planet or planet-satellite. The more massive member of the two bodies is referred to as the *primary* mass and the other as the *secondary* mass.

When the approximate description of an orbit provided by the orbital elements is insufficient, it is customary to specify the elements for a given epoch and to give the rates at which these elements change with time.† Departures from motion in conic sections can be caused by the attraction of bodies other than the primary, or by the oblateness or lack of spherical symmetry of the primary and secondary. In nature, these are the principal causes of departure from conic-section orbits, but theoretically any resisting medium or any force other than that of mutual gravitation will cause similar deviations.

The orbital elements are:

a, semimajor axis,　　　　　　　　ω, argument of perihelion,

e, eccentricity,　　　　　　　　　P, sidereal period,

i, inclination,　　　　　　　　　　T, time of perihelion passage.

Ω, longitude of ascending node,

---

* This will be demonstrated in Chapter 4.
† A discussion of this point will follow in Chapter 5.

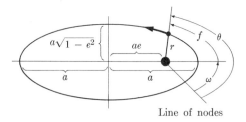

FIG. 2–1. Graphic representation of some of the terms used in describing an elliptic orbit.

The elements a and e define the size and shape of the orbit (see Fig. 2–1). When the orbit is not closed, one uses q, the distance of closest approach of object to sun, or perihelion distance, instead of a.

For objects whose primary is the sun, the semimajor axis is usually given in astronomical units, a unit which is very nearly equal to the earth's orbital semimajor axis. The exact definition of the astronomical unit will be given in the discussion of Kepler's laws in Section 2–2. The term "mean distance" is also employed when speaking of the element a. Astronomers sometimes refer to the line defined by the major axis of an orbit as the "line of apsides."

The eccentricity, e, measures the ellipticity of the orbit, and its value is less than unity for elliptical orbits, is unity for parabolic orbits, and is greater than unity for hyperbolic orbits. The relationships of e to the major and minor axes in an elliptical orbit are shown in Fig. 2–1. Instead of e, the quantity ϕ defined by $e = \sin \phi$ is occasionally used in orbit calculations as a measure of ellipticity.

The inclination, i, is the angle between the orbital plane and a reference plane. In planetary orbits, the reference plane is taken as the plane of the ecliptic, which in turn is the orbital plane of the earth. Hence, by definition, the value of i for the earth's orbit is zero. For a satellite, the reference plane used may be the orbital plane of the parent planet, or its equatorial plane, or the so-called *proper plane*. The latter, which is used for those satellites whose motion is sensibly affected by the sun, takes into account the precession effects of the orbital plane of the object. It is perpendicular to the pole of the precession and passes through the center of mass of the primary. The inclination is sometimes taken to be such that it is always less than 90°, and the orbital motion is then defined as *direct* if it has the same sense as that of the earth, or as *retrograde* if its sense is opposite. It is customary and convenient to let the inclination vary from 0° to 180°, and to consider the orbital motion always as direct.

The inclination, i, and the longitude of the ascending node, Ω, define the position of the orbital plane in space, where Ω is the *angle measured*

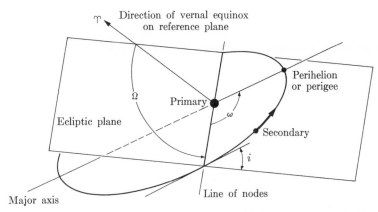

FIG. 2–2. The elements used to define the space orientation of an elliptic orbit.

toward the east in the reference plane from the direction of the vernal equinox to the direction of the ascending node. This element is shown in Fig. 2–2.

The element ω, or *argument of perihelion, is the angle in the orbital plane from the ascending node to the major axis at the point of closest approach,* as shown in Fig. 2–2. It serves to denote the orientation of the major axis in the orbital plane. The *longitude of perihelion,* which equals the sum of ω and Ω, is often used instead of ω for designating the perihelion orientation. The term "perihelion" is used for objects whose primary is the sun. When the earth is the primary, "perigee" is used. For the other planets a word derived from the planet's name may be used, for example, "perijovian" or "perimartian" distance.

In order to locate the object in the orbit at a given time, two more elements are required, the period of revolution, P, and the exact moment when the object passes through some reference point in the orbit. The time of perihelion passage, T, may be used for the latter purpose. Often instead of P the mean rate of motion on the orbital plane n is given, expressed in degrees per day, in which case $n = 360°/P$.

The period referred to so far is more precisely called the *sidereal period,* which is defined as the time interval between successive passages of an object across a line from the primary to the fixed star field. In practice, successive crossings of the same right-ascension circle are taken to be equivalent to the sidereal period. Then there are the *anomalistic* period, which is the interval between successive passages through perihelion, and the *nodical* period, which is the interval between successive passages through the ascending node. Precession effects and orbital perturbations cause these periods to be different. The term "period" by itself usually means the sidereal period.

The mean anomaly, M, at a given date may be used instead of T. The mean anomaly is the angle which would be swept out by the planet's radius vector since the last passage through perihelion if it moved at the uniform rate given by the mean motion n. It should not be confused with the true anomaly, f, defined in Fig. 2–1. It will be shown in Chapter 4 how it is possible to derive the position in the orbit at other epochs when the orbital elements are known.

Orbital elements for the major planets and their variations are listed in the *American Ephemeris*. Table V in the appendix presents a summary for the epoch September 23.0, 1960, Universal Time. The rates of variation of the less constant elements are included, the figures representing variations for 100-day periods. Positive variations mean that the elements are increasing with time at the given epoch. The unit used in the variations is unity in the last significant figure quoted for a given element. These variations are presented only to give the reader an idea of their nature. In general, the rate of variation varies with time, and the table of planetary elements from the current year in the *American Ephemeris* should be consulted for the most accurate possible set of elements. The numbers are given with the precision required if theoretically derived right ascensions and declinations of the planets based on these elements are to have an accuracy of the order of $0''.01$.

In Table V the longitudes of perihelion quoted are measured from the vernal equinox. Periods in tropical years are included. In each case these have been derived by dividing $360°$ by the quoted daily motion n. The result gives the sidereal period of revolution in ephemeris days. The periods have then been converted to tropical years by using the relationship

$$365.24219879 \text{ ephemeris days} = 1 \text{ tropical year.}$$

Table V also includes synodic periods. *The synodic period is defined as the time of orbital revolution of a planet with respect to the sun-earth line.* The relationship between synodic period (S) and sidereal period (P) of a planet is

$$\frac{1}{S} = \pm \left(\frac{1}{P} - \frac{1}{E} \right), \tag{2-1}$$

where E is the earth's sidereal period. The negative sign is to be used for the *superior* planets, i.e., the outer planets from Mars to Pluto, and the positive sign for the *inferior* planets, i.e., Mercury and Venus.

With respect to the earth, some positions of the planets in their orbits have special designations. These are shown in Fig. 2–3, where the orbits are projected on the same plane. The *elongation* is the angle between the lines earth-to-planet and earth-to-sun.

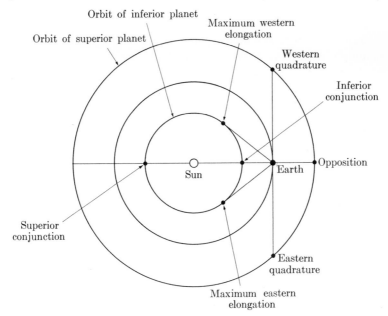

Fig. 2–3. Planetary configurations.

The orbital elements of the planets exhibit certain regularities worth summarizing.

(1) From Mercury to Uranus the values of the semimajor axes increase in such a way that they are given approximately by the formula

$$a = \tfrac{1}{3}(2^{n-2} + 1), \tag{2–2}$$

where $n = 1$ for Mercury, 2 for Venus, and so on. The value predicted by the formula for $n = 5$ is occupied by the asteroid belt instead of by one of the planets. A similar relationship known as Bode's law played a very important role in the discovery of the asteroid belt.

(2) The orbital eccentricities are usually small. In a drawing of the solar system in which the mean distance of Pluto is made equal to one foot, the planet's orbit cannot be distinguished from a circle. The position of the sun in the drawing, however, is off center by 3 in.

(3) With the exception of Pluto's orbit, the planetary orbits lie remarkably close to the ecliptic plane, as shown by the small values of the orbital inclinations. This is in contrast to the parabolic orbits among the comets, which will be discussed in Chapter 6.

(4) The direction of revolution of the planets about the sun and of the majority of the satellites about the planets is the same, that is, eastward or counterclockwise as seen from the north pole of the ecliptic.

The dynamics of the motions of the planets in their orbits will be considered in Chapter 4.

2–2 Kepler's laws of motion. In the seventeenth century Johannes Kepler announced three laws of planetary orbital motion which were derived empirically from observations of the planet Mars made by Tycho Brahe. The laws are known as Kepler's first, second, and third laws. Each of them embodies a result of motion under gravitational attraction, and hence these laws are obeyed by all members of the solar system. Analytical derivations for them are given in Chapter 4. Kepler's laws are:

First law: *Each planetary orbit is elliptical, and the sun is located at one of the foci.* The polar equation of the elliptical orbit in terms of symbols defined previously (see Fig. 2–1) is

$$r = \frac{a(1 - e^2)}{1 + e \cos (\theta - \omega)}. \tag{2–3}$$

Second law: *The line joining the centers of mass of the sun and a planet covers an area that increases at a constant rate as the planet moves in its orbit.* Another way of saying this is that the radius vector sweeps over equal areas in equal periods of time, regardless of where the planet is in its orbit. Since the radius vector is shortest when the planet is at perihelion, this means that it will move fastest then.

Third law: *The square of the sidereal period P divided by the cube of a planet's mean distance a from the sun forms a ratio that is the same for all planets, that is, $P^2/a^3 =$ constant.* If a is expressed in astronomical units and P in tropical years, the constant is approximately unity. It is shown in Chapter 4 that the ratio does vary slightly, and that it is a function of the sum of the masses of the sun and the planet in question, including its satellites. The planetary masses, however, are so small compared with the mass of the sun that, unless the most exact results are desired, the ratio P^2/a^3 may be taken as constant.

The precise statement of Kepler's third law is

$$\frac{P^2}{a^3} = \frac{4\pi^2}{k^2(m_1 + m_2)}, \tag{2–4}$$

where k^2 is the gravitational constant used in expressing the force of attraction between two bodies of masses m_1 and m_2 separated by a distance r. This force law is

$$F = -k^2 \frac{m_1 m_2}{r^2}. \tag{2–5}$$

In astronomy, the fundamental unit of mass is that of the sun, and the unit of length is the astronomical unit (A.U.). The mean distance of a given planet from the sun is found in astronomical units from Eq. (2–4) when unity is used for m_1 and the exact* value is used for k, i.e.,

$$k = 0.01720209895,$$

the period P being expressed in ephemeris days. This value of k is called the "gaussian constant." In astronomical units the earth's mean distance from the sun is (according to the foregoing definition) 1.00000003, the mass m_2 being taken as 1/329390 solar mass units, which includes the moon. The planetary masses used by the Nautical Almanac Office in the determination of the mean distances presented in Table V are slightly different from the most accurate values available at present, but these differences produce only negligible changes in the distances quoted. In Section 2–3 some of the most recent determinations of planetary masses are presented.

Kepler's third law also applies to the motion of satellites about their parent planets. In the case of the earth's artificial satellites, it is convenient to express a in kilometers and P in minutes. Remembering that the mass of such a satellite is negligible, and taking the earth's mass by itself as 1/333432 solar mass units, we find from Eq. (2–4)

$$a = 332.3P^{2/3}. \tag{2–6}$$

In the derivation of this result, the astronomical unit in kilometers has been taken (see Section 2–3) as 1.4953×10^8.

Since P can be readily determined for an artificial satellite, this formula provides a quick method for finding a. If from a one subtracts the radius of the earth, one obtains the mean height above sea level of the satellite. For example, the Sputnik I carrier rocket (whose sidereal period on October 9.4, 1957, Universal Time, was quoted as 96.046 min) had a mean distance of 6,968 km according to Eq. (2–6). The earth's equatorial radius is 6,378 km. Therefore the mean height above equatorial sea level was 590 km. Since at that time the orbital eccentricity was 0.05, the closest approach to equatorial mean sea level was initially (see Fig. 2–1) $[6968(1 - e) - 6378]$ or 242 km. The farthest distance from equatorial mean sea level can be computed in a similar manner, and the result is 938 km. These values refer to the initial period of the rocket carrier and not to later epochs when the period had diminished.

Because of the effects produced by the oblateness of the earth, Eq. (2–6) provides only approximate results for a close earth satellite. The formula holds only for an idealized earth possessing spherical symmetry.

---

* See Section 4–3 for a more detailed specification of the constant k.

2–3 The masses of the planets. Table VI in the appendix presents the physical characteristics of the planets and lists the probable errors wherever these data are available from observations. Many of the values in Table VI depend on the masses and radii of the planets. Astronomical methods yield masses in terms of the solar mass. To reduce the results to grams one must use the ratio of the mass of the earth to that of the sun, together with the mass of the earth in grams. The latter is based on physical experiments, while the ratio of the sun's mass to that of the earth can be derived astronomically.

The masses of the planets and of other fundamental astronomical constants have been discussed by Newcomb [1], De Sitter [2], and Brouwer [3]. Since these reviews were made, new observations as well as more accurate computations based on the older observations have become available. The masses for the planets presented here are based primarily on new information published since De Sitter's study.

The mass is calculated from Kepler's third law for a planet that has satellites or from the perturbations that the planet causes in the motions of other objects of the solar system. If, as usual, we use in Kepler's third law the astronomical unit and mean solar days, the mass is obtained in solar mass units. The resulting value of the mass depends on how well the period and the semimajor axis a are known. Periods can be very exactly evaluated when the observations cover many revolutions. The observation of a is much more difficult. For a planetary satellite, a is obtained in seconds of arc, and the level of precision of past observations gives probable errors of $\pm 0''.02$ to $\pm 0''.05$. As a result, the fractional error in the quantity a^3 which is used in Kepler's law is rarely less than $0.06/a''$, an error which is carried over into the computed mass. The derivation of the mass for an oblate planet must allow for the perturbing effects in the satellite's orbit due to the equatorial bulge of the parent planet. In this case, Kepler's third law is used with a small correction $(1 - \sigma)$ (see Section 5–9), namely,

$$\frac{P^2}{a^3} = \frac{4\pi^2}{k^2(m_1 + m_2)} (1 - \sigma), \qquad (2\text{–}7)$$

σ being a function of the planet's oblateness and of the distance of the satellite from the planet given by

$$\sigma = \left(\frac{A}{a}\right)^2 \left(f - \frac{1}{2} m\right), \qquad (2\text{–}8)$$

where A is the planet's equatorial radius, a the satellite's semimajor axis in the same units, f the flattening or oblateness of the planet [see Eq. (2–19)], and m the ratio of centrifugal force to the gravitational force on the planet's equator. If the solar attraction produces a significant effect, one should then subtract from σ the quantity $\frac{1}{2}(n_1/n)^2$, where n_1 and n are the mean rates of orbital motion of the planet and of the satellite. In a study of the orbital motions of Mars' satellites, Woolard [4] presents the data required to derive from Eq. (2–7) the mass of Mars:

$$f = \frac{1}{191.8},$$

$$m = \frac{1}{218},$$

$$a \text{ (Phobos)} = 12''.895,$$

$$a \text{ (Deimos)} = 32''.389,$$

$$\frac{4\pi^2}{P^2} \text{ (Phobos)} = 388.17065 \text{ (rad/day)}^2,$$

$$\frac{4\pi^2}{P^2} \text{ (Deimos)} = 24.77067 \text{ (rad/day)}^2,$$

where the a's are angular mean distances as observed from a distance of one astronomical unit. The solar perturbation term in σ is negligible when compared with the uncertainties in a mentioned below. Thus

$$\sigma \text{ (Phobos)} = 0.00074,$$

$$\sigma \text{ (Deimos)} = 0.000012.$$

A value of $A = 4''.59$ may be obtained from Section 2–4. For the mass of Mars in solar mass units, we then obtain, using Eq. (2–7),

$$\text{from Phobos:} \quad \frac{1}{3,108,000(\pm 15,000)},$$

$$\text{from Deimos:} \quad \frac{1}{3,091,200(\pm 6000)}.$$

In solving for the mass one finds that the greatest uncertainty comes from the factor a^3. When a probable error in a of $\pm 0''.02$ for each of the satellites is used, the errors quoted are obtained. The reciprocal of the mass equals $3,095,400 \pm 6000$ if the result for Deimos is given three times the weight of that for Phobos. An independent derivation of the mass of Mars is described further on.

For the earth, Spencer Jones [5] gives a computation in which the oblateness is taken into account by substituting for the gravitational constant k^2 a function of the equatorial acceleration due to gravity and of the oblateness. This approach is equivalent to the previous one. Spencer Jones's value for the earth's mass in solar units is

$$m_\oplus = \frac{\pi_\odot^3}{2.26444 \times 10^8},$$ (2–9)

where $\pi_\odot$ is the sun's *equatorial horizontal parallax* in seconds of arc. This is the angle at the center of the sun subtended by the equatorial radius of the earth. The evaluation of $\pi_\odot$ permits us to convert the length of the astronomical unit into linear units, such as kilometers, by means of the relation

$$\frac{\pi_\odot}{206265} = \frac{R_\oplus}{1_{\text{A.U.}}},$$ (2–10)

where 206,265 is used to change seconds of arc into radians, and $R_\oplus$ is the earth's equatorial radius.

The solar parallax has proved to be a very difficult quantity to determine with the desired accuracy. Yet it is required in the conversion of the astronomical determinations of planetary masses and radii to cgs units. At the time of writing, the most reliable of the various ways in which $\pi_\odot$ has been determined are those based on observations of the asteroid Eros. The observations of Eros lead to two methods for obtaining $\pi_\odot$. In one method, the distance between the earth and Eros, determined trigonometrically, is used; in the other, the solar parallax is derived as part of a general solution involving the perturbing effects of all the planets on Eros. The first method yields, in essence, a value in centimeters for a distance which can be computed in astronomical units from Kepler's third law when the orbital periods of the asteroid and the earth are used. In the second method, equations relating the theoretical perturbing effects to the position of Eros in the sky are employed. The masses of the planets as well as the solar parallax can be introduced as unknown quantities. A simultaneous solution for the unknowns is then made such that the right-ascension and declination observations of Eros are satisfied. Since Eros passed relatively close to the earth in 1930–31, the determinations of sky positions were very sensitive to the adopted values of the unknowns. Spencer Jones [5] has determined $\pi_\odot$ from trigonometric calculations, while E. Rabe [6] has made a study based on the perturbing effects. As a result of Rabe's work, values were obtained for $\pi_\odot$ as well as for the masses of Mercury, Venus, Mars, the earth, and the moon. Because these results are based on the planet's dynamical effects and form a self-consistent group of quantities, they are the best determinations available for astronautical applications. According to Rabe's determination,

$$\pi_\odot = 8''.79835 \pm 0''.00039,$$ (2–11)

while Spencer Jones's trigonometric result is $8''.790 \pm 0''.001$. Rabe's determination is adopted here.*

In order to evaluate the astronomical unit in kilometers, we may use Rabe's value for $\pi_\odot$ and the earth's equatorial radius of 6,378,150 m derived from the most recent geodetic data [7].† The result is

$$1 \text{ A.U.} = 149,527,000 \pm 8000 \text{ km}, \tag{2-12}$$

which is the value adopted here. If one uses the equatorial radius of the earth as adjusted by Jeffreys [3] and quoted by Spencer Jones, namely 6,378,084 m, and Rabe's value for $\pi_\odot$, a value is found that is within the probable error indicated above.

Observations by radar-echo techniques of the distance between the earth and Venus or other planets are currently being conducted or planned. The reduction of these observations must allow for any retardation effects that may exist when the radar beam traverses a planet's atmosphere. One must also include corrections for the fact that the radar-echo method yields a distance between the earth's surface and the surface of the planet under study, whereas astronomical methods yield center-to-center distances. These corrections may be greater than the probable error of the astronomical unit quoted in Eq. (2-12). The question of the proper distance between an observed planet's center and the effective reflecting surface must also be answered. Preliminary results obtained for Venus [8] yield a value for the solar parallax of $8''.8020 \pm 0.0005$. The disagreement among the various solar parallax determinations is disturbing. The differences are usually appreciably larger than those expected from the probable errors indicated by the individual workers.

Returning to the problem of the earth's mass, we obtain from Eqs. (2-9) and (2-11) the result

$$m_\oplus = 332,473 \pm 45 \tag{2-13}$$

in *reciprocal solar mass units;* this figure excludes the mass of the moon. For the earth-moon system Rabe finds from the perturbations on Eros the reciprocal mass value $328,452 \pm 43$, and for the ratio mass of earth: mass of moon a value of 81.375 ± 0.026 from which we obtain $m_\oplus = 332,489 \pm 43$. For the mass of the earth we adopt here the mean of these two determinations,

$$m_\oplus = 332,480. \tag{2-14}$$

---

* Since this was written, $\pi_\odot$ has been determined with high precision from observations of the artificial planetoid Pioneer V. The result: $\pi_\odot = 8''.7974 \pm 0.0008$ is in agreement with Rabe's determination. (See M. Melin, "Pioneer V and the Scale of the Solar System," *Sky and Telescope,* Vol. XX, No. 6, p. 337, December, 1960.)

† See Section 3-1 for a discussion of the earth's radius.

To convert planetary masses to cgs units we need the mass of the earth in grams. This has been determined by measuring the acceleration due to gravity g on the earth's surface. At the equator the determination adjusted by Jeffreys [5] yields $g_e = 978.0384 \pm 0.0021$ in cgs units. The relationship between g_e and the earth's mass is given by [see Section 5–9, Eq. (5–73)]

$$g_e = \frac{Gm_\oplus}{A^2}\left(1 + f - \frac{3}{2}\,m\right),\qquad(2\text{--}15)$$

where A, f, and m are the quantities defined previously and G is the universal constant of gravitation in cgs units. Since G is known with less accuracy than g_e, it is the principal cause of uncertainty in the mass derivation. From laboratory measurements [9]

$$G = 6.670 \pm 0.005 \times 10^{-8}\text{ dynes·cm}^2\text{·gm}^{-2}.\qquad(2\text{--}16)$$

The values of f and m for the earth are $\frac{1}{297}$ and 0.003461, respectively [10]. Using the earth's equatorial radius and the values of G and g quoted previously, we obtain

$$m_\oplus = 5.976 \pm 0.004 \times 10^{27}\text{ gm}.\qquad(2\text{--}17)$$

The masses quoted in Table VI for Mercury and Venus are taken from Rabe's [6] work. For the earth, the value in Eq. (2–14) is used. The mass of Mars is an average of Rabe's value [6] and the mass derived from the orbital motions of Mars' satellites according to the data discussed previously. These two determinations agree within their probable errors. The masses of Jupiter and Uranus are taken from a list by Clemence [11], that quoted for Saturn is due to Hertz [12].* The mass listed for Neptune is due to Van Biesbroeck [13] and is based on observations of the satellite Nereid. No value is quoted for the mass of Pluto. Determinations of Pluto's mass from perturbations of the planet Neptune yield a value between 0.25×10^{-5} and 0.35×10^{-5} solar masses, nearly equal to the mass of the earth [14]. However, the use of this mass in conjunction with the observed radius yields an excessive mean density of about 50 gm/cm^3 for Pluto. New observations of the perturbations of Neptune by Pluto, now in progress, may eventually resolve this problem.

2–4 The radii and shapes of the planets. The radii of the planets, excepting that of the earth, are usually computed from observed angular

---

* Hertz's value for Saturn's mass may require a slight revision according to G. Clemence (*Astron. J.*, **62**, 21, 1960) whose work indicates a mass of 1/3499.7 solar units.

diameters. The observed diameter α is usually expressed in seconds of arc and is related to the radius ρ and to the distance D by

$$\rho = \frac{1}{2} \frac{\alpha D}{206{,}265} .$$ (2–18)

In practice the measurement of angular diameters is difficult because of the unsteadiness and poor quality of the telescopic image, which suffers from telescopic aberrations and diffraction effects as well as from atmospheric turbulence. In addition, physiological phenomena of the eye vary from observer to observer and for a given observer with time.

Measurements may be made at the telescope by means of *micrometers* or *disk meters*. Photographic determinations have also been made. *Filar micrometers* in which two parallel wires, one movable with respect to the other, are placed tangent to opposite limbs of the planet's telescopic image, were used in older observations that now date back well over a hundred years. Another variation is the *double-image micrometer* in which two images of the planet are formed which can be moved with respect to each other. In this case the limbs of the two images are brought in contact. The errors that can be made in micrometer measurements (discussed by W. W. Campbell [15]) always tend to make the measured diameter too large except in the case of accidental errors or errors caused by human factors. The probable error obtained in one of these observations is rarely less than $\pm 0''.5$. These figures refer to internal consistency. A newer variation of the double-image micrometer which uses a birefringent filter to separate the two images has been described by Dollfus [16]. Some of the errors arising in this case as well as in the older version of the double-image micrometer, have been investigated by Camichel [17], who concludes that at present the best measurements available for objects with small angular diameters are those obtained with disk meters.

Using the *disk meter* [18], one forms on the focal plane of the telescope the image of an artificial disk of variable diameter, whose color and brightness can be made to match those of the planet's image. Eye estimates are made of the equality of size of the two images. Laboratory experiments show that the eye can, under ideal conditions, detect diameter differences of the order of one per cent. In telescopic observations the variations experienced in different trials show a mean error of about five per cent of the measured diameter [19].

Photographic observations measure the size of the image. The observational errors in this method, whose principal source is the incorrect exposure of the limb of the planet's disk (see Plate I), have been discussed by Ross [20]. The probable error of photographic determinations of diameter is somewhat larger than that of filar-micrometer observations.

The measurements usually include the atmosphere of the observed planet if it has an appreciable atmosphere. The telescopic disk in this case is larger than if the planet were devoid of an atmosphere. Diameter measurements with light of different colors yield different results; for example, for Venus and Mars, diameters observed in yellow light are smaller than those observed in blue light. As a result, the diameters of the planets quoted in Table VI usually include part of their atmospheres. This is particularly true for Venus, Jupiter, Saturn, Uranus, and Neptune, whose actual surfaces have never been observed.

Systematic differences are found when a planet's diameter is measured with different instruments. Of the various methods of observation, the disk meter in theory yields results much less affected by the earth's atmospheric turbulence than do the other methods. In preparing a list of planetary radii from available observations, one could attempt to weight the published observations according to the systematic errors of each observer and then find an average. This is a difficult task and was last attempted by W. Rabe [21] for the observations available up to 1926. His results for Jupiter and Saturn may still be regarded as the best micrometric measurements available. For Jupiter, the angular diameter has also been determined by Sampson [22] from the time interval in which the four bright satellites are occulted by the planet. The results indicate a smaller value (by $0''.25$) than that found by micrometric measurements. Because of the precision with which the occultation observations can be made, Sampson's equatorial diameter for Jupiter is adopted here. For the other planets, more recent observations, or observations with newer techniques, have been made which show that systematic errors were present in the older determinations. These are more serious for the planets with small angular diameters. The radius quoted for Mercury is that obtained by Muller [23], and the radius given for Venus was computed by De Vaucouleurs [24]. For Mars the solid surface diameter determined by Trumpler [25] is adopted. Recent determinations by Camichel [26] seem to bear out this measurement. Trumpler's and Camichel's determinations are based on studies of the motion of surface markings across the disk as the planet rotates. The path followed by a marking as the planet rotates is a portion of an ellipse whose size and orientation depend on the radius of the planet and the direction in space of its axis of rotation. The observations by Dollfus [27] with a birefringent micrometer are in fair agreement with these results. The specific results obtained by Rabe, Trumpler, and Dollfus for Mars' angular diameter illustrate some of the difficulties encountered in the determination of planetary diameters:

$$
\begin{array}{ll}
\text{W. Rabe [21]} & 9''.47 \pm 0''.05, \\
\text{R. J. Trumpler [25]} & 9''.178 \pm 0''.015, \\
\text{A. Dollfus [27]} & 9''.25 \pm 0''.03.
\end{array}
$$

For Uranus we adopt the observations of Camichel [18]; for Neptune [28] and Pluto [29] we adopt Kuiper's results. The data for the last three planets are based on disk-meter observations.

There has been some disagreement about the observed diameter of Pluto. Alter [30] argues that if the planet reflects sunlight specularly (in the manner of a polished sphere), the observed diameter would be smaller than the actual diameter. However, no other object in the solar system is known to reflect sunlight in this manner. In addition, photometric observations show that the planet varies in brightness as it rotates [31], indicating that it cannot be considered as a uniformly polished sphere.

Table VI lists the angular diameters reduced to distances from the earth that are approximately the true distances. The linear values given are based on the solar parallax and the earth's diameter previously adopted in this chapter.

The oblateness, or flattening, is defined as the ratio

$$f = \frac{A - B}{A}, \qquad (2\text{-}19)$$

where A and B are respectively the equatorial and polar radii. It may be derived from direct observations or from the effects that the equatorial bulge has on the orbital motion of a satellite. The oblateness of the earth can also be evaluated from the theory of the observed precession of the earth's axis, from gravity measurements, or from artificial satellite observations.

The dynamical methods for evaluating the oblateness make use of the differences (discussed in Chapter 5) in gravitational potential between an oblate planet and a spherically symmetric planet. The equivalence of the dynamical and geometrical evaluations of f is a result of the fact that the surface of the planet may be assumed to be in equilibrium in the sense that a plane tangent at any point is always perpendicular to the resultant of the gravitational acceleration due to the planet's mass and the centrifugal acceleration due to the planet's rotation. In theory, the planet's figure is an ellipsoid of revolution about the minor axis, that is, an oblate spheroid. As a result of oblateness, the poles of the orbital planes of its satellites precess about the direction of the pole of rotation of the planet. The more exact evaluations of oblateness from satellite motions take into account the internal distribution of mass in the planet. Satellite observations therefore yield information about the internal structure of the planets.

Reliable geometrical and dynamical determinations of f are available for Jupiter and Saturn; the determinations by Struve [32] are given in Table VI. For Mars, a dynamical determination has been given by Woolard [4]. For the earth, various methods have been used which are reviewed in Sections 3–1 and 5–9, where the theoretical relationship between the flattening and the gravitational and rotational properties of a planet are discussed in more detail. For Uranus, a dynamical study [33] yields $\frac{1}{20}$ for f, whereas direct observations yield $\frac{1}{14}$. In this case, both determinations are based on difficult observations. For Neptune, no definite observational evidence of flattening exists, but the inner satellite Triton shows a strong precession effect from which Jackson [34] has derived the oblateness quoted in Table VI.

The mean radius of an oblate spheroid is

$$\bar{r} = \frac{2A + B}{3}$$

since, out of three mutually perpendicular radii, two will lie in the equatorial plane and one will point toward a pole.

2–5 Periods of rotation and axial inclinations. In addition to the mass, radius, and oblateness, another important observationally determined characteristic of a planet is its period of rotation. The periods of rotation that will be discussed and which are presented in Table VI (appendix) are sidereal periods. The *axial inclination is the angle between the plane of the orbit and the plane of the planet's equator.* Sidereal rotational periods and axial inclinations have been derived in a variety of ways.

For Mercury, surface markings have been used. According to Antoniadi [35], Mercury rotates about its axis with the periodicity of its orbital revolution, always presenting the same face to the sun. However, the observations are very difficult to make because the observed markings are extremely faint (see Section 2–11), and Antoniadi's conclusions about the rotation are not generally accepted.

Photographic observations of Venus in ultraviolet light show bandlike markings which by their orientation suggest that Venus rotates about an axis inclined from the orbital poles by about $32° \pm 2°$ [36]. That the planet rotates is also suggested by the similarity of temperatures measured on the dark and the illuminated hemispheres of Venus, as discussed in Section 2–7. On the other hand, several attempts at measuring the period of rotation of Venus by spectroscopic methods indicate that the period cannot be short.

If a spectrogram of a planet is obtained while the slit of the spectrograph is oriented along the planet's equator, the Doppler effect produced by the rotation causes the spectral lines to be tilted in the spectrogram with respect to the direction of dispersion. The tilt is a function of the velocity of rotation and of the angle between the sun, the planet, and the earth. In the case of Venus, the equator has not been established definitely, but the observations have allowed for this uncertainty. Richardson [37] has recently made a study of this nature, and he has summarized the results of previous studies. A slow retrograde rotation is indicated, but the probable error of the measured equatorial rotational velocity is rather high. Richardson concludes (1) that the probability is 0.50 that the rotation is retrograde with a period between 8 and 46 days, or (2) that the probability is 0.94 that the direct period is longer than 14 days or the retrograde period longer than 5 days. In other words, although a rapid rotation is suggested by Venus' ultraviolet markings and by its uniformity in temperature, the spectroscopic method suggests a slow rotational period, possibly retrograde.

The Martian period of rotation and axial inclination are very accurately known from observations of surface markings [38]. Since these observations have extended over many years, the period is known with high precision. An error in the period would mean that, in n rotations, the timing of the crossing of a given Martian feature across the line earth-Mars or any other reference line would be off by n times the error. Conversely, the mean error with which the period is known (± 0.0026 sec) is equal to the mean error in the timing in the oldest reliable observation (1877) of the passage of a Martian feature across any convenient reference line, divided by the number of revolutions that have occurred since.

The period of rotation of Jupiter is the shortest among the planets. Seen through a telescope of moderate size, this planet shows cloud belts on which semipermanent markings or spots may be observed. The angular motions of these markings about the planet's axis indicate that the periods are variable, the shortest being at the planet's equator. The observations of Jupiter's markings have long been the task of amateur astronomers, who have compiled a wealth of data related to the rotation of this planet [39]. Unlike the sun, whose rotational period is found to vary smoothly with solar latitude, Jupiter shows zonal variations of rotational periods that, although more complex, are reminiscent of the earth's atmospheric circulation belts. The rotational period observed within $10°$ of the equator is about $9^{h}50^{m}$, while periods observed outside of this zone vary about a mean of $9^{h}55^{m}$.

Saturn also exhibits belts of cloudiness, but semipermanent markings of the kind observed in Jupiter are very rare. A few such markings have been observed, and these yield a period of $10^{h}14^{m}$ at the equator, $10^{h}21^{m}.4$ at latitude $12°30'$, and $10^{h}38^{m}$ at latitude $36°$ [40]. Spectroscopic observations by Moore [41] yield a period of $10^{h}2^{m} \pm 4^{m}$ at the equator, two per cent shorter than the visually determined period.

For the planet Uranus, the spectroscopic method [42] has yielded a rotational period of $10\frac{3}{4}$ hr. A periodic variation of the brightness of Uranus with a similar period has also been noticed [43]. Because of the smaller angular diameter of this planet, the tilt of the lines in the spectrogram is more difficult to measure than in the case of Saturn. Markings similar to those commonly observed in Jupiter and rarely seen in Saturn have not been found, although faint bands with too little detail to permit determination of the rotational period have been observed. The axis of rotation of Uranus lies almost in the planet's orbital plane. In the solar system, the vastly preponderant direction of planetary and satellite revolutions and rotations is such that, as seen from the north pole of the ecliptic, the motion is counterclockwise. The axis of Uranus is so inclined that what should be its north pole according to the "counterclockwise" convention actually lies south of the ecliptic. For this reason the axial inclination is usually quoted as 98° instead of 82°. The orientation of this axis can be determined from the orientations of the orbital planes of the satellites of Uranus, which do not show the orbital perturbations caused by the planet's equatorial bulge. These would be appreciable if the satellite orbits lay outside the equatorial plane. Since the orbital period of the planet is about 84 years, alternate poles of Uranus are observed from the earth every 42 years. In 1945 the south pole of the planet was facing towards the sun.

For Neptune, the spectroscopic method has yielded a rotational period of 15.8 ± 1 hr [44]. However, in the derivation of this period the older value for the radius of the planet was used, a value that has been revised downward by about 10% in more recent observations. This 10%-error is reflected directly in the period which, accordingly, should be taken as 14.2 ± 1 hr. Variations in the brightness of Neptune have been observed which suggest a period of $12^h43^m.5$ [45]. In view of the difficulties in the observation, the two results cannot be considered to be in serious disagreement.

Photometric observations of the planet Pluto show that its brightness varies periodically, indicating the rotational period of 6.39 days presented in Table VI [31].

Radio observations provide a possible method for determining the periods of rotation of the major planets. Detection of radio noise originating during storms in the atmospheres of Jupiter and Saturn has been useful for this purpose. Certain areas on these planets appear to be more active in originating radio noise than the rest of the planet, and extended observations may be used to determine the period of rotation [46]. For Jupiter a rotational period of $9^h55^m28^s.8$ has been obtained. According to the investigators, this may be the rotation period of Jupiter's solid surface. Initial results for Saturn suggest a period of 10.37 hr.

2–6 Densities, velocities of escape, and accelerations of gravity. These quantities can all be derived from the observational data described in the previous sections.

(a) *Densities:* The volume of an oblate spheroid is $\frac{4}{3}\pi A^2 B$, where A and B are, respectively, the equatorial and polar radii. The densities quoted in Table VI are derived from the masses in grams presented in that table and from the volumes computed with this formula. It is interesting that use of the mean radius, $\bar{r}$, to compute approximate volumes, $\frac{4}{3}\pi\bar{r}^3$, leads to an error of only 0.1% in the volume of Saturn, the most oblate of the planets. The inner planets, Mercury to Mars, have densities two or more times those of the major planets, Jupiter to Neptune. The density of Pluto is at present unknown, but the known densities of the larger satellites of the major planets, which, in size, are roughly comparable with Pluto, indicate that the density may be about 2.0.

(b) *Velocity of escape:* This quantity, v_e, also called *parabolic velocity*, determines the minimum initial velocity that a projected body would require to reach infinity, starting from the surface of a planet. Assuming spherical planets, we may compute the escape velocity by equating the initial kinetic energy, $\frac{1}{2}mv^2$, of the projected body, to the change in potential energy,

$$\int_R^\infty \frac{GMm}{r^2}\, dr = \frac{GMm}{R}, \tag{2–20}$$

that it will undergo as it is carried against the gravitational field from the surface of a planet of radius R and mass M to infinity. We find

$$v_e = \left(\frac{2GM}{R}\right)^{1/2} \tag{2–21}$$

or, in terms of the acceleration of gravity g,

$$v_e = (2g_e R)^{1/2}. \tag{2–22}$$

The escape velocities quoted in Table VI were derived from Eq. (2–22). The gravitational acceleration at the equator, g_e, used in Eq. (2–22) includes the centrifugal acceleration component.

The escape velocities play an important role in the theoretical determination of whether a planet can retain an atmosphere. Any molecule in the upper layers of the atmosphere of a planet where collisions are negligible would be lost to the planet if its velocity equaled or exceeded the escape velocity. Since at a given temperature there are always some fast-moving molecules, even at kinetic temperatures that are lower than can be derived from the escape velocity, a planet's atmosphere will suffer losses. The above arguments, applied quantitatively to Mercury as well as to the moon, lead to the conclusion that these bodies are practically devoid of any atmosphere.

(c) *Acceleration due to gravity:* According to the first-order theory developed by Clairaut in 1743 and discussed in Section 5–9, the gravitational acceleration, g_e, at the equator of a rotating oblate planet is given by Eq. (2–15). At geocentric latitude β, the acceleration is

$$g = g_e[1 + (\tfrac{5}{2}m - f) \sin^2 \beta], \qquad (2\text{--}23)$$

where m and f are the quantities defined in Eq. (2–8). In this relationship, the component due to the centrifugal acceleration caused by the planet's rotation is included. Equatorial accelerations due to gravity computed according to Eq. (2–15) are presented in Table VI. Gravitational acceleration is often expressed in *gal* units. One gal equals 1 cm·sec^{-2}.

2–7 Albedos and temperatures. The *albedo* of a planet is the ratio of the total amount of radiant energy reflected by the planet in all directions to the amount it receives from the sun. This is also called *Bond's albedo* to distinguish it from other albedo definitions which have been proposed in the literature [47].

Let S be the rate at which solar energy is received per cm^2 at a distance of 1 A.U. from the sun. A planet with a mean radius $\bar{r}$ located at a A.U. from the sun will then intercept per unit time the solar radiation given by

$$\frac{S\pi\bar{r}^2}{a^2}.$$

Let $P(O)$ be the brightness of the planet, or rate at which the reflected radiation is received per cm^2 by an observer at distance Δ (cm) from

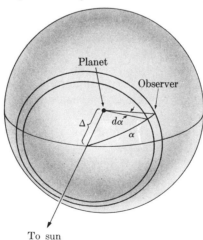

FIG. 2–4. Computation of the total brightness of a planet.

the planet, assuming the observer to be in the line sun-planet, that is, looking at the full sunlit face of the planet. If the planet shone in all directions with equal intensity, its total brightness would be $4\pi \Delta^2 P(0)$. For an observer at the same distance Δ from the planet but looking in the direction of phase angle α (sun-planet-observer), let the observed brightness be $P(\alpha)$. The phase function is then defined as

$$\phi(\alpha) = \frac{P(\alpha)}{P(0)}. \tag{2–24}$$

The total rate at which the planet reflects radiation in all directions is (see Fig. 2–4)

$$2\pi \Delta^2 P(0) \int_0^\pi \phi(\alpha) \sin \alpha \, d\alpha.$$

Bond's albedo, A, then is defined by the relationship

$$\frac{S\pi \bar{r}^2}{a^2} A = 2\pi \Delta^2 P(0) \int_0^\pi \phi(\alpha) \sin \alpha \, d\alpha,$$

or

$$A = \frac{\Delta^2 a^2 P(0)}{\bar{r}^2 S} 2 \int_0^\pi \phi(\alpha) \sin \alpha \, d\alpha.$$

Since one usually observes $P(\alpha_0)$ instead of $P(0)$, where α_0 is a given fixed phase angle, we may combine the last expression with Eq. (2–24) and obtain

$$A = \frac{\Delta^2 a^2}{\bar{r}^2 S} \frac{P(\alpha_0)}{\phi(\alpha_0)} 2 \int_0^\pi \phi(\alpha) \sin \alpha \, d\alpha. \tag{2–25}$$

For the inner planets and for the moon, $\phi(\alpha)$ can be determined since $P(\alpha)$ is observable over a large range of α-values. Values of the quantity

$$\frac{2}{\phi(\alpha_0)} \int_0^\pi \phi(\alpha) \sin \alpha \, d\alpha$$

are presented by Russell [47] for the moon, Mercury, and Venus. These are plotted in Fig. 2–5, together with the corresponding quantities for a diffusely reflecting sphere obeying Lambert's law. In this case, $\phi(\alpha)$ is given by

$$\phi(\alpha) = \frac{1}{\pi} [\sin \alpha + (\pi - \alpha) \cos \alpha]. \tag{2–26}$$

For Mars, $P(\alpha)$ can be observed up to $\alpha = 47°$ and, according to Russell, this is sufficient for a reliable determination of its albedo. For Jupiter and the other outer planets, $P(\alpha)$ is observed only at small values

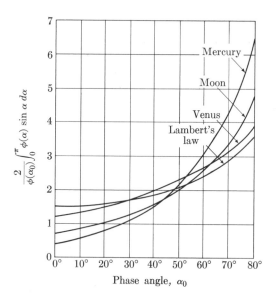

FIG. 2–5. Reflecting properties of some planets.

of α, and hence their phase functions are unknown. Jupiter and Saturn show, within the observable range of α's, a lesser decrease of $P(\alpha)$ with phase angle than does Venus, which suggests that Lambert's law may be used for the outer planets without serious error. This assumption has been made in deriving the albedos of the outer planets presented in Table VI. The values presented in that table are quoted from Kuiper's list [48]. The earth's albedo has been determined by observing the earthlight falling on the areas of the moon not exposed to direct sunlight.

For the theoretical determination of planetary temperatures, the albedos should be defined in terms of light of all wavelengths. Fortunately, since one is measuring reflected sunlight, measurements made over a relatively broad spectral band about the visual wavelengths give a good estimate of the total albedo.

A theoretical temperature for the reflecting surface of a planet may be derived from the blackbody radiation laws. Stefan's law states that the total radiant energy in ergs/sec/cm^2 emitted by a black surface is given by σT^4, where $\sigma = 5.672 \times 10^{-5}\,\mathrm{erg \cdot cm^{-2} \cdot sec^{-1} \cdot deg^{-4}}$ and T is the absolute temperature. It is known that the rate of solar energy received per cm^2 outside the earth's atmosphere and normal to it is about 1.94 calories/min or $1.35 \times 10^6\,\mathrm{ergs \cdot sec^{-1} \cdot cm^{-2}}$. This quantity is called the *solar constant*. The principal difficulty in evaluating the solar constant is the problem of allowing for the ultraviolet energy absorbed by the earth's atmosphere. The value quoted is due to Aldrich and Hoover [49], who

have estimated the absorption losses by means of rocket observations.[*]
From the solar constant we may obtain a blackbody temperature for the
sun. The solar constant multiplied by $4\pi D^2$, where D is one astronomical
unit in centimeters, is equal to the total rate of solar energy output. This
total rate divided by the surface area of the sun, $4\pi R_\odot^2$, then equals $\sigma T_\odot^4$.
The temperature derived in this manner is called an effective temperature,
and, in the case of the sun, the best data available yield 5778°K with a
mean error of $\pm 10°$ due to the lack of precision in the solar constant.

Letting a be the semimajor axis of a planet's orbit, $\bar{r}$ its mean radius,
and $R_\odot$ the solar radius, we obtain for the rate of radiant energy received
by a planet the quantity

$$\sigma T_\odot^4 \, \frac{R_\odot^2}{a^2} \, \pi \bar{r}^2. \tag{2-27}$$

If A is the planet's total albedo, then the rate of absorption of radiant
energy by the planet is

$$\sigma T_\odot^4 \, \frac{R_\odot^2}{a^2} \, \pi \bar{r}^2 (1 - A). \tag{2-28}$$

The absorbed energy is reradiated by the planet, since it must be in
equilibrium in the sense that it emits as blackbody radiation the radiant
energy that it absorbs from the sun. If we suppose the planet to be a
perfect conductor of heat or if it rotates so rapidly that it may be assumed
to radiate equally from its dark and sunlit sides, then the planet's effective
temperature is given by

$$T^4 = \frac{T_\odot^4}{4} \frac{R_\odot^2}{a^2} (1 - A), \tag{2-29}$$

or

$$T = \frac{T_\odot}{\sqrt{2}} \left(\frac{R_\odot}{a}\right)^{1/2} (1 - A)^{1/4}. \tag{2-30}$$

The last expression can be obtained by equating $4\pi \bar{r}^2 \sigma T^4$ to our previous
result. If the planet radiates preponderantly from its sunlit side, a better
approximation of its effective temperature is obtained by using $2\pi \bar{r}^2 \sigma T^4$
instead of $4\pi \bar{r}^2 \sigma T^4$. The result is

$$T = \frac{T_\odot}{(2)^{1/4}} \left(\frac{R_\odot}{a}\right)^{1/2} (1 - A)^{1/4}. \tag{2-31}$$

---

[*] A discussion of possible variations in the solar constant is given in Section 1-11.

Finally, if the planet is such a poor conductor that each unit of sunlit surface area radiates the same energy that it absorbs, we have

$$T = T_{\odot} \left(\frac{R_{\odot}}{a} \right)^{1/2} (1 - A)^{1/4}. \qquad (2\text{-}32)$$

This would be the effective temperature at the subsolar point of a poorly conducting planet. Values for the three effective temperatures just defined are presented in Table VI. In the computations the solar radius was taken as 6.96×10^5 km, and the mean distances given in Table V were used for a. These formulas neglect the well-known "greenhouse" or blanketing effects of planetary atmospheres, which may be appreciable. To illustrate these atmospheric effects, let us consider the earth, where the blackbody formulas predict a maximum temperature of 347° K at the subsolar point, a temperature that is much higher than the observed one of about 30°C or 303°K. The blackbody temperature predicted by Eq. (2–30) is 245°K, while the actual average surface temperature is 15°C or 288°K. Most of the radiation that escapes into space is radiated from the higher levels of the atmosphere.

Measurements of the infrared radiation received from several of the planets have been used to estimate their blackbody temperatures. Pettit and Nicholson [50] have described the method of observation and reduction. The radiation received on the earth from a planet consists of two parts: (1) the reflected sunlight which is mostly in the wavelength range from 0.3 to 5μ and (2) the planetary heat emitted by the planet after absorption which reaches the earth mostly through the water-vapor transmission band between 8 and 14μ. Very little planetary heat is radiated in the wavelengths where the reflected sunlight is received, and little reflected sunlight is contained in the transmission band at 8 to 14μ. The observational technique used by Pettit and Nicholson consisted of exposing a thermocouple to the radiation collected by a telescope. The response of the thermocouple was carefully calibrated in absolute units. Measurements of the thermocouple response were made with, and without, filters that favored transmission in one of the two spectral bands. The filters employed were either glass or a water cell, both of which are transparent to the 0.3 to 5μ band, and opaque to the 8 to 14μ band. By means of the filters, the energy of the planetary heat may be separated from that of reflected sunlight. The rate at which the telescope receives the energy component that we have called planetary heat may be used to compute the total rate of energy emitted in all directions by the observed body, and from the latter quantity the effective temperature may be derived. For planet Mercury, Pettit and Nicholson [51] find a temperature of

600°K for the subsolar point, a result that is in fair agreement with the temperature computed from Eq. (2–32).

Mercury always faces the sun, has no atmosphere (see Section 2–9), and if its surface is composed of silicates like that of the earth, it is a poor conductor. Therefore, Eq. (2–32) would be the most appropriate one to use in deriving Mercury's temperature. For Venus, Pettit and Nicholson [52] find that both the sunlit and the dark sides have nearly the same temperature, about 237°K, a result verified by Strong and Sinton [53]. These values are somewhat lower than expected from the theory presented here. That the difference is due to atmospheric effects in Venus has been pointed out by Adel [54]. Adel finds that absorption and emission bands of CO_2 affect the quality of the infrared radiation that we receive from this planet. The presence of CO_2 and other constituents in the atmosphere of Venus causes the infrared radiation that is received from the planet to be selective and banded, and in some cases these bands are absorbed by our own atmosphere. For Mars, Strong and Sinton [55] find a temperature of 303°K for the subsolar point and 203°K at the equator at sunrise. The temperatures of Jupiter, Saturn, and Uranus have been observed by Menzel, Coblentz, and Lampland [56], with results that are in rough agreement with the temperatures calculated theoretically.

Planetary temperatures may also be measured by means of radio-astronomical observations. Thermal radiations from Venus, Mars, and Jupiter at radio wavelengths have been detected [57]. Observations of Venus near conjunction made at centimeter wavelengths show temperatures of about 550°K with a probable error of approximately 10% of this value. This temperature is much higher than that indicated by previous estimates. Since at radio wavelengths the thermal radiation from the surface is not absorbed appreciably by the planet's atmosphere, we may conclude that this high temperature is that of the planet's surface. The lower temperatures obtained from observations at optical wavelengths must apply to layers above the clouds. A very strong greenhouse effect is indicated by these results. For Mars a temperature of 218° ± 50°K was reported during the opposition of 1956. For Jupiter the observations give variable results, suggesting that the observed radiation is nonthermal. How much of the observed radiation from all of these planets is of purely thermal origin is in question [58].

Physical theory permits the evaluation of temperatures from the observation of the intensities of certain spectroscopic features. Chamberlain and Kuiper [59] report for Venus a temperature of 285°K obtained in this manner. This result was determined from the intensities of the molecular bands of CO_2, and the temperature is appropriate for the levels above the clouds of Venus where the absorption of light by this gas takes place.

2–8 The brightness of the planets. As a measure of brightness astronomers use *magnitudes*. This unit originated in ancient times, when about 20 of the brightest stars were arbitrarily assigned to first magnitude and the faintest visible stars to sixth magnitude. A first-magnitude star is about as bright as a candle seen at a distance of one kilometer. Experiments with light sources placed at various distances were made by John Herschel in 1830 to show that the average first-magnitude star was 100 times brighter than one of sixth magnitude, and that light from stars differing by one magnitude had a constant ratio of intensity equal to about 2.5, regardless of magnitude. The constancy of the ratio is a result of Fechner's law, according to which the contrast sensitivity of the eye, or its ability to detect slight differences in brightness, is nearly constant over a wide range of illumination. Herschel's experiments led to a more precise definition of astronomical magnitudes by Pogson in 1850. A precise definition of magnitudes must specify the following quantities:

(a) The value of the illumination in absolute units or in terms of an arbitrary constant source of light that shall be called magnitude x. This defines the *zero point* of magnitudes.

(b) The change in illumination that corresponds to a change of one magnitude. This defines the *scale* of magnitudes.

Pogson proposed that the zero point should be such that magnitude 6.0 would agree with stars of the sixth magnitude in the *Bonner Durchmusterung*, a catalogue and atlas of stars to about the ninth magnitude, compiled in Germany in the last century. For a scale, Pogson proposed that a star should be of magnitude 1.0 if it is exactly 100 times brighter than one of magnitude 6.0. The ratio of the intensity of light for stars that differed by one magnitude was to be taken as $\sqrt[5]{100} = 2.5119$. The Pogson proposal is now in general use. It harmonizes the more exact modern observations with the older ones. A magnitude difference $m_1 - m_2$ between two stars is related to their light intensity ratio I_1/I_2 by

$$m_1 - m_2 = -K \log \frac{I_1}{I_2}, \tag{2–33}$$

where K is a positive constant. The negative sign on the right side of the equation arises from the fact that magnitudes *increase* for fainter stars. The Pogson scale definition may be used to evaluate K. Using logarithms to base 10 and remembering that the ratio in brightness between a first- and a sixth-magnitude star is 100, we have

$$-5 = -K \log 100. \tag{2–34}$$

Hence $K = 2.5$ exactly, and Eq. (2-33) may be rewritten as

$$m_1 - m_2 = -2.5 \log \frac{I_1}{I_2}.$$ (2-35)

According to the inverse-square law, objects of the same intrinsic brightness appear to have different magnitudes at different distances. If the magnitude of an object is known at distance r_1, this law permits the computation of the magnitude at distance r_2,

$$m(r_1) - m(r_2) = -2.5 \log \left(\frac{r_2}{r_1}\right)^2,$$ (2-36)

a relationship that assumes no absorption in the intervening space.

To distinguish between the observed brightness and the intrinsic brightness of stars, the terms *apparent* and *absolute* magnitudes are used. The magnitude discussed above is the apparent magnitude of a star. The absolute magnitude of a star is, by definition, the apparent magnitude it would have if observed from a standard distance of 10 parsecs, where 1 parsec equals 206,265 A.U.

For planets it is customary to list the apparent magnitude at some standard position of the planet with respect to the earth and the sun. For the inner planets the magnitude at elongation is usually listed, and for the outer planets the magnitude at opposition. The magnitude observed at other relative positions is a function of the distances sun-planet and planet-earth, as well as of the angle between sun and earth as seen from the planet, that is, the phase angle. It is also a function of the side of the planet which is exposed to sunlight.

The observed brightness of an object depends on the spectral sensitivity of the instrument used for measurement in addition to the other factors mentioned previously. An apparent magnitude measured by an instrument whose spectral sensitivity approximately matches the eye is called an *apparent visual magnitude*. *Apparent photographic magnitudes* are those obtained from photographic plates that have ordinary blue-sensitive emulsions.

In Table VI, apparent visual magnitudes at elongation and opposition are given. For magnitudes at other planetary positions, the reader is referred to Allen's compilation of planetary apparent-magnitude formulas [60]. The apparent magnitudes quoted are taken from Allen's list. As seen from the sun, the visual apparent magnitude of the earth would be about -3.8.

The most exact current method for the measurement of astronomical magnitudes uses photoelectric cells. A mean error of one hundredth of a magnitude is not uncommon in photoelectric photometry.

The difference between the apparent magnitudes obtained for two separate regions of the spectrum can be easily determined by photoelectric means. Such a difference is called a *color index*. The color index, defined as

(magnitude in blue light) — (magnitude in yellow light),

is commonly quoted in the literature [61]. The color index of a planet compared to the color index of the sun provides some information about the nature of the planetary atmosphere or surface.

2–9 Planetary atmospheres. According to the kinetic theory of gases, the mean speed of a gas molecule is a function of the temperature T and the mass m of the molecule and is given by

$$v = \sqrt{\frac{3kT}{m}}, \tag{2–37}$$

where k is Boltzmann's constant, 1.38×10^{-16} erg·deg^{-1}, and where T and m should be in degrees kelvin and grams, respectively, for v to be expressed in cm·sec^{-1}. Whether a planet can retain an atmosphere depends on the relationship between v and the velocity of escape, v_e. Equality of these velocities clearly means that the atmosphere will diffuse rapidly into space. Even if $v < v_e$, given sufficient time, a planet may lose its atmosphere. Estimates of the rate at which gases can be lost under different conditions have been made. Since the order of magnitude of the age of the earth is 10^9 years, computations can be made to determine the fraction of the atmosphere left at the end of this period of time for various values of the ratio v/v_e. A convenient criterion resulting from the computations [62] is that half of the original atmosphere will be lost in 10^9 years if $v = 0.2v_e$.

With Eq. (2–37) and the foregoing criterion, we may compute the temperature necessary for different objects in the solar system to retain certain molecules in their atmospheres. The results are given in Table 2–1. We may compare these temperatures with the planetary temperatures discussed in Section 2–7.

The proper temperatures to consider, however, are those of the layers in the atmosphere from which the molecules are most likely to escape. The *critical level* is the layer in an atmosphere from which a particle moving outward to infinity will have a probability of e^{-1} of suffering no collisions on its flight. Escaping molecules in an atmosphere will come from somewhere near the critical level. The atmosphere above this level is called the *exosphere*. In the case of the earth, the exosphere starts somewhere about 500 km from the surface, and in it the temperature is constant or isothermal, and equal to about 1500° to 2000°K [63].

TABLE 2–1

PLANETARY TEMPERATURES, IN °K, REQUIRED FOR
ESCAPE FROM PLANETARY ATMOSPHERES

	H_2	H_2O	N_2	O_2	CO_2	Xe
Mercury	56	500	779	889	1,220	3,650
Venus	339	3,030	4,720	5,390	7,410	22,100
Earth	403	3,600	5,610	6,400	8,810	26,300
Mars	83	739	1,150	1,310	1,800	5,390
Jupiter	10,700	95,500	148,000	170,000	233,000	696,000
Saturn	3,470	31,000	48,200	55,000	75,700	226,000
Uranus	1,480	13,200	20,500	23,400	32,300	96,200
Neptune	2,000	17,900	27,900	31,800	43,800	131,000

Table 2–1 indicates that it would be unlikely for Mercury to have kept any but the heavier elements present in any original atmosphere that it may have had. The temperature for molecular hydrogen listed for the earth explains the relative scarcity of this element. Hydrogen is the most abundant element found in the sun and in the universe at large. Little or no molecular hydrogen would be expected in the atmosphere of Mars. The major planets appear able to keep all the molecules listed.

This analysis assumes that at no time in the past were the temperatures so much above present-day temperatures that the rate of escape of the atmospheric constituents would have been radically different from the present one. The analysis also neglects the possibility of chemical reactions between the planet's surface and its atmosphere. These reactions could modify the composition, especially of the chemically more active constituents such as oxygen.

The determination of the composition of planetary atmospheres is a very complex problem. Spectroscopic observations and the available physical theory bearing on this subject have been reviewed by Kuiper [48] and by Urey [64]. Spectroscopic studies can be used to identify and to determine the abundance of some of the constituents of a planet's atmosphere. In these studies, one encounters a number of difficulties. In any observation made from the earth's surface, the absorption of the earth's atmosphere must be taken into account. The high content of water vapor and oxygen in the earth's atmosphere makes it difficult to detect small amounts of these constituents in other planetary atmospheres. When a sufficiently high dispersion is used in the spectrograph, the Doppler shift resulting from the relative radial velocity of earth and planet may permit some separation of the planet's absorption features from those originating in the earth's atmosphere. The transmission characteristics of the earth's atmosphere are such that only certain spectral regions are readily observable from the earth's surface. Some gases, like nitrogen and argon, do not absorb in the observable regions of the spectrum. Observations from artificial satellites or from high-altitude balloons will, of course, eliminate or minimize this difficulty.

Another problem in spectroscopic studies is the derivation of abundances from the observed strength of the absorption features. For this purpose, comparisons are usually made of the strength of certain absorption lines with that of similar ones produced in the earth's atmosphere or in laboratory transmission tubes. However, the radiation observed from a given planet has been *in toto* or in part back-scattered in the planet's atmosphere, and the net path length within the atmosphere may not be known. If most of the observed light is known to have been reflected from the planet's surface, as one may assume in the case of Mars, then the comparisons are simplified.

The observational techniques and the results obtained in high-dispersion studies of planetary spectra at the Mt. Wilson Observatory and elsewhere have been reviewed by Dunham [65] and in reference 66. Infrared spectroscopic observations of the planets and some satellites have been described by Kuiper [67]. Spectroscopic observations of Venus from a high-altitude balloon have been described by Strong [68].

Mercury appears to be devoid of an atmosphere. No bands or lines have been detected in its spectrum that differ from those observed in sunlight. No refraction of the light of stars grazing the limb of the planet's disk has been detected. This is the case also for the moon. Mercury and the moon have similar albedos and color indices. Upper limits for the densities of the atmospheres of these two bodies are 3×10^{-3} and 10^{-9}, respectively, in terms of the atmospheric density at the surface of the earth. Patches of haze have been reported on Mercury by various observers. This is a puzzling phenomenon if it is real. If Mercury always presents the same hemisphere to the sun, its dark side must be very cold. Low temperatures on the dark side are indeed indicated by the observed integrated planetary heat at various phase angles. This function agrees closely with that expected from a nonrotating and nonconducting smooth sphere whose absolute temperature at the surface away from the sun is zero [51]. As suggested by Kuiper [29], in such a case one would expect any vestiges of atmospheric gases to be either frozen on the permanently cold side or to have escaped from the hot side.

Absorption bands characteristic of CO_2 have been detected in the spectrum of Venus. Since the surface of Venus is not observed, these bands must originate in the upper layers of Venus' atmosphere, above the opaque, cloudlike layers. The precise amount of CO_2 is difficult to determine; different amounts are derived from the intensities of different absorption bands [69]. According to Van de Hulst [70], this is due to multiple scattering which causes the weak bands to appear stronger relative to the strong bands than is the case in a laboratory transmission tube. When spectral features are weak, the light received from the planet has penetrated deeper into its atmosphere and suffered more scattering than the light observed in strong lines or bands. Weak features are, as a result, enhanced relative to the strong features by the longer total absorption path inside the atmosphere. If the CO_2-bands observed on Venus are assumed to be formed in light reflected from the top of Venus' cloud layer, then above this layer, Venus contains the equivalent of a layer of CO_2 about 10^5 cm thick at normal temperature and pressure. By contrast, the entire CO_2-content in the earth's atmosphere could be reduced to a layer 220 cm thick, and the entire earth's atmosphere to a layer 8×10^5 cm thick at normal temperature and pressure. It thus appears that CO_2 is likely to be the main constituent in Venus' atmosphere.

Evidence for the presence of nitrogen has been found in the spectrum of the *dark* side of Venus, where emissions that correspond to the aurora and the nightglow on the earth (see Section 6–2) are observed. The composition of the clouds of Venus has been the source of much controversy. The yellow color of the planet suggests that the clouds which permanently cover the surface of the planet are not formed from water droplets. Such clouds are almost perfectly white.

Recently Strong [68] has described the results of spectroscopic observations of Venus made from a balloon at a height of 80,000 ft. During these observations, the 11,300-A bands of H_2O were detected. We quote: "The measured Venus water is about four times more water than lies in our own stratosphere (4 microns precipitable water above 48,000 feet over the British Isles), and is about the same as lies above high-level clouds on the earth. We presume there is much more water in the Venusian atmosphere below the cloud level."

Carbon dioxide has also been detected in Mars' atmosphere, but indirect evidence indicates that nitrogen (N_2) is the main atmospheric constituent [71, 72]. The polar caps have been identified as ice. White clouds are observed on Mars from time to time, and the light reflected from them is polarized to the same extent as that produced by clouds of ice crystals, such as cirrus clouds. Yellow clouds, probably dust clouds, have also been observed.

In all the major planets, the presence of methane or marsh gas (CH_4) has been detected spectroscopically. The presence of ammonia (NH_3) has been detected by similar means in Jupiter and in Saturn. Indirect evidence indicates the presence, in the major planets, of considerable hydrogen (H_2), nitrogen (N_2), and helium (He).

The *structure* of planetary atmospheres has been the object of some theoretical study. By structure is meant the relationship between temperature, T, pressure, p, and density, ρ, at each height, h, above the surface. The determination of atmospheric structures is so complex that very few definitive results are available. Since this is a field of research that is likely to be advanced by means of observations from outside the earth's atmosphere, we describe now some of the present lines of inquiry.

All theoretical studies of atmospheric structure are based on the *equation of hydrostatic equilibrium*. Imagine an atmospheric column with unit cross-sectional area. An increase in height from h to $h + dh$ in such a column will cause a drop of pressure, dp, which must equal the weight of the material (in the column) that is found between h and $h + dh$. If g is the acceleration due to gravity, the weight mentioned must be $g\rho\, dh$. Hence

$$dp = -g\rho\, dh, \tag{2-38}$$

the equation of hydrostatic equilibrium. As a first approximation to the theory of atmospheric structure, g may be taken as a constant throughout the atmosphere. In the case of the earth, g decreases by only 3% over the first 100 km of height.

The ideal gas law is

$$p = \rho\, \frac{R}{\mu}\, T, \tag{2-39}$$

where R, the universal gas constant, equals 8.314×10^7 ergs·gm^{-1}·deg^{-1}, and μ is the mean molecular weight in atomic mass units in which the oxygen atom equals 16. Combining Eqs. (2–38) and (2–39), we have

$$\frac{dp}{p} = -g\, \frac{\mu}{RT}\, dh,$$

which, when integrated, yields

$$p = p_0 e^{-\int_0^h (\mu g/RT)\, dh}. \tag{2-40}$$

Combining this result with Eq. (2–39), we obtain

$$\rho = \rho_0\, \frac{T_0}{T(h)}\, e^{-\int_0^h (\mu g/RT)\, dh}, \tag{2-41}$$

where μ is assumed constant and T_0 and $T(h)$ denote the temperature at the surface and at height h.

The quantity $RT/\mu g$, which has the dimension of length, is called the *scale height*. In the earth's atmosphere its value on the surface is about 8 km, but it depends on the amount of water vapor in the air. If the scale height is assumed to be constant, simple logarithmic relationships for the variation with height of the pressure and of the product density times temperature follow from Eqs. (2–40) and (2–41), i.e.,

$$\ln \left(\frac{p}{p_0}\right) = \ln \frac{\rho T(h)}{\rho_0 T_0} = -\frac{h}{H}, \qquad (2\text{–}42)$$

where H is the scale height.

Present knowledge permits only rough estimates of scale heights and of p_0 and ρ_0 in the atmospheres of the other planets. In the determination of scale heights, fairly accurate values of g are available, as described in Section 2–6. Approximate estimates of the temperatures can be made by the methods outlined in Section 2–7 or from other thermodynamic considerations which may yield the rate of variation of temperature with height. For example, in a turbulent atmosphere the *adiabatic lapse rate* may be assumed to hold. This is the quantity dT/dh that occurs when no loss or gain of heat by radiation or conduction is suffered by the volume elements of the gas as they are compressed or expanded in their turbulent motion. At present, the determinations of μ and p_0 are the most difficult problems to solve before one can make use of the foregoing relationships. The available estimates of μ are based on complex considerations about the origin and past history of the planets [48, 64]. The difficulties explain our imperfect understanding of how the earth's atmosphere was formed. Yet this is the only planet for which we have precise information about atmospheric composition, and where we can observe at least some of the factors that may produce variations in the composition.

If either p_0 or ρ_0 is known, the other can be determined from Eq. (2–39) provided T and μ are also known. Several determinations of p_0 for the planet Mars have been reviewed by De Vaucouleurs [71]. As an illustration of the methods used to obtain p_0, we outline one of the derivations by Dollfus [73].

Dollfus has estimated the ratio, B_a/B_s, of the brightness of the atmosphere of Mars to that of its surface, or the proportion of visual radiant energy scattered by the atmosphere to that reflected by the surface. In the computations, use is made of a number of approximate relationships. It is assumed that Mars' atmosphere is plane-parallel, optically thin, and composed of particles that scatter light isotropically according to Rayleigh's theory of molecular scattering, i.e., the scattering is proportional to λ^{-4}. Certain empirical formulas based on laboratory experiments are also used. The equations relate the variation of polarization across the disk at various phase angles to B_a/B_s. The polarization is

$$P = \frac{I_\perp - I_{||}}{I_\perp + I_{||}}, \qquad (2\text{–}43)$$

where $I_\perp$ and $I_{||}$ are respectively the intensities of polarized light, with electric field vectors perpendicular and parallel to the plane sun-planet-earth. The observed polarization at various points on the disk and at different wavelengths is then used to derive a value of about 0.028 for B_a/B_s. Various combinations of the observational data and formulas, however, yield values for this ratio in the range 0.018 to 0.036. From measurements of the integrated brightness of Mars (surface plus atmosphere), the brightness of Mars' atmosphere is then obtained. Finally, Dollfus makes an estimate of what the brightness of the earth's atmosphere would be if placed on Mars, and he concludes that the Martian atmosphere has $\frac{1}{4}$ the mass of that of the earth. Taking into account the lesser gravitational acceleration on Mars, a pressure, p_0, of 90 millibars is obtained.

The rigorous solution of such problems as the variation of brightness with phase angle, the observed polarization of the planet's light, the variation of brightness across the observed disk, and the formation of absorption lines in a planetary atmosphere requires the integration of an *equation of transfer* which is now presented. In a field of radiation, if we wish to specify the rate at which energy is transmitted along a beam of radiation, we must first define a unit area perpendicular to the central ray of the beam and a unit solid angle surrounding this ray. The radiant energy flowing per unit time across the unit area in directions confined within the unit solid angle is called the *specific intensity* and is usually denoted by I. If we wish to talk about radiation within a given range of frequency, then the specific intensity should also be defined for a unit frequency interval. It is then denoted by I_ν.

In studying the behavior of radiation in a volume filled with absorbing, scattering, or emitting particles, we must take into account the interaction of matter and radiation. For this purpose an absorption and an emission coefficient are defined as follows.

Numerous experiments have shown that absorption and scattering reduce the specific intensity of a beam according to

$$dI_\nu = -\kappa_\nu I_\nu \rho \, ds, \tag{2-44}$$

where dI_ν represents the loss of specific intensity along the distance ds in a medium of density ρ. The value of the *mass absorption coefficient*, κ_ν, depends on the nature of the absorbing matter. The energy loss may be due to scattering or to *true absorption*. In the latter case, the absorbed energy will result in atomic or molecular excitation or ionization, or molecular dissociation.

The specific intensity may, on the other hand, be increased as the beam traverses ds, by energy emitted in the right directions. This energy may arise from *true emission*, i.e., from de-excitation or recombination, or from

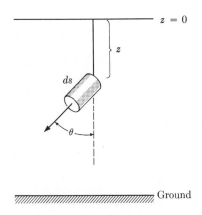

FIG. 2–6. Definitions of z and θ in a plane-parallel atmosphere.

radiation scattered into the beam from other directions. The *emission coefficient*, j_ν, is used to define the gain in intensity,

$$dI_\nu = j_\nu \rho \, ds. \qquad (2\text{–}45)$$

According to the principle of energy conservation, the net change in specific intensity along the axis of an elementary cylinder of unit base area and length ds is given by

$$dI_\nu = -\kappa_\nu I_\nu \rho \, ds + j_\nu \rho \, ds. \qquad (2\text{–}46)$$

This is called the *equation of radiative transfer* or commonly the *equation of transfer*. In connection with this equation, it is usual to define a *source function*, J_ν, by

$$\frac{j_\nu}{\kappa_\nu} = J_\nu. \qquad (2\text{–}47)$$

In astronomy, Eq. (2–46) arises in problems related to the flow of radiation through an atmosphere. It is convenient, then, to use the linear depth, z, below some arbitrary atmospheric level, and also to specify the direction of a given beam by the angle θ, its angular distance from the vertical (see Fig. 2–6). Then, assuming the atmosphere to be plane-parallel, we have

$$ds = \frac{dz}{\cos \theta}.$$

Furthermore, the *optical thickness*, τ_ν, from $z = 0$ to z is defined by $d\tau_\nu = \kappa_\nu \rho \, dz$ or by

$$\tau_\nu = \int_0^z \kappa_\nu \rho \, dz. \qquad (2\text{–}48)$$

Thus, from Eq. (2–44) the total attenuation of intensity across z is $e^{-\tau}$.

By means of Eqs. (2–47) and (2–48), the equation of transfer may be written

$$\mu \frac{dI}{d\tau} = -I + J, \qquad (2\text{–}49)$$

where $\mu = \cos \theta$. The subscripts ν have been dropped since the equation also holds without loss of generality for radiation of all frequencies.

If the level $z = 0$ is taken as that above which absorption and emission are negligible, then for a planet, at $\tau = 0$, and for $\cos \theta < 0$, Eq. (2–49) has the boundary condition $I = I_0$, which is the intensity of the incident solar illumination.

The emergent intensity at a given point on a planet could be obtained by integration of the equation of transfer if one knew the source function at different optical depths, the total optical depth of the atmosphere, and the reflecting characteristics of the planet's surface. In planetary atmospheres, absorption and emission are produced principally by scattering. The derivation of the source function in the case of multiple scattering has proved difficult. The most thorough analyses of radiative transfer in planetary atmospheres have been made by Chandrasekhar [74] and by Van de Hulst [75]. A convenient summary of the results has been presented by Van de Hulst [70].

The computations are based on the assumption of some suitable analytical form for the phase function of the scattering particles and for the reflectivity of the planetary surface. The theory is easier in the case of a very opaque atmosphere, such as that of Venus. The mathematical analog of an opaque atmosphere is called a *semi-infinite* atmosphere. For semi-infinite atmospheres, Chandrasekhar's solutions of Eq. (2–49) yield the variation of intensity across the planet's disk for any given phase angle. The solutions are limited to simple scattering processes, such as isotropic scattering and Rayleigh scattering. In the latter* case, Chandrasekhar's theory yields the variation of polarization across the disk. The theory, however, appears too simple to account for the polarization observed on Venus [70]. On the other hand, for the earth's atmosphere, the theory has been very successful in explaining the observed distribution of intensity and polarization in the sunlit sky.

Reflection from the ground, which is important on Mars, has also been approximately taken into account in the computations. Van de Hulst [70] has presented the modifications required in the analysis. A comparison has been made by Blackadar [76] of the intensity variation across Mars' disk with the predictions of the theories of Chandrasekhar and Van de

---

* The phase functions of scattering by a single particle for the two planes of polarization defined in relation to Eq. (2–43) are in this case

$$\phi_{||} = \tfrac{3}{4} \cos^2 \theta, \qquad \phi_{\perp} = \tfrac{3}{4}.$$

Hulst. On the assumption of perfect isotropic scattering, the comparisons yield a total optical thickness of 0.10 to 0.15 for the atmosphere of Mars in blue light. In the earth's atmosphere (see Table 3–2) the corresponding optical thickness is about 0.30.

In the interpretation of polarization studies, some of the theoretical difficulties may be overcome by using the empirical results of Lyot [77]. Lyot finds that the polarization, P, observed in scattering media in the laboratory is mostly due to radiation that has been scattered once. Multiple scattering results in little additional polarization. Except at grazing reflection angles, the observed polarization is not sensitive to the orientation of the scattering surface and, in practice, depends only on the phase angle of the observed object.

Lyot made use of these conclusions to show that the polarization observed on Venus at visual wavelengths, in light from the entire planet, varies with phase angle in a manner resembling that expected if the scattering is produced by water droplets with a diameter of 2.5μ. However, Kuiper [67], who observed the polarization in Venus' atmosphere at wavelengths of 1 and 2μ rules out water droplets as the scattering medium; he finds a better fit for the observations with the polarization expected from polymerized carbon suboxide (C_3O_2).

2–10 Satellites of the planets: Saturn's rings. No satellites of Mercury and Venus are known. According to Kuiper's [78] theory of the origin of the solar system, satellites were formed in a flat disk of material that surrounded some of the protoplanets. If the periodicity of rotation of Mercury is equal to that of its revolution, it is unlikely that this planet ever had sufficient angular momentum of rotation to form the disk from which satellites might form. Therefore, it is only by a rather unlikely accidental capture of an asteroid (see Chapter 6) that Mercury could acquire a satellite.

A similar argument applies to Venus if the periods of rotation and revolution are identical, a point that has not yet been settled. If a satellite of Venus brighter than visual magnitude 12 existed, it would very probably have been discovered already. Assuming the albedo of a hypothetical Venus satellite to be the same as that of Mercury or the moon, we can estimate how large a satellite could be and remain undiscovered. The result is that no satellite of Venus larger than 100 km in radius can exist.

The moon is extraordinary in being the largest (relative to its parent planet) known satellite. The view we have of its surface is the clearest of any object in the solar system, apart from the earth. Table VII of the appendix contains the data describing the orbital and physical characteristics of the satellites. A detailed account of the nature of the moon's surface is presented in Chapter 3.

The satellites of the planets are designated either by name or by Roman numerals. The latter follow either the order of discovery of the satellites (Jupiter), or the order of distance from the planet (Saturn), or both (Uranus).

Several excellent tables of the known orbital and physical properties of the satellites are available, for example, in references 79, 80, and 81. In addition, the *American Ephemeris* contains some orbital information as well as predictions of the positions of the satellites in their orbits as functions of time.

The mean distances and periods of the satellites have been derived from micrometric or photographic observations extending over many years. The figures presented in Table VII have been subjected to reductions which are based on the planetary parameters presented in Table VI. The number of significant digits gives an indication of the accuracy with which the various quantities are known. The orbital elements of the satellites in general suffer changes due to the ellipticity of the parent planet and to mutual attractions among the satellites themselves. Hence the quoted figures are to be taken as averages. In a number of cases, definitive studies have been made which permit the determination of the elements at any given epoch. References to these definitive studies, where available, are given in Table VII. The mutual attraction of the satellites permits the calculation of masses for a number of them. Where available, these results are included in the table. The mass of Rhea is not very accurately known. Original papers on satellite masses are listed in reference 14.

The radii of a number of satellites have been determined directly by measuring the angular diameter that they show in telescopic observations. The results quoted by Camichel [18] are included in the table. Jupiter I and II (Io and Europa) are only slightly smaller than the moon. Ganymede and Saturn's Titan are about equal to Mercury, while Callisto is somewhat smaller. Where radii and masses are available, density estimates can be made. Also, the radii and the brightness of satellites permit the computation of albedos. Some indirect estimates of satellite sizes can be made from plausible assumptions about the albedo and observations of the brightness. These estimates are not presented in the table, but are given below whenever the size is of interest.

Mars' two known satellites are the smallest known natural satellites in the solar system. The diameters of Phobos and Deimos, as estimated from their brightnesses, are 15 and 10 km, respectively. Phobos, the inner satellite, has an orbital period shorter than Mars' period of rotation, a unique case in the solar system, if man-made satellites are excluded. This causes Phobos to rise in the west and set in the east as viewed from the planet.

Jupiter has four major satellites (the Galilean satellites) and eight known minor ones. The major satellites and the inner satellite, Jupiter V (JV), have small orbital inclinations to the planet's equator and also small eccentricities. Photometric observations [82, 83] suggest that the major satellites always turn the same face toward Jupiter. The distribution of mean distances of the satellites of Jupiter shows three distinct groups. The Galilean satellites and JV have mean distances less than 2 million kilometers. A second group consisting of satellites JVI, JX, and JVII has mean distances somewhat less than 12 million kilometers. Finally, satellites JXII, JXI, JVIII, and JIX are grouped at about 23 million kilometers. The orbits in the two outer groups are appreciably eccentric and exhibit high inclinations to the planet's equatorial plane. As seen from the ecliptic north pole, all satellites in the outer group have retrograde motions instead of the counterclockwise revolution which is usually found throughout the solar system. Inclination values above 90° are used to indicate retrograde orbital motion.

These satellites are subject to strong solar perturbations. For satellites JVIII and JIX, the eccentricities have been found to vary between 0.25 and 0.50 during a single revolution, while the semimajor axes vary by nearly 10% in the same interval. Thus the orbits of these satellites are not even approximately closed ellipses, and the orbital elements quoted can be regarded as only averages. The outer satellites, in the course of their irregular motion, can get so far away from Jupiter that the solar attraction will predominate. If this should occur while a satellite is in conjunction, the satellite may become lost to Jupiter. It appears likely that the outer group of satellites consists of transient members of the Jovian family which were captured from the asteroid belt (see Chapter 6) and may be lost again. Diameters between 20 and 30 km have been estimated for satellites JIX, JX, and JXI [84].

The infrared spectra of JII and JIII show a drop in intensity with increasing wavelength at 1.5μ. This is a characteristic also of the spectra of H_2O-frost and snow and indicates that these satellites may be covered, at least in part, by these substances [67].

Saturn has nine known satellites in addition to the well-known system of rings. Their orbits, except for SVII and SIX, are appreciably circular. The two outer satellites show slight inclinations to the equatorial plane, and the outermost satellite has a retrograde orbital motion.

According to Kuiper [85], the spectrum of Titan shows bands of methane, demonstrating the presence of an atmosphere. Kuiper's survey included all large satellites in the solar system as well as Pluto. Apart from Titan, negative or inconclusive results were obtained in attempts to find an atmosphere on these other bodies.

Saturn's rings (see Plate II) are composed of separate particles which orbit independently in the plane of the planet's equator, a fact shown clearly by the period of revolution in different parts of the ring. Observations by spectroscopic methods show that the periods agree with those expected from Kepler's third law. There are three separate rings: a faint outer ring, generally known as Ring A, whose outer and inner diameters respectively are 271,000 and 238,000 km; a middle bright ring (Ring B) with corresponding dimensions of 233,000 and 180,000 km; and a faint inner one (Ring C), also called the "Crape Ring," measuring 121,500 and 107,500 km in outer and inner diameters. *Cassini's division* separates rings A and B, and in addition to the space separating rings B and C, several faint divisions have been observed in rings A and B. The satellite nature of the rings is confirmed by the fact that these divisions fall at mean distances where orbiting particles would have periods amounting to simple fractions of the periods of Saturn's inner satellites. A particle orbiting in a division would suffer strong perturbations which would remove it to a more stable mean distance. When the rings are seen edgewise, they become practically invisible, a fact indicating that the orbiting particles occupy a very flat volume of space.

The satellites of Uranus revolve in the planet's equatorial plane. Their revolution is in the same sense as that of the planet's rotation, which, as the reader will recall, has an inclination of axis equal to 98°. No mass determinations for the satellites of Uranus are available.

Neptune's inner satellite, Triton, is sufficiently massive to cause this planet to show an appreciable wavy path along its orbit. This effect is produced by the revolution of Neptune about the center of gravity of the system. Observations of this effect by Alden [86] have permitted a derivation of the ratio of the mass of Triton to that of Neptune. Triton's orbit is retrograde. Triton's brightness indicates that it has a large diameter, estimated at about 4500 km. The orbit shows perturbations due to the equatorial bulge of the planet. Observations of the orbital motion permit the evaluation of the oblateness of the planet. The outer satellite, Nereid, has the most eccentric satellite orbit known in the solar system, the eccentricity being 0.749. The orbital motion is direct.

2-11 Surface features on the planets Mars and Mercury. The only planets, apart from the earth, on which surface details have been observed are Mars and Mercury. Concise summaries of the observations made of these planets are given in references 35, 79, 87.

(a) Mars (see Plate III): Light and dark areas of a more or less permanent nature have been observed on Mars. The planet as a whole has an orange color. Through the telescope this appears to be the color of the lighter areas. The darker areas cover about one third of the surface and show seasonal variations in intensity. They are usually more prominent during the spring for the particular hemisphere in which the region may lie. Comparisons of photographs taken in blue and in infrared light show that there are no seas on Mars. Thus the dark regions are not due to surface waters.

Seasonal variations in the color of the dark areas (yellowish green to bluish green in the spring, and brown in the winter) have been reported by many observers. However, other observers report neutral shades. It is possible that the adjacent bright orange areas cause the observer to see complementary hues in the darker region. Nevertheless, the reported color changes have long been considered suggestive of the existence of plant life on Mars.

Kuiper [88] has compared the infrared reflection spectrum of the dark areas of Mars in the region of 0.4 to 2.4μ with that of the higher forms of plant life (spermatophyta and pteridophyta). These plants show absorption bands characteristic of chlorophyll and liquid water. The bands are absent in the spectra of Mars' dark areas.

Sinton [89] has studied the reflection spectrum of the dark areas at wavelengths ranging from 2.7 to 3.8μ. He finds absorption bands at 3.43, 3.56, and 3.67μ which are absent in the spectrum of the brighter areas. These bands fit very closely those found in the spectra of the lichen Physcia and the alga Cladophora, and have been tentatively identified as due to carbohydrates. Although one cannot conclude from this evidence that lichens and algae of terrestrial type exist on Mars, Sinton's result points strongly to the existence on Mars of some form of plant life.* The precise nature of this plant life remains unknown for the present. Higher forms of plant life similar to those found on the earth are excluded not only by Kuiper's observations but also by the rigors of the Martian climate.

McLaughlin [90] finds that the dark markings on the southern equatorial belt quite often suggest by their shapes and orientations that they are deposits of material which originate at discreet points and are spread in a systematic fashion by winds. He concludes that volcanic activity is present on Mars, and that the dark areas may be areas covered by volcanic ash. Furthermore, McLaughlin suggests that these deposits of ash may be the most fertile areas on Mars and the only places capable of supporting plant life.

In 1877, fine straight or slightly curved lines on the surface of Mars were detected by Schiaparelli, who called them *canals*. Numerous independent observers have reported seeing the canals, but their descriptions exhibit striking agreements as well as disagreements. Some of the most

---

* An alternative interpretation has recently been advanced by C. C. Kiess, S. Karrer, and H. K. Kiess (*Publ. Astron. Soc. Pacific*, **72**, 256, 1960). These authors point out that the absorption bands observed by Sinton may be caused by N_2O_4. Furthermore, many of the observed Martian phenomena may be interpreted as the result of reactions involving nitrogen compounds.

experienced observers do not agree that these fine features are lines. For example, F. G. Pease, who observed Mars for several years with the 100-in. telescope at Mt. Wilson [91], writes: "There are times when a geometrical structure may appear on the planet and even be photographed. When the seeing [see Section 3–3] improves, however, this structure disintegrates into the familiar forms prevalent in nature. . . . The fine threadlike lines to the south [of Solis Lacus], and others extending to the north and east, were meandering in form, but poor seeing turned them into an apparently curved line." Barnard [92], another experienced observer, reported similar results with the 60-in. telescope at Mt. Wilson. The results of Antoniadi [93] and Dollfus [94], both noted observers of Mars, are similar.

The white polar caps of Mars and their seasonal variability are the most easily seen phenomena on Mars' surface. During the winter a polar cap may reach as far as latitude 60°. As spring advances, the cap diminishes in extent and may break up into several distinct patches. Near the end of summer, the cap is very small, but it does not usually vanish. As the cap breaks up, some of the rifts and the white patches appear in the same location in different years. This is indicative of irregularities in the topography of the polar areas.

The composition of the polar caps has been studied by Kuiper [48] by means of infrared spectra. The spectra of terrestrial frost and snow show a drop in intensity with increasing wavelength at 1.5μ. The spectra of the polar caps of Mars show the drop in intensity with increasing wavelength at 1.5μ characteristic of terrestrial frost and snow. The spectrum of frozen CO_2, which has also been considered as a possible constituent, does not show this drop in intensity. Kuiper concludes that the polar caps of Mars are almost certainly composed of H_2O-frost at temperatures much below 0°C.

The bright areas of Mars appear to be rather dusty regions. The yellow clouds which are observed from time to time can be explained as dust carried aloft from these areas. The polarization characteristics in the reflected light of Mars have been studied by Dollfus [73]. For the bright areas of Mars, Dollfus finds that the surface must be covered by dust.

De Vaucouleurs [71, 87] has reviewed the infrared spectrophotometric and radiometric observations in the bright areas of Mars. The presence of silica compounds is indicated by these observations. The available data are consistent with the hypothesis that the bright areas of Mars are desertlike regions.

The International Astronomical Union [95] adopted in 1958 a nomenclature for the more permanent Martian features. Many of these features have been assigned proper names according to the customary usage in the past. These names often consist of the Latin or Greek forms of various terrestrial geographical names. Smaller features are to be designated by longitude and latitude rather than by proper names. Longitudes and latitudes for Mars are defined in the same manner as for the earth. Zero longitude is a meridian that passes through one of the permanent dark features appropriately called *Sinus Meridiani*. Among the more prominent darker areas in Mars are: Syrtis Major (longitude 290°, latitude +10°), Sinus Sabaeus (340°, −8°), Mare Erythrarum (40°, −25°), and Mare Acidalum (30°, +45°).

(b) Mercury: The surface markings on Mercury have been described by Antoniadi [35] whose description, quoted below, is in agreement with the observations of other experts. Dark and light areas are observed which resemble those on the moon. "The dark areas were in general very pale and difficult to distinguish, but they were real, and their coloration appeared grayish, like that of the lunar seas.... "The dark and light areas observed [independently] by me appeared fixed with respect to the terminator through intervals of several hours in the same day even though the air in the afternoon was in general less calm than in the morning, and they accompanied from one day to the next, the terminator in its constant movement to the left in the telescopic image." Since Mercury is so close to the sun, observations are best made during daylight hours. From his observations, Antoniadi was able to conclude that Mercury always presents the same side to the sun.

Canals have also been reported on Mercury, but Antoniadi rejects their reality.

References

1. S. Newcomb, *The Elements of the Four Inner Planets and the Fundamental Constants of Astronomy.* Washington: U.S. Government Printing Office, 1895.

2. W. de Sitter, "On the System of Astronomical Constants," *Bull. Astron. Inst. Netherlands,* **8,** 213, 1938.

3. Colloques Internationaux du Centre National de la Recherche Scientifique, *Constantes Fondamentales de l'Astronomie,* p. 3. Paris: C.N.R.S., 1950.

4. E. W. Woolard, "The Secular Perturbations of the Satellites of Mars," *Astron. J.,* **51,** 33, 1944.

5. H. Spencer Jones, "Dimensions and Rotation," *The Earth as a Planet,* p. 21, G. P. Kuiper, editor. Chicago: University of Chicago Press, 1954.

6. E. Rabe, "Derivations of Fundamental Astronomical Constants from the Observations of Eros During 1926–1945," *Astron. J.,* **55,** 112, 1950.

7. A. H. Cook, "Developments in Dynamical Geodesy," *Geophys. J. Roy. Astron. Soc.,* **2,** 238, 1959.

8. J. V. Evans and G. N. Taylor, "Radio Echo Observations of Venus," *Nature,* **184,** 1358, 1959.

9. P. R. HEYL, "A Redetermination of the Constant of Gravitation," *J. Research Natl. Bur. Standards*, **5**, 1243, 1930.

10. See p. 8 of reference 5.

11. G. M. CLEMENCE, "On the System of Astronomical Constants," *Astron. J.*, **53**, 173, 1948.

12. H. C. HERTZ, "The Mass of Saturn and the Motion of Jupiter 1884–1948," *Astronomical Papers Prepared for the Use of the American Ephemeris and Nautical Almanac*, **15**, 188, 1953.

13. G. VAN BIESBROECK, "The Mass of Neptune from a New Orbit of its Second Satellite Nereid," *Astron. J.*, **62**, 272, 1957.

14. G. P. KUIPER, "Titan: A Satellite with an Atmosphere," quotation of a letter by D. Brouwer, *Astrophys. J.*, **100**, 380, 1944.

15. W. W. CAMPBELL, "A Determination of the Polar Diameter of Mars," *Astron. J.*, **15**, 145, 1895.

16. A. DOLLFUS, "Le Diamètre des Satellites de Jupiter, de Titan, et de Neptune, Déterminé par le Micromètre Biréfringent Avec un Grand Pouvoir Séparateur," *Compt. Rend.*, **238**, 1475, 1954.

17. HENRI CAMICHEL, "Erreur Systématique Sur la Mesure des Diamètres des Petits Astres avec le Micromètre à Double Image," *Annales d'Astrophysique*, **21**, 217, 1958.

18. HENRI CAMICHEL, "Nouvelle Méthode de Mesure des Diamètres des Petits Astres et ses Résultats," *Annales d'Astrophysique*, **16**, 41, 1953.

19. G. P. KUIPER, "The Diameter of Pluto," *Pub. Astron. Soc. Pacific*, **62**, 133, 1950.

20. F. E. ROSS, "Photographs of Venus," *Astrophys. J.*, **68**, 83, 1928.

21. W. RABE, "Untersuchungen über die Durchmesser der grossen Planeten," *Astron. Nachr.*, **234**, 153, 1926.

22. R. A. SAMPSON, "Theory of the Four Great Satellites of Jupiter," *Memoirs Roy. Astron. Soc.*, **63**, 258, 1921.

23. P. MULLER, "A Propos du Diamètre de Mercure," *L'Astronomie, Bull. Soc. Astron. France*, **69**, 82, 1955.

24. G. DE VAUCOULEURS, "Remarks on Mars and Venus," *J. Geophys. Research*, **64**, 1739, 1959.

25. ROBERT J. TRUMPLER, "Observations of Mars at the Opposition of 1924," *Lick Obs. Bull.*, **13**, 19–45, 1927.

26. H. CAMICHEL, "Détermination Photographique du Pôle de Mars, de son Diamètre et des Coordonnées Aréographiques," *Bulletin Astronomique*, **18**, 83–174, 1954.

27. A. DOLLFUS, "Un Micromètre à Double Image Permettant un Grand Dédoublement," *Compt. Rend.*, **235**, 1479, 1952.

28. G. P. KUIPER, "The Diameter of Neptune," *Astrophys. J.*, **110**, 93, 1949.

29. G. P. KUIPER, "Planetary and Satellite Atmospheres," *Reports on Progress in Physics*, Vol. XIII, p. 247, 1950. The Physical Society, London.

30. DINSMORE ALTER, GEORGE W. BUNTON, and PAUL E. ROQUES, "The Diameter of Pluto," *Pub. Astron. Soc. Pacific*, **63**, 174, 1951.

31. MERLE F. WALKER and ROBERT HARDIE, "A Photometric Determination of the Rotational Period of Pluto," *Pub. Astron. Soc. Pacific*, **67**, 224, 1955.

32. H. STRUVE, "Beobachtungen der Saturnstrabanten am 30-zölligen Pulkowaer Refraktor," *Publications de l'Observatoire Central Nicolas, Ser. 2*, **11**, 233, 1898.

33. Ö. BERGSTRAND, "Sur la Figure et la Masse de la Planète Uranus, Déduites des Mouvements des Deux Satellites Intérieurs," *Compt. Rend.*, **149**, 333–336, 1909.

34. J. JACKSON, "Note on the Figure and Rotation Period of Neptune," *Monthly Notices Roy. Astron. Soc.*, **86**, 295, 1926.

35. E. M. ANTONIADI, *La Planète Mercure*, p. 35. Paris: Gauthier-Villars, 1934.

36. G. P. KUIPER, "Note on the Determination of the Pole of Rotation of Venus," *Astrophys. J.*, **120**, 604, 1954.

37. ROBERT S. RICHARDSON, "Spectroscopic Observations of Venus for Rotation Made at Mount Wilson in 1956," *Pub. Astron. Soc. Pacific*, **70**, 251, 1958.

38. J. ASHBROOK, "A New Determination of the Rotation Period of the Planet Mars," *Astron. J.*, **58**, 145, 1953.

39. B. M. PEEK, *The Planet Jupiter*. New York: Macmillan, 1958.

40. W. H. WRIGHT, "An Account of Some Photographic Observations of the Bright Spot on Saturn, and General Remarks on Saturn's Rotation," *Pub. Astron. Soc. Pacific*, **45**, 236, 1933. See also: H. CAMICHEL, "Mesure de la Dureé de Rotation de Saturn d'après une Tâche," *Bulletin Astronomique*, Paris, **20**, 141, 1956.

41. J. H. MOORE, "Spectroscopic Observations of the Rotation of Saturn," *Pub. Astron. Soc. Pacific*, **51**, 274, 1939.

42. PERCIVAL LOWELL, "Spectroscopic Discovery of the Rotation Period of Uranus," and V. M. SLIPHER, "Detection of the Rotation of Uranus," *Bull. Lowell Obs.*, **2**, no. 53, 1912.

43. L. CAMPBELL, "Variability of Uranus," *Astron. Obs. Harvard Coll., Circulars*, no. 200, 1917.

44. J. H. MOORE and D. H. MENZEL, "Preliminary Results of Spectroscopic Observations for Rotation of Neptune," *Pub. Astron. Soc. Pacific*, **40**, 234, 1928.

45. O. GÜNTHER, "Der Rotationslichtwechsel des Neptun," *Astron. Nachr.*, 282, 1, 1955.

46. HARLAN J. SMITH and J. N. DOUGLAS, "Observations of Planetary Nonthermal Radiation," *Paris Symposium on Radio Astronomy*, p. 53, R. N. Bracewell, editor. Stanford: University Press, 1959. See also: T. D. CARR, A. G. SMITH, R. PEPPLE, and C. H. BARROW, "18-Megacycle Observations of Jupiter in 1957," *Astrophys. J.*, **127**, 274, 1958.

47. H. N. RUSSELL, "On the Albedo of the Planets and Their Satellites," *Astrophys. J.*, **43**, 173, 1916.

48. G. P. KUIPER, "Planetary Atmospheres and Their Origin," *The Atmospheres of the Earth and Planets*, 2nd ed., p. 306, G. P. Kuiper, editor. Chicago: University Press, 1952.

49. L. B. ALDRICH and W. H. HOOVER, "The Solar Constant," *Smithsonian Contributions to Astrophysics*, **3**, 23–24, 1958.

50. EDISON PETTIT and SETH B. NICHOLSON, "Lunar Radiation and Temperatures," *Astrophys. J.*, **71**, 102, 1930.

51. EDISON PETTIT and SETH B. NICHOLSON, "Radiation from the Planet Mercury," *Astrophys. J.*, **83**, 84, 1936.

52. EDISON PETTIT and SETH B. NICHOLSON, "Temperatures on the Bright and Dark Sides of Venus," *Pub. Astron. Soc. Pacific*, **67**, 293, 1955.

53. JOHN STRONG and WILLIAM M. SINTON, "Radiometry of Mars and Venus," *Science*, **123**, 676, 1956.

54. ARTHUR ADEL, "The Importance of Certain Carbon Dioxide Bands in the Temperature Radiation from Venus," *Astrophys. J.*, **93**, 397, 1941.

55. WILLIAM M. SINTON, "Taking the Temperatures of the Moon and Planets," *Astron. Soc. Pacific Leaflets*, **7**, no. 345, 1958.

56. D. H. MENZEL, W. W. COBLENTZ, and C. O. LAMPLAND, "Planetary Temperatures Derived from Water-Cell Transmissions," *Astrophys. J.*, **63**, 177, 1926.

57. CORNELL H. MAYER, "Planetary Radiation at Centimeter Wave Lengths," *Astron. J.*, **64**, 43, 1959.

58. See p. 77 of reference 46.

59. JOSEPH W. CHAMBERLAIN and GERARD P. KUIPER, "Rotational Temperature and Phase Variation of the Carbon Dioxide Bands of Venus," *Astrophys. J.*, **124**, 399, 1956.

60. C. W. ALLEN, *Astrophysical Quantities*, p. 158. London: University Press, 1955.

61. H. L. JOHNSON, "Magnitude and Magnitude Systems," *Proc. Nat. Sci. Found., Astronomical Photoelectric Conference*, p. 14, J. B. Irwin, editor. 1953.

62. J. J. JEANS, *Dynamical Theory of Gases*, Chapter 15. Cambridge: University Press, 1916. See also reference 64.

63. L. SPITZER. See Chapter 7 of reference 48.

64. HAROLD C. UREY, "The Atmospheres of the Planets," *Handbuch der Physik*, vol. 52, p. 363. Berlin: Springer, 1959.

65. T. DUNHAM, JR., Chapter XI of reference 48.

66. Colloque International d'Astrophysique, Liége, 1956: *Les Molécules dans les Astres*. Institute d'Astrophysique Cointe-Scléssin, Belgium, 1957.

67. G. P. KUIPER, "Infrared Observations of Planets and Satellites," *Astron. J.*, **62**, 245, 1957.

68. J. STRONG, "Water in Venus' Atmosphere," *Sky and Telescope*, **XIX**, 343, (April) 1960.

69. G. HERZBERG, "The Atmospheres of the Planets," *J. Roy. Astron. Soc. Canada*, **45**, 100, 1951.

70. H. C. VAN DE HULST, Chapter III of reference 48.

71. G. DE VAUCOULEURS, *Physics of the Planet Mars*, Chapter 4. London: Faber & Faber, 1953.

72. J. GRANDJEAN and R. M. GOODY, "The Concentration of Carbon Dioxide in the Atmosphere of Mars," *Astrophys. J.*, **121**, 548, 1955.

73. A. DOLLFUS, "Étude des Planètes par la Polarisation de leur Lumière," *Suppléments aux Annales d'Astrophysique*, no. **4**, 1957.

74. S. CHANDRASEKHAR, *Radiative Transfer*. Oxford: Clarendon Press, 1950.

75. H. C. VAN DE HULST, "Scattering in a Planetary Atmosphere," *Astrophys. J.*, **107**, 220, 1948.

76. A. K. BLACKADAR, "On the Scattering of Blue Light in the Martian Atmosphere," 1950 *Report of "Project for the Study of Planetary Atmospheres,"* Lowell Obs., Flagstaff.

77. B. LYOT, "Recherches sur la Polarisation de la Lumière des Planètes et de Quelques Substances Terrestres," *Annales de l'Observatoire de Meudon,* Paris VIII, 1, 1929.

78. GERARD P. KUIPER, "On the Origin of the Solar System," *Astrophysics,* p. 404, J. A. Hynek, editor. New York: McGraw-Hill, 1951.

79. H. N. RUSSELL, R. S. DUGAN, and J. Q. STEWART, *Astronomy,* Vol. I., rev. ed., appendix. Boston: Ginn & Co., 1945.

80. W. E. FORSYTHE, *Smithsonian Physical Tables,* p. 735. Washington: Smithsonian Institution, 1954.

81. *Handbook of the British Astronomical Association.* Oxford: Vincent Baxter Press, published yearly.

82. P. GUTHNICK, "Die veränderlichen Satelliten des Jupiter und Saturn. Als planetarisches Analogon der Veränderlichen vom δ Cephei-Typus betrachtet," *Astron. Nachr.,* **198,** 233, 1914.

83. JOEL STEBBINS, "The Light Variations of the Satellites of Jupiter and their Application to Measures of the Solar Constant," *Pub. Astron. Soc. Pacific,* **38,** 321, 1926.

84. SETH B. NICHOLSON, "Discovery of the Tenth and Eleventh Satellites of Jupiter and Observations of These and Other Satellites," *Astron. J.,* **48,** 132, 1939.

85. G. P. KUIPER, "Titan: A Satellite with an Atmosphere," *Astrophys. J.,* **100,** 378, 1944.

86. HAROLD L. ALDEN, "Observations of the Satellite of Neptune," *Astron. J.,* **50,** 110, 1943.

87. G. DE VAUCOULEURS, *The Planet Mars.* London: Faber & Faber, 1950.

88. See Chapter 12 of reference 48.

89. WILLIAM M. SINTON, "Further Evidence of Vegetation on Mars," *Science,* **130,** 1234, 1959.

90. DEAN B. McLAUGHLIN, "A New Theory of the Surface of Mars," *J. Roy. Astron. Soc. Canada,* **50,** 193, 1950.

91. F. G. PEASE, "Planetary and Lunar Notes I," *Pub. Astron. Soc. Pacific,* **36,** 347, 1924.

92. *Annual Report of the Director of the Mount Wilson Solar Observatory of the Carnegie Institution of Washington, 1911,* p. 198.

93. E. ANTONIADI, *La Planète Mars,* Chapter 5. Paris: Librairie Scientifique, Hermann et Cie., 1930.

94. A. DOLLFUS, "Étude Visuelle de la Surface de la Planète Mars avec un Pouvoir Séparateur 0".2," *L'Astronomie, Bull. Soc. Astron. France,* **67,** 85, 1953.

95. *Transactions of the International Astronomical Union,* **10,** 259, 1958. See also J. ASHBROOK, "The New IAU Nomenclature for Mars," *Sky and Telescope,* **18,** 23, 1958.

CHAPTER 3

THE EARTH-MOON SYSTEM

3–1 Dimensions and figure of the earth. The characteristics of the earth and of the other planets have been described in Chapter 2. In the present section we review in more detail how information about the dimensions and figure of the earth is obtained. This constitutes the science of geodesy.

The International Union of Geodesy and Geophysics adopted in 1924 as a convenient mathematical model for the earth's surface the figure of an ellipsoid of revolution with a polar radius of 6,356,912 m and an equatorial radius of 6,378,388 m. These radii correspond to an oblateness of 1/297.0. This figure is called the Hayford Ellipsoid of 1909, or the International Ellipsoid, and is based on measurements made in the United States. It is used as a reference figure from which the departures of the actual surface are to be computed.

Direct determinations of the size and shape of the earth are based on geodetic triangulations. The basic problem of geodesy is to determine the separation, in linear units, between points whose latitudes and longitudes are well known. Triangulation technique makes it possible to limit the actual linear measurements to relatively short base lines. For example, if in Fig. 3–1 the distance AB is known, the triangulation network shown could be used to determine the distance along the earth's surface between O and A. For this purpose, besides knowing the length of the reference line AB with sufficient accuracy, one would have to determine the various angles shown in the figure. Differences in height from point to point would

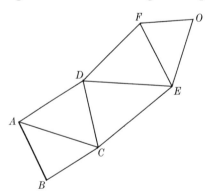

Fig. 3–1. Triangulation network.

78

also have to be determined. For the practical problems involved in this procedure, the reader should consult references 1 and 2.

The most difficult part of the measurements is precise determination of the reference base line. For this purpose the most accurate device developed is the iced-bar apparatus, in which a bar of standardized length is packed in melting ice to prevent errors due to thermal expansion. Direct use of this technique in the field is slow and costly; hence one usually employs Invar tapes or wires which have been calibrated by the iced-bar technique. In the field the tapes or wires are stretched between stakes driven at intervals of 25 to 50 m. Tension must always be the same, and shielding against wind and solar radiation must be provided. Base-line measurements with Invar tapes in California in 1924 for Professor Michelson's determination of the velocity of light, which were carried out with utmost care, yielded an accuracy of one part in about 12,000,000 [3].

Triangulation is classed as first, second, or third order, according to the accuracy attained. First-order triangulation should yield distances which are correct to within one part in 25,000; the results of second-order and third-order triangulations should be correct to within one part in 10,000 and 5000, respectively. The average length of the sides of the triangles in first-order triangulation is about 30 km. Because of atmospheric effects, the mean error with which angles can be measured is not less than approximately $0''.3$. Sides of triangles of about 30 km can be measured to within 6 cm, but accuracy decreases for the triangles farther away from the standard base line.

Reduction of the triangulation observations must allow for the effects of variations of height above sea level from point to point and also for the earth's spheroidal shape. First, the geodetic measurements are reduced to mean sea level. The triangulation network may then be compared with a set of triangles with straight-line sides resting on an ellipsoid. No two triangles will lie on the same plane, a fact that must be allowed for in the reductions of the observations. A knowledge of the figure of the earth is thus essential in the computations, and the International Ellipsoid is one convenient model that can be used for this purpose.

Although triangulation is principally used in surveying and mapping, the procedure permits computation of the geographic coordinates at other points when the geographic latitude and longitude of one point in the network are known. Comparisons of such calculated coordinates with astronomically determined ones yield the data necessary for a better determination of the size and figure of the earth. At present, the scalar dimensions of the earth can be obtained most accurately by triangulation, but the oblateness, or degree of flattening, of the spheroid can be found with more precision from observations of the gravitational field of the earth. This will be discussed in Section 3–2.

In addition to the triangulation procedure described above, several other techniques have been employed in geodetic surveys. Distances can be determined by electronic range finders, called *geodimeters* or *tellurometers*, which utilize the principle of radar. Flares dropped from high altitudes and observed simultaneously from different locations against the stellar background can also be used for geodetic purposes. Observations of the moon's position with respect to the stars have been employed by the U.S. Army Map Service for geodetic surveying. Since in this case the large distance to the moon makes it an unfavorable triangulation point, the times of occultation of certain stars along the moon's path in the sky have been used to improve the accuracy of the observations. These times may be accurately determined if the observations are made photoelectrically. In practice, the results have been affected by poor knowledge of the details of the lunar surface at the apparent lunar edge which occults the star.

From geodetically determined geographic coordinates, one can compute the direction of the vertical with reference to the stars at a given time. On the other hand, the vertical axis of a properly leveled theodolite also yields a direction of the vertical. The difference between the two results, called the *deflection of the vertical*, is due to irregularities in the gravitational attraction of the earth's crust, such as those caused by mountains.

Attempts at computing local deviations of the vertical by estimating the gravitational attraction of nearby mountains yield values that are much higher than the observed deviations. The discrepancies are explained by the principle of *isostatic compensation*, according to which the material under a mountain has a lower density than elsewhere.

Errors in astronomically determined latitudes and longitudes caused by a local deviation of the vertical are the *station errors* discussed in Section 1–2. In the absence of station errors astronomical observations could readily yield a position on the earth's surface within 5 m. Average station errors, however, cause an uncertainty of about 100 m.

The accuracy with which the International Ellipsoid represents the actual figure of the earth has been the object of a great deal of research using the different methods mentioned so far. The mean surface of the oceans has the property that the direction of the acceleration due to gravity is everywhere perpendicular to it. The shape of the earth which is based on the assumption that an idealized smooth surface covers the entire earth, coincides in the oceans with mean sea level, and is everywhere perpendicular to a plumb line, is called the *geoid*. The theory of the gravitational field of the earth, outlined in Section 5–9, indicates that the geoid departs slightly from an ellipsoid even when the earth is assumed to have no irregularities in its distribution of mass. At latitude 45° where this departure is at a maximum, the theoretical difference is, however, only about 4 m. Irregularities in the gravitational field that cause deviations of the vertical result in larger differences between the geoid and the ellipsoid, but these two surfaces are always within 100 m of each other [4].

In a discussion of geodetic data in 1948, Jeffreys [5] reported for the geoid an equatorial radius of 6,378,099 ± 116 m and a value for the reciprocal of the oblateness of 297.10 ± 0.36.

The results from data obtained since Jeffreys' analysis have been summarized by Cook [6]. These indicate an equatorial radius of 6,378,150 ± 50 m and a reciprocal of the oblateness of 298.2 ± 0.03. The oblateness has been derived from observation of artificial satellites. The equatorial radius is based on a reduction of geodetic data which include recent measurements along two major arcs, one in North and South America and one extending from Finland to the southern tip of Africa. The second of these arcs is about 95° long and is more than three times as long as the previous geodetic arcs used for this purpose. In the reduction of the geodetic data, the oblateness obtained from observations of artificial satellites was used. Heiskanen [7] has found from gravimetric observations that the mean-sea-level surface has local undulations that may rise or sink up to 40 m above or below an ideal ellipsoidal surface. Analysis of the motion of the Vanguard satellite, 1958-β_2, has indicated [8] that the geoid is 15 m higher at the north pole than at the south pole, so that the earth is very slightly pear-shaped. At the equator, the International Ellipsoid is thus somewhat larger than the earth, and the reciprocal of the oblateness is uncertain in the third digit. For most practical applications the International Ellipsoid is sufficiently accurate.

Assuming the mean-sea-level surface to be equivalent to the International Ellipsoid, we may derive several useful relationships. A meridional cross section of the earth is then an ellipse, and can be represented by

$$\frac{x^2}{A^2} + \frac{y^2}{B^2} = 1 \tag{3–1}$$

in a system of coordinates with origin at the earth's center, x-axis in the equatorial plane, and y-axis pointing toward the north pole. The symbols A and B denote the equatorial and polar radii, respectively. The relationship between A and B and the eccentricity of the ellipse is

$$B = A(1 - e^2)^{1/2}. \tag{3–2}$$

Equation (3–1) can be changed, after insertion of Eq. (3–2), into polar form, that is,

$$r^2 = \frac{A^2}{\cos^2 \beta + \sin^2 \beta/(1 - e^2)}, \tag{3–3}$$

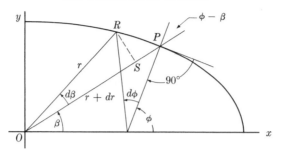

FIG. 3-2. In the elementary triangle SPR, $SP = dr$, $RS = r\,d\beta$, and the angle $PRS = \phi - \beta$. Therefore $\tan(\phi - \beta) = -(dr/r\,d\beta)$. Since r decreases as β increases, $dr/d\beta$ is negative. Equation (2-8) follows since $(\phi - \beta)$ is a small quantity.

in which r is the distance from the earth's center to a point with geocentric latitude β on the surface of the ellipsoid. The relationship between the oblateness f and the eccentricity e may be found by combining Eqs. (2-19) and (3-2):

$$(1 - f)^2 = (1 - e^2). \tag{3-4}$$

Hence, in terms of the oblateness,

$$r^2 = \frac{A^2}{\cos^2\beta + \sin^2\beta/(1 - f)^2}. \tag{3-5}$$

Using the values of A and f for the International Ellipsoid in Eq. (3-5) will permit the calculation of r as a function of β. Equation (3-5) may be recast in the form

$$r = A\left[\frac{1}{(1 + \xi)^{1/2}}\right], \tag{3-6}$$

where

$$\xi = \left[\frac{1}{(1 - f)^2} - 1\right]\sin^2\beta.$$

Using a binomial expansion, and keeping terms to the order of f^2 only, we obtain

$$r = A\left[1 - \left(f + \frac{3f^2}{2}\right)\sin^2\beta + \frac{3}{2}f^2\sin^4\beta\right],$$

which can be reduced to

$$r = A[1 - f\sin^2\beta - \tfrac{3}{8}f^2\sin^2 2\beta]. \tag{3-7}$$

In practice it is more convenient to express this relationship in terms of the geographic latitude ϕ. From Fig. 3-2 we have

$$\beta = \phi + \frac{dr}{r\,d\beta}. \tag{3-8}$$

We may evaluate $(1/r)(dr/d\beta)$ from Eq. (3–7). For this purpose we need to retain only the terms to the first power of f, since in the conversion to geographic latitude the higher-order terms will be multiplied by terms in f^2, and hence can be neglected. Thus we have

$$\beta = \phi - f \sin 2\beta. \tag{3–9}$$

Since $\phi - \beta$ is always a small quantity, a Taylor expansion of the function $\sin^2 \phi$ to the second term yields

$$\sin^2 \phi = \sin^2 \beta + 2 (\sin \beta \cos \beta)(\phi - \beta).$$

Combining this with Eq. (3–9) and neglecting terms in f^2, we then obtain

$$\sin^2 \beta = \sin^2 \phi - f \sin^2 2\beta.$$

We insert this result in Eq. (3–7) and disregard the difference between β and ϕ for terms in f^2. The result is

$$r = A[1 - f \sin^2 \phi + \tfrac{5}{8}f^2 \sin^2 2\phi]. \tag{3–10}$$

For the International Ellipsoid, r in km is given by

$$r = 6378.388[1 - 0.0033670 \sin^2 \phi + 0.0000071 \sin^2 2\phi]. \tag{3–11}$$

The difference between geographic and geocentric latitudes given by Eq. (3–9) is accurate to about one second of arc. If the terms in f^2 are included, a derivation analogous to that used for Eq. (3–9) yields, after conversion to geographic latitude,

$$\phi - \beta = 11'35''.66 \sin 2\phi - 1''.17 \sin 4\phi \tag{3–12}$$

to an accuracy of $0''.01$. Expressions for r and for $\phi - \beta$ which include higher-order terms are presented in reference 9.

The perpendicular distance s between mean sea level and the earth's axis of rotation is given by

$$s = r \cos \beta. \tag{3–13}$$

This is the distance to be used for computations of the centripetal acceleration due to the earth's rotation at a given point. The centripetal acceleration is given by

$$\alpha = 1.3294 \times 10^{-9}s, \tag{3–14}$$

the units being in km·sec^{-2} if s is in km. The component of this acceleration along the vertical is $\alpha \cos \phi$ if no local deviations of the vertical occur. This is the component that aids a rocket during vertical take-off.

3-2 **The earth's atmosphere.** The composition and structure of the earth's atmosphere have been subject to intensive study for years. In the present section we shall summarize briefly the current state of knowledge, and the reader is referred to the extensive literature for more details [10 through 16].

There are four principal components of dry air that are regarded as permanent in the sense that their proportions in the lower atmosphere remain approximately constant with time. These components and their fractional volume concentrations are [12]: molecular nitrogen (N_2), 0.78084; molecular oxygen (O_2), 0.20946; argon (A), 0.00934; and carbon dioxide (CO_2), 0.00033. These fractions are all somewhat uncertain in the last digit. In addition, there are traces of other permanent gases whose proportion by volume in each case is less than 20 parts per million. In order of abundance these are: neon (Ne), helium (He), methane (CH_4), krypton (Kr), molecular hydrogen (H_2), nitrous oxide (N_2O), and xenon (Xe). Up to a height of 70 km the atmosphere has a constant composition of the permanent gases. Above this height, the dissociation of molecular components by the action of solar radiation and the gravitational separation (or diffusion) of the components according to their molecular weights become increasingly important. Although carbon dioxide has been listed as a "permanent" component, a secular increase in its content amounting to 13% between 1900 and 1950 has been reported [12]. The increase can be explained as a result of the human use of fuels in this period. The proportion quoted above for this gas is for the year 1950.

Excluding water vapor, the components of the atmosphere that show variable concentration are: ozone (O_3), sulphur dioxide (SO_2), and nitrogen dioxide (NO_2). In addition, ammonia (NH_3) and carbon monoxide (CO) have been suspected, and probably are of industrial origin. The fractional volume concentration of these components is in each case less than 10^{-7}.

Ozone is formed by the dissociation of molecular oxygen under ultraviolet radiation of solar origin and the subsequent recombination of atomic and molecular oxygen. Near the ground, the amount of ultraviolet radiation is too small, while at great heights the combinations of atomic and molecular oxygen are too infrequent, for the effective production of ozone. As a result, ozone is mostly confined to a layer in the stratosphere. Maximum concentration occurs between 15 and 40 km above sea level. The concentration of ozone shows diurnal, seasonal, and possibly geographic variations, being highest during daylight hours and during the summer, as expected from the relationship of ozone to solar radiation. Ozone absorbs radiant energy very effectively in the spectral region extending from wavelength 2100 A to 3200 A, and, in a diminishing extent, to $\lambda = 3690$ A. This absorption is principally responsible for the cut-off in atmospheric transparency on the short-wavelength side of the visible spectrum.

Water vapor has an irregular geographic distribution. In the equatorial regions of highest humidity, 30 gm of H_2O-vapor per kgm of air may be found. The water-vapor concentration decreases rapidly with height; in the stratosphere each kilogram of air contains at most only about 1/100 gm of vapor. Water vapor may condense by cooling, as air from the lower levels is carried aloft to form clouds which consist of water droplets of 5 to 70μ in diameter, the most frequent size being 18μ. Ice crystals of comparable size may also result from the condensation of water vapor at a sufficient height. The presence of hygroscopic nuclei of condensation is required for the formation of droplets or crystals. Sea spray, smoke, dust raised by storms, and minute meteoritic particles have been considered as sources of condensation nuclei.

At its Brussels meeting in 1951, the International Union of Geodesy and Geophysics suggested several schemes of nomenclature for the various sections of the atmosphere. These are illustrated in Fig. 3–3. The simplest division is into two parts, according to chemical composition. These are the *homosphere*, or lower section, where the composition may be regarded as constant with height, and the *heterosphere*, or upper section, where dissociation and diffusive separation are important. These two sections are divided by a layer called the *homopause*, which will be found somewhere above 70 km. In the recommended nomenclature, the designations of the dividing layers have in general the suffix *pause*, while the main sections have the suffix *sphere*.

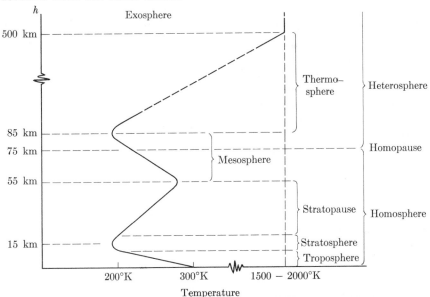

FIG. 3–3. Atmospheric nomenclature and temperature variation.

Temperature distribution with height is such that the atmosphere can be divided into six principal sections. The *troposphere*, whose total thickness varies with latitude and with the seasons from about 7 to 18 km, is thickest at the equator or during the summer. It contains 90% of the total mass of the atmosphere, and in it the temperature decreases with height to a minimum of about 200°K. The layer between the troposphere and the stratosphere is called the *tropopause*. The *stratosphere* is approximately isothermal, and its thickness may vary from zero to about 20 km. Above it there is a region, the *stratopause*, where temperature increases with height and, at a height from 50 to 60 km, may reach a value similar to that on the ground. Above the stratopause lies the *mesosphere*, in which temperature decreases with height and, at 80 to 90 km, reaches a value similar to that in the stratosphere. The temperature-height gradient then changes in sign at the *mesopause;* going higher, one finds the *thermosphere*, where again the temperature increases with height up to the *thermopause*. Above the thermopause lies the outer, isothermal region called the *exosphere* (see Section 2–9) whose base is at a height of about 500 km, and where the temperature ranges from 1500 to 2000°K.

In the troposphere the decrease of temperature with height occurs at an average rate of about 0.65°C/100 m. The actual rate may be as high as 0.8°C/100 m, and *inversions*, or layers through which the temperature increases with height, may frequently occur under certain meteorological conditions. For example, at the ground, on a clear, windless night, radiation cooling of the surface may result in the formation of an inversion layer. The general trend of a decrease in temperatures with height in the troposphere is a result of the absorption of solar radiation by the ground. Convective motions set up by radiation heating of the ground and other factors that contribute to the turbulence formed in the lower layers result in the establishment of an approximate adiabatic lapse rate (see Section 2–9) in the troposphere. A mass of air moving vertically will obey an adiabatic relationship, according to which the changes in temperature and pressure for dry air are given by

$$c_p \frac{dT}{T} = \frac{R \, dp}{J \mu p}, \tag{3–15}$$

where c_p is the specific heat at constant pressure, R is the gas constant, $J = 4.185 \times 10^7$ ergs/cal is the mechanical equivalent of heat, μ the mean molecular weight, and T, p are the temperature and pressure, respectively. This equation may be derived readily from the first law of thermodynamics and the gas law, Eq. (2–39), by assuming that an element of air mass displaced vertically does not change its total heat content. This assumption is plausible, since the gains or losses by conduction or radiation of such an element are negligible. Combining Eq. (3–15) with the equation of hydrostatic equilibrium [Eq. (2–38)], we obtain for the lapse rate

$$\frac{dT}{dz} = \frac{-g}{c_p J},$$ (3–16)

where z is a measure of height. For dry air, $c_p = 0.239 \text{ cal·gm}^{-1}\text{·deg}^{-1}$, and the expected lapse rate is $0.98°C/100 \text{ m}$, a quantity called the *dry adiabatic lapse rate*. Atmospheric water vapor, whose effect has been neglected so far, can be condensed as a given air mass is displaced upward, releasing the latent heat of condensation which will increase the temperature of the rising air above the values expected from the dry adiabatic lapse rate. Hence for moist air, a smaller decrease of temperature with height is expected than for dry air.

The existence of a region of increasing temperature in the stratopause was first indicated by the anomalous propagation of sound waves. Cannon fire in France during World War I was often heard in England, due to the refraction of sound waves in the stratopause. The velocity of sound in air is given by

$$v^2 = \gamma \frac{RT}{\mu},$$ (3–17)

where γ is the ratio of specific heats c_p/c_v (1.4 for dry air), and μ is the molecular weight (=28.9 for air). An increase in temperature with height will, therefore, cause sound waves to be refracted downward. Thus the sound may be heard at a distance from the source even if one or more zones of silence occur between the source and the area of audibility. Warming in the stratopause is due to the absorption of solar ultraviolet light by ozone and its subsequent radiation according to the radiation laws.

The theory accounting for the temperature distribution observed above the ozone layer is still imperfect. Above this layer in the mesosphere, the temperature drops to about 200°K at 85 km, according to data obtained with rockets. This temperature decrease with height was suspected as early as 1929 because meteors detonating at an altitude of more than 60 km are not usually heard, indicating that above this level the temperature decreases with height so that sound waves originating above it are refracted upward.

The top of the mesosphere, at about 85 km, is indicated by the phenomenon of *noctilucent clouds* [17]. These appear occasionally in the west, after sunset, and reflect sunlight even after the sun, for an observer on the ground, is well below the horizon. Their location in the sky viewed from different ground locations permits a derivation of their height. The results average about 82 km. According to Vestine [17], noctilucent clouds are composed of dust particles; but they may be composed of ice crystals [15]. The increase in temperature above 85 km is evidenced by the existence of these clouds, which may be formed out of particles trapped at the base of the thermosphere where little convection can exist.

Above 85 km important changes in the composition of the atmosphere occur. The dissociation of molecular oxygen and photochemical reactions involving molecular and atomic oxygen in the presence of solar radiation probably cause the increase of temperature with height in the thermosphere. Dissociation of nitrogen becomes important at a height greater than that at which the dissociation of oxygen occurs. Finally, at the critical level (see Section 2–9), the isothermal conditions of the exosphere prevail, at the height where collisions between the constituent particles become statistically negligible. The exosphere is composed mostly of particles which, as a result of collisions near the critical level, travel outward in free flight and fall back after a few minutes [11].

The increase in temperature above 85 km is required by the theory of atmospheric tides. It has been recognized for a long time that solar tides in the atmosphere are much more pronounced than lunar tides, contrary to what is expected from an analogy with surface water tides. As a result, atmospheric pressure varies markedly with a 12-hr period. This may be explained as a resonance effect if the atmosphere is assumed to have a 12-hr natural period of oscillation. Another mode of oscillation with a period of 10.5 hr was indicated by a study of the speed of propagation of sound around the world at the times of the Krakatoa explosion in 1883 and of the Siberian meteor explosion of 1908. These modes of oscillation of the atmosphere are satisfied by an atmospheric temperature distribution such as that described above.

The high temperatures of the upper thermosphere and exosphere are also indicated by the high particle concentration found in the *ionosphere*. The ionosphere is the region in the atmosphere where free electrons exist in sufficiently large numbers to alter appreciably the velocity of propagation of electromagnetic radiation. Free electrons may originate from the ionizing action of ultraviolet solar radiation or in sufficiently energetic collisions of the air particles with particles traveling inward from the sun. Several separate ionospheric layers have been identified by means of radio signals transmitted vertically at different frequencies. The signal reflected from an ionized layer is subsequently observed. The refractive index n of the ionized gas is given by

$$n^2 = 1 - \frac{4\pi N e^2}{m\nu^2},\qquad (3\text{--}18)$$

where ν is the frequency of the electromagnetic radiation, m and e are the mass and charge of the electron, and N is the number of electrons per cm^3.

Reflection of a radio signal occurs when n is imaginary. When $n = 0$, Eq. (3–18) yields a frequency ν_p called the *penetration frequency* [18]:

$$\nu_p^2 = \frac{4\pi N e^2}{m}.$$ (3–19)

If the frequency of the signal is gradually increased, reflection will occur until the frequency ν_p is attained. At higher frequencies penetration occurs. Observation of the penetration frequency can yield an estimate of N, the concentration of electrons, while the time interval between signal emission and echo return can be used to determine the height of the ionized layer.

The data indicate that three principal ionized regions exist: the so-called E-layer at about 100 km, the F_1-layer at about 200 km, and the F_2-layer at about 300 km. The E-layer lies on top of an extended region of ionization reaching downward to a height of 60 km above the ground, which is called the D-layer. Irregular cloudlike regions of abnormally high ionization are found embedded in the E-layer; these form the *sporadic E*-layer. Diurnal and seasonal statistics of the heights of these layers and of their electron densities indicate a correlation with solar radiation. This relationship will be discussed in more detail in Section 6–2. The existence of several layers of ionization may be explained by differences in the ionization mechanisms which are most effective at different atmospheric heights.

If the temperature and the mean molecular weight are assumed constant with height, a simple atmospheric structure yields, according to Eq. (2–42), at a height of 300 km, a density of about 10^{-6} that near mean sea level. In terms of the number of particles per cm^3, this means something like 1000 particles at 300 km. However, the electron concentration in the F_2-layer is of the order of 10^6 per cm^3, and we may infer that the actual number of particles per cm^3 is appreciably higher because not all atoms are necessarily ionized. Since this simple atmospheric model fails, we must conclude that the scale height $H = RT/\mu g$ varies with elevation above sea level. Below the thermosphere, we know that T is never high compared with the temperature at the ground, and that μ, the mean molecular weight, is sensibly constant. Hence the increase in the scale height occurs principally in the thermosphere, and is due partly to a decrease in μ resulting from dissociation of oxygen and nitrogen and partly to diffusion. The decrease in μ, however, can scarcely be expected to amount to much more than a factor of 2 (the case for full dissociation of O_2 and N_2). Thus, in order to explain the high particle concentration in the F_2-layer, a marked increase in temperature must be assumed to occur in the thermosphere. The actual variation of μ and T with height in the upper atmosphere cannot be readily derived from the available data which only indicate an increase in the scale height with elevation.

TABLE 3–1

STRUCTURE OF THE EARTH'S ATMOSPHERE*

Altitude, km	Pressure, millibars	Density, gm/m³	Temperature, °K
Surface (1.22 km)	881	1054	291
10	279	422	231
20	57.0	93.0	213
30	12.5	18.7	232
40	3.14	4.17	262
50	0.90	1.16	271
60	0.25	0.35	252
80	0.012	0.021	205
100	0.0006	0.00086	240

* Adapted by permission of the University of Chicago Press and Dr. G. P. Kuiper from Table 2, p. 500 of *The Earth as a Planet,* G. P. Kuiper, ed. Chicago: University of Chicago Press. Copyright 1954 by the University of Chicago.

The variation of pressure with height may be derived directly from gauges carried aloft by balloons or by rockets. Knowledge of the pressure-height relationship can be used to derive T/μ as a function of height by means of Eq. (2–40). The velocity of sound, which may be readily measured as a function of height by rocket experiments, can also yield values of T/μ [Eq. (3–17)]. In order to derive the temperature-height relationship from these data, μ must be known. Since the latter is reliably known only below 80 km, the two variables T and μ cannot be separated at greater altitudes.

From rocket and balloon observations made in New Mexico (latitude 33°), as summarized by Whipple [10], the structure of the atmosphere below 100 km has been derived with the results shown in Table 3–1. Data obtained from rockets and artificial satellites indicate that air densities at 200 km are of the order of 10^{-6} to 10^{-7} gm/m³ [19]. At extreme heights, the available data indicate that the density is quite variable and intimately related to solar activity. These observations are discussed more extensively in Section 6–2.

3–3 Observing through the atmosphere. When an astronomical object is observed from the earth's surface, the light of the object is subject to a change of direction caused by atmospheric refraction. While traversing the atmosphere, the light also suffers a loss of intensity called *extinction*. In addition, any atmospheric inhomogeneities, such as small regions whose

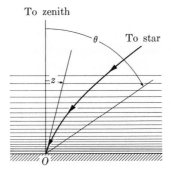

FIG. 3–4. Atmospheric refraction.

refractive indices are slightly different from those of their surroundings, will introduce a combined effect known as *scintillation*, or "seeing." Refraction, extinction, and scintillation are the subject of this section.

Let z be the zenith distance (see Fig. 3–4) at which the astronomical body is observed from the ground, and let θ be the zenith distance of the light beam before it enters the atmosphere. In this case Snell's law may be rewritten as

$$\sin \theta = n \sin z, \tag{3–20}$$

where n is the index of refraction of the air at the point of observation O [20]. Let $\theta - z = r$, the *angle of refraction*. In terms of r, Eq. (3–20) becomes

$$\sin (z + r) = n \sin z, \tag{3–21}$$

or

$$\sin z \cos r + \cos z \sin r = n \sin z.$$

Because r is a small angle (for $z < 75°, r < 4'$), we write

$$\sin z + r \cos z = n \sin z.$$

Hence,

$$r = (n - 1) \tan z. \tag{3–22}$$

This is the basis of a number of empirical formulas such as Comstock's formula for refraction in which the quantity $(n - 1)$ is given as an empirical function of atmospheric pressure p and temperature T, at the observer's point. In seconds of arc, according to Comstock's formula, r is given by

$$r = \frac{983p}{460 + T} \tan z, \tag{3–23}$$

where p is expressed in inches of mercury and T is in degrees fahrenheit. If p is in millibars (1 millibar = 0.02953 in. = 0.7502 mm of mercury) and T in degrees celsius, the constants in the numerator and denominator of Eq. (3–23) are 52.25 and 886, respectively, for r in seconds of arc. Except under extreme pressure and temperature conditions, this formula gives r within $1''$ for $z < 75°$. A set of tables based on a theory that takes into account the curvature of the earth as well as astronomical and meteorological data has been presented by Garfinkel [21]. With these tables r can be estimated even at zenith distances greater than $90°$, which may be required when the observer is at an appreciable height above sea level. At $z = 90°$, Garfinkel's tables list the value $r = 2206''$ for standard atmospheric conditions for an observer at sea level.

Special care must be taken when allowance for refraction is made in the determination of the space coordinates of a rocket observed against the stellar background immediately after take-off. If the stars are used in deriving the coordinates, one must allow for the fact that their light suffers the full effect of atmospheric refraction, whereas the light of a rocket during its flight at an altitude of less than about 40 km does not. The difference in refraction angles will be minimized by making the observations of the rocket as close to the zenith as possible.

The extinction of light in the atmosphere is caused by molecular scattering, by scattering and absorption by water droplets and dust particles, and by selective absorption of the atmospheric gases in certain spectral regions. In Eq. (2–44) we have presented the relationship for the loss of intensity suffered by a beam of radiation traversing a scattering and absorbing medium of thickness ds. The product of the mass absorption coefficient and the density is often called the *extinction coefficient*. We may thus write

$$dI = -I\beta\, ds,$$

where I and β may refer to a wavelength interval of any desired extension. The light intensity reaching the observer after the beam traverses a total thickness s is

$$I = I_0 e^{-\tau}, \tag{3–24}$$

where τ, as defined in Section 2–9, is the optical thickness. For the atmosphere it is usually impossible to derive the value of τ from the extinction coefficient. However, astronomical observations yield the optical thickness. The change in the apparent magnitude of a star produced by the extinction may be obtained by combining Eqs. (2–35) and (3–24). The result is

$$\Delta m = 1.085\tau. \tag{3–25}$$

If we assume a plane-parallel model for the atmosphere and neglect refraction, the total path traversed by a beam of light approaching the observer from zenith distance z is (Fig. 3–5) sec z times the path length

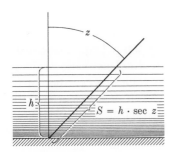

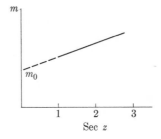

FIG. 3–5. Total air path as a function of z.

FIG. 3–6. Determination of apparent magnitude outside of atmosphere from surface observations.

traversed by a beam approaching the observer from the zenith.   This holds well for $z < 75°$, in the earth's atmosphere.

If we let τ_0 be the optical thickness of the atmosphere at $z = 0$, the change in magnitude for a star at zenith distance z will be

$$\Delta m_z = 1.085\tau_0 \sec z.$$

Therefore, the observed magnitude will be

$$m_z = m_0 + 1.085\tau_0 \sec z \qquad (\sec z \geq 1). \tag{3–26}$$

A plot of the observed m_z at several zenith distances against the corresponding $\sec z$ will be a straight line (for $z < 75°$). The intercept on the m_z-axis (arbitrarily setting $\sec z = 0$) will be the magnitude m_0 of the star as it would be observed outside the atmosphere (Fig. 3–6).   From the value of m_0 and any observed m_z, the value of τ_0 can be found from Eq. (3–26).

This method for eliminating the effects of atmospheric extinction in stellar photometry is known as Bouguer's method.   It may be extended to zenith distances greater than 75° by taking into account the curvature of the earth. Bemporad's table [22] is useful for this purpose. It shows, for example, that when the actual total atmospheric mass is 3 atmospheres, the plane-parallel model yields $\sec z = 3.057$.

TABLE 3–2

OPTICAL THICKNESS $(1.085\tau_0)$ AS A FUNCTION OF WAVELENGTH

λ, A	3530	4220	4880	5700	7190
$1.085\tau_0$	0.61	0.29	0.18	0.14	0.06

Table 3–2 shows the quantity $1.085\tau_0$ obtained at the Mt. Wilson Observatory (elevation 1742 m) during clear weather for the visible region of the spectrum in band widths of the order of 500 A [23]. These quantities may be used in Eq. (3–26) to estimate total extinctions. Van de Hulst [24] has compared these data with the values of τ_0 expected from the theory of scattering, and concludes that even under clear conditions the extinction is somewhat higher than expected for pure air. Near ground level, Van de Hulst finds that the extinction is produced mostly by haze.

Momentary disturbances in the index of refraction of the air, causing the phenomenon of scintillation, are responsible for the limit in resolution attainable with the telescopes used in astronomical research. Theoretically, without this effect, a telescope's resolving power is given according to Rayleigh's criterion [25] by

$$R = \frac{1.22\lambda}{a}, \tag{3-27}$$

where R is in radians, λ is the wavelength, and a is the aperture of the telescope expressed in the same units as λ. R is the smallest angular distance that may separate two stars and still permit an observer to distinguish them as two separate objects through a telescope. In seconds of arc, for white light, with a in inches, we have

$$R = \frac{4''.56}{a}, \tag{3-28}$$

which is called "Dawes's formula." If F is the focal length of a telescope, the theoretical diameter of the telescopic image of a star is

$$d = \frac{F}{206,265} \frac{4''.56}{a}, \tag{3-29}$$

with F and d in the same units and a in inches.

Atmospheric turbulence causes rapid variations in the brightness and position of an image in the focal plane of a telescope. As a result, the observed image is spread out, especially when a prolonged photographic observation is made. The actually observed diameter of the image, δ, may be used as a measure of the "seeing," and the range s in seconds of arc over which the direction of the light beam is disturbed is given by

$$s = 206,265 \frac{\delta}{F}, \tag{3-30}$$

where F is again the focal length and s is referred to as the "seeing." Statistics on the value of s have been collected [26]. Baum [27] has presented the relative frequency of s observed at Mt. Palomar (see Fig. 3–7).

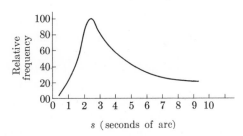

FIG. 3–7. Frequency distribution of "seeing" at Mt. Palomar according to Baum.

The most commonly observed value of s is about $2''.5$. If this number is inserted in Eq. (3–30), we may draw comparisons between the theoretical image diameters given by Eq. (3–29) and the actually observed diameters. For example, at the prime focus of the 200-in. telescope ($F = 660$ in.), the theoretical diameter is 10^{-4} in., while the seeing diameter (under average atmospheric conditions) is 0.01 in. It should be mentioned that, even in the absence of an atmosphere, a telescope used in photographic observations may be limited in resolution by the resolving power of the photographic emulsion. The usual emulsions employed in astronomical observation have resolving powers of somewhat less than 100 lines/mm.

An observer can see visually more details than can be photographed, because the eye can readily select the moments of atmospheric steadiness. Photographic observation, however, can reach very faint stars and, in addition, provide a permanent record of the observations. The effects of seeing can be compensated for to some degree in photographic observations by making the exposures as short as possible. Since this will limit the total light energy collected in the photographic emulsion, some means must be used to increase the efficiency of the process. Image intensification by electronic means offers a possible solution [28].

Observations from a balloon carrying a telescope above the main scintillation-producing atmospheric layers have also been made. Beautiful photographs of the solar photosphere have been obtained in this manner [29, 30] (see Plate VII). Balloon observations, however, still suffer from the ultraviolet absorption of the upper atmosphere. The problem of making observations in a wide range of frequencies, to the limit of resolution of the telescope and of the photographic emulsion (or of another sensing device employed with the telescope), can be completely solved only by making observations above the atmosphere.

3–4 The moon's motion, mass, and size. In this section certain dynamical and geometrical aspects of the moon and its motion are described.

A more mathematical treatment of the lunar motion will be covered in Section 5–11. General references on the topics discussed here are 9, 31, and 32.

The orbital motion of the moon displaces it eastward among the stars by about 13° every day. The phase of the moon is a function of its *elongation*, or angular distance from the sun, as seen from the earth. When the elongation is 90°, the moon, whose phase then is either first or last quarter, is said to be in *quadrature*. New and full moons occur when the moon is in *syzygy*, that is, in an approximately straight line with the sun and earth. The path of the moon among the stars shows that the moon's orbit is inclined about 5° to the plane of the ecliptic. Hence we may define a line of nodes, an ascending node, and a descending node in a manner analogous to the case for planetary orbits (see Section 2–1). Because of the 5°-inclination, a solar or lunar eclipse can occur only when the moon is very close to a node and the line of nodes points approximately toward the sun.

The *nodical*, or *draconitic*, month is defined as the interval between successive passages of the moon through a given node and is a very important period of time in eclipse theory.

The *month* may also be defined in terms of other reference directions: the *sidereal month* in terms of the stars; the *tropical month* in terms of the vernal equinox; the *anomalistic month* in terms of perigee; and the *synodic month* in terms of the sun. The mean lengths of these months for the year 1900 were:

	Days	d	h	m	s
Sidereal	27.321661	27	07	43	11.5
Nodical	27.212220	27	05	05	35.8
Tropical	27.321661	27	07	43	04.7
Anomalistic	27.554551	27	13	18	33.2
Synodic	29.530589	29	12	44	02.9

These mean values vary by only a few hundredths of a second per century, but in a given lunation appreciable changes may occur because of perturbations of the lunar motion due to the sun. For example, the sidereal month may vary by 7 hr and the synodic month by 13 hr from the mean values quoted. The perturbations cause the inclination to vary from 4°58′.8 to 5°18′.6.

As a result of solar attraction, the orbital plane of the moon precesses in such a way that the nodes are continually displaced westward, a motion known as the *regression of the nodes*. A complete sidereal revolution of the ascending node takes place in 6798.3 days (approximately 18.6 years). An interesting effect occurs when the regression of the nodes is measured with respect to a line connecting the earth and the sun. The interval between successive passages of the sun through a lunar node is 346.62 days. Nineteen such crossings will occur in 6585.8 days. Now, 223 synodic periods (or lunations) equal 6585.3 days (18 years, 11 days). This period is called the *saros*. If, at a given new moon, the sun is sufficiently close to the line of nodes for a solar eclipse to occur, then one saros later the same relative positions of the earth, sun, and moon will be very closely reproduced, and a solar eclipse will again take place. Lunar eclipses will be repeated in a similar manner. Since the saros is not an exact multiple of a solar day, and since it is not quite equal to 19 solar crossings of the node, the successive eclipses will not be observable from the same location on the earth.

The eccentricity of the moon's orbit averages 0.0549, but, because of perturbations, it varies from 0.044 to 0.067. As a result, the moon's apparent diameter varies from 33′31″ to 29′21″. The line of apsides of the moon's orbit rotates eastward with a period of nearly nine years. A discussion of the perturbations of the orbital elements of the moon from the point of view of celestial mechanics is given in Section 5–11.

Among the many perturbations of the moon's motion there is one that is related to the solar parallax, which can be used to determine the ratio of the distances from earth to moon and earth to sun. This perturbation is called the *parallactic inequality*. It is caused by the fact that the solar attraction suffered by the moon varies slightly, depending on whether the moon is located on the half of its orbit facing the sun or on the other half. According to Brouwer [33], this inequality causes the moon's ecliptic longitude D to be displaced by

$$-(124''.97 \pm 0''.04) \sin D.$$

The theoretical value of the coefficient of sin D has been investigated by Brown [34]. De Sitter [35] has given a value of the coefficient based on Brown's theory which we may equate to the observational value

$$49{,}853''.2 \, \frac{1 - \mu}{1 + \mu} \, \frac{\pi_\odot}{\pi_{\mathbb{C}}} = 124''.97, \tag{3–31}$$

where μ is the ratio of the mass of the moon to that of the earth, and $\pi_\odot$, $\pi_{\mathbb{C}}$ are the solar and lunar parallaxes, respectively. Theoretically this expression yields one of the three quantities μ, $\pi_\odot$, $\pi_{\mathbb{C}}$ when two of these are known. However, it is a difficult task to carry out observational determinations of the parallactic inequality with the required precision.

The best observational data consist of occultations of stars by the moon, and here the irregularities in the moon's limb affect the results. With an improvement of observational technique, Eq. (3–31) will doubtless become a more useful relationship.

The angular velocity of the moon in its orbit varies according to Kepler's second law and because of perturbations. On the other hand, since the moon *rotates* once in a sidereal month, the same face is turned toward the earth, and the angular velocity of this rotation is very nearly uniform. These effects lead to a *libration in longitude* which permits us to see a maximum of 7°45′ around each of the eastern and western limbs of the moon. Furthermore, the poles of the moon's axis of rotation differ by about 6.5° from the poles of the orbit. This causes a *libration in latitude* which tips the moon's north pole toward the earth once a month; and half a month later the south lunar pole is tipped toward the earth. The maximum value of this libration is 6°44′. Without these librations one could also see around the edge of the moon's disk up to 1° by observing the moon from different locations on the earth. This effect is called the *diurnal libration*. The librations mentioned so far are collectively called *geometrical* librations. Small irregularities in the velocity of rotation of the moon cause a *physical libration* amounting at most to $3\frac{1}{2}$ min of arc. These oscillations are due to the fact that the moon has a slight bulge in the direction of the earth which is not always directed toward the earth's center because of the geometric librations in longitude and latitude. The earth's attraction on the bulge then causes pendulumlike oscillations of the moon which constitute the physical libration. As a result of all librations, 59% of the moon's surface had been observed before the epoch-making Russian observation of the other side of the moon.

The mean distance from the earth to the moon has been determined both trigonometrically and by radar. The measurements, which agree within 1 km, yield 384,405 km or 60.268 times the earth's equatorial radius. The variations of the moon's orbital elements cause the actual distance to vary between 406,697 and 356,410 km. The mean velocity of the moon in its orbit is 1.023 km/sec.

The linear diameter of the moon can be derived by combining the distance observations with those of the angular diameter. The result is 3476 km. In Section 2–3, Rabe's determination of the ratio of the mass of the earth to that of the moon (81.375 ± 0.026) was discussed. This ratio, together with the mass of the earth, Eq. (2–17), yields 7.344×10^{25} gm for the mass of the moon. A mean density of 3.34 gm/cm^3 follows from these figures. Since this does not differ greatly from the density of the outer layers of the earth (3.3 to 3.8 gm/cm^3 at a depth of 35 km [36]), it is probable that the moon is rocky throughout and does not have a dense core such as that found in the earth.

The departure of the moon's figure from a perfect sphere can be determined observationally in the following way. The times of occultation of stars by the moon provide a measure of the circularity of the moon's visible disk. Such observations indicate only the random variations from circularity that are expected from the irregular features on the moon's surface [32]. The departure from circularity of the lunar meridian directed toward the earth is more difficult to determine. Librations cause displacements of the features of the moon from the center of the disk. The observed displacement of a given point on the moon's surface depends on its distance from the center of the moon. Baldwin [37] has analyzed the displacements measured by several observers and concludes that in the center of the lunar disk, the tidal bulge amounts to about 2 km in excess of the mean radius of the moon as observed in the plane of the sky.

The amplitude of the moon's physical libration may also be used to make a theoretical estimate of the size of the moon's bulge. From a discussion of these observations by Jeffreys [5], one can obtain 650 m for the height of the central bulge above the mean radius of the moon. The determination of the bulge elevation by the first of the two methods outlined above is based on observations that are difficult to make; this may account for the disagreement.

If the moon were completely adjusted to the earth's attraction and to its own centrifugal force, its figure would be such that the semiaxis pointing in the earth's direction would be 38 m above the moon's mean radius. The polar radius would be 27 m less and the equatorial radius in the plane of the sky 11 m less than the moon's mean radius. These figures are based on Jeffreys' analysis of the shape that the moon would have if its surface were an equipotential surface [5]. The bulge expected from this theory is much smaller than that indicated by observations, and one must conclude that the interior of the moon is sufficiently rigid to support the excess of the bulge above that given by hydrostatic considerations. Jeffreys has interpreted the difference by assuming that the observed bulge represents a frozen tide that was acquired by the moon in the past, when the moon was much nearer to the earth. The argument for the moon's relative proximity in the past is based on the conservation of angular momentum in the earth-moon system. Because the tides that the moon raises on the earth are slowing down the earth's angular rotation (see Section 1–7), the moon must be receding slowly from the earth. According to Jeffreys, the theory of an equipotential surface at the time when the moon was 90,000 km from the earth would lead to a bulge equal to that determined by the dynamical method mentioned previously. If it is further assumed that at that time the moon changed from a body having sufficient plasticity to adjust to the tidal forces to one with a rigid structure, it follows that the old tide would remain frozen on the moon.

3–5 The surface of the moon. *Selenography* is the study of the surface features of the moon, which we shall consider in this section. The multiplicity of these features and the clues that they present regarding their origin make a fascinating subject. An excellent treatise on this topic is Baldwin's *The Face of the Moon* [37], which contains an extensive bibliography. The nomenclature of the lunar surface features in general use is that proposed by the International Astronomical Union [38], which has published lists of names—usually those suggested by the selenographer who first called attention to a given feature. Many lunar maps have been prepared. One readily available is Andel's Lunar Map [39], which is reproduced at the rear of the book. *The Moon* by W. Goodacre [40] is an excellent older book devoted to the description of lunar features.*

To the unaided eye the moon's surface appears to be divided into light and dark areas (see Plate IV). The dark area can in turn be divided into *maria* (singular, *mare*), the so-called "seas of the moon." Through a telescope, the maria appear as relatively smooth plains which often have a roughly circular perimeter, or a perimeter made up of circular arcs. The largest mare is Oceanus Procellarum which occupies an extensive part of the eastern half of the moon. Procellarum joins on its western side with a group of maria (Imbrium, Nubium, Humorum) which may be regarded as bays of Procellarum if we carry along the analogy with the earthly seas. The principal maria on the eastern half (Vaporum, Serenitatis, Tranquilitatis, Nectaris, Foecunditatis, and Crisium) are all fairly well demarcated. The diameters of those maria that are roughly circular vary from 1100 km for Imbrium to 270 km for Nectaris. Telescopic observations also show that the maria contain isolated mountain peaks, mountain chains, and numerous craters. Craters found in the maria may be very rugged in appearance, as for example, Copernicus (diameter 90 km, one of the most conspicuous lunar craters), or may appear only as a faint trace as if the material from which the maria were formed had almost covered them. The majority of the maria craters, however, are small pits whose angular diameter is near the limit of telescopic resolution. These features suggest that the maria were formed from flowing material. It has been assumed by many students of the moon that the maria represent extended lava flows.

The maria are often bordered by high mountain ranges, such as the Apennine Mountains on the south side of Mare Imbrium, which rise higher than 6000 m above the level of this mare. Cracklike formations, called *rills*, are often found near the borders of the maria, but they may also occur elsewhere.

---

* *A Photographic Lunar Atlas*, G. P. Kuiper, editor. Chicago: University of Chicago Press, 1960. This has a collection of the best available lunar photographs reduced to a common scale of 1:1,370,000.

The light-colored areas of the moon's surface show a profusion of craters ranging in size from the limit of visibility (of the order of 1 km) to formations such as crater Clavius, which has a diameter of 200 km. Out of the many thousand craters visible on the earth side of the moon, only about 150 are larger than 80 km. In the lighter areas the most conspicuous crater is Tycho, near the south pole, with a diameter of 87 km. Tycho is associated with very extended *rays* or splashlike light-colored streaks which can be traced to distances of over 1500 km from Tycho (see Plate IV).

That the craters differ in age is evident from a comparison of Copernicus, which evidently was formed later than the mare in which it is located, with those craters which appear partly covered by the maria. In many instances, series of craters can be found in which the rim of one crater is overlapped by another, and so on (see Plate V). Where such an overlap between two craters exists, the one with the unbroken rim is of more recent origin. Also, if the ray from a given crater crosses over another crater and is visible in the bottom of the second crater, the crater from which the ray radiates is younger than the second one. Gold [41] has compared the sharpness of features in groups of craters for which age sequences can be established in this manner. The result is that the younger a crater, the sharper are its features. According to Gold, this indicates that erosion of the lunar surface material has taken place. Gold suggests various mechanisms for the erosion, such as meteoritic bombardment and the effects of electrostatic forces on dust particles in the presence of electrons that bombard the moon from the sun. He concludes that even the maria may be deposits of eroded material rather than lava flows.

It is a straightforward procedure [40] to estimate the height of a lunar mountain or crater rim above the surrounding plain by measuring the lengths of the shadows they cast at a given moment. From observations of this nature, the heights of many crater rims above the surrounding terrain, as well as the depths of the crater depressions, are known. The measurements indicate a relationship known as Schröter's rule, according to which the volume of the elevated rim is approximately equal to the volume of the central depression when the latter is referred to the level of the crater's surroundings. As shown by Baldwin [42], these observations indicate that the diameters and depths of the craters are related according to

$$D = 0.1083d^2 + 0.803d + 0.62, \tag{3-32}$$

where

$D = $ log diameter in meters,

$d = $ log depth in meters measured from the top of the rim.

The height of a crater rim above the surrounding plain is also related to the diameter, according to

$$E = -0.097D^2 + 1.442D - 1.56, \qquad (3\text{--}33)$$

where

$$E = \log \text{ rim height in meters.}$$

These relationships are shown by Baldwin to be satisfied also by terrestrial craters of known meteoritic origin, as well as by bomb and shell craters.

Baldwin's relationships are based on plots that show appreciable scatter. That the scatter is not entirely due to errors of observation is confirmed by a comparison of Tycho and Copernicus which, although of equal diameter, have rim heights of 2400 and 1000 m, respectively. The nature of the lunar surface at the points where these craters were formed is undoubtedly an important factor in crater formation. As mentioned previously, Tycho is found in the light-colored lunar highlands, whereas Copernicus is in Oceanus Procellarum.

The inferences that one can draw regarding the origin of the lunar features have been summarized by Baldwin [37], Urey [32], Kuiper [43], and Gilvarry [44]. These writers present references to the numerous research papers in this field.

In the nineteenth century it was commonly believed that the craters and maria of the moon had a volcanic origin. A study by G. K. Gilbert in 1893 of the sizes of lunar craters and maria showed that their diameters or depths do not follow the pattern of terrestrial volcanic craters and lava flows. Gilbert favored the hypothesis that the lunar features were caused by meteoric collisions. The fact that the craters are circular instead of showing more or less elongated shapes depending on the angle of incidence of the meteor was explained by A. C. Gifford to be due to the explosive nature of meteoric impacts. Schröter's rule and Baldwin's relationships [Eqs. (3–32), (3–33)] strongly support the explosive nature of crater formation.

Gilvarry and Hill [45] have studied the energy requirements for the production of an explosion during a meteoric impact. They conclude that explosive impacts will occur for meteorites traveling with speeds that are astronomically plausible. The hypothesis of a meteoric origin for the majority of lunar features thus rests today on very strong observational and theoretical arguments.

It has been suggested that the maria were formed as a result of the fall of relatively large bodies on the moon. During or after the explosion, lava may have surged from the moon's interior and formed extensive lava flows. Alternatively, the lava may have originated in the fusion of the surrounding terrain during the explosion, a view advanced by Urey [32].

As mentioned previously, an alternative hypothesis about the nature of the maria is Gold's eroded-dust theory. Gilvarry [44] has shown that one can discriminate between a rocky and a dusty lunar surface by observing the rate of cooling when the sunlight is suddenly cut off during a lunar eclipse. Radiometric measurements of lunar surface temperatures as a function of time during an eclipse were made by Pettit and Nicholson [46] for an area in the lunar highlands, and by Pettit [47] for an area which included Mare Vaporum. The technique used is described in Section 2–7. Gilvarry's analysis of these data indicates that in the maria as well as in the highlands, the thermal behavior of the surface can only be explained by assuming that it consists of dust exposed to a vacuum.

Gilvarry also compared the observed cooling rates in Mare Vaporum during eclipse with theoretical rates for a model surface consisting of rock covered with a thin layer of dust and also with those for a deep-dust model (suggested by Gold's hypothesis). A slightly better fit is found for Gold's model, but the observations are not sufficient to warrant a definite conclusion.

Observations of the polarization of moonlight have been described by many investigators. A summary of these observations and bibliographical data have been presented by Dollfus [48]. Relationships between the amount of polarization and the phase angle (sun-moon-earth), on the one hand, and albedo, on the other, have been established. Dollfus and Lyot [49] have found that terrestrial volcanic ash has optical properties resembling the surface of the moon. Baldwin [37], however, points out that in the production of polarization effects the particular form of the dust particles is more important than their composition.

Thermal radiation from the moon at radio frequencies has been investigated in detail [50]. Variations of radiation during a lunation indicate the low thermal conductivity that can only be explained if we assume the lunar surface material to be composed of dust. Good agreement of the observations with a model consisting of a dust layer at least 2 cm deep is obtained.

From numerous visual and photographic observations of the moon, Kuiper [51] concludes that there are two surface types in the bright areas of the moon. One type is found in those bright areas that border on dark ones. There the surface appears to be covered with material ejected from the dark regions when these were formed. The other surface type is found in certain regions located mostly in the southwest quadrant of the moon. Apparently these are ancient areas which have suffered little rearrangement by major meteoric impacts since the lunar crust was formed. A thick layer of interplanetary dust may have been accumulated in these areas. Kuiper states that this type of surface, having the consistency of crusty snow, may prove to be the softest for landing purposes.

Elsmore [52] has observed an occultation by the moon of the Crab nebula, a cosmic radio source. The observed duration of the occultation at a wavelength of 3.7 m was 0.4 ± 0.26 min longer than the duration expected on the basis of the moon's motion across the nebula. The difference can be ascribed to refraction of the radio signal at the moon's limb if a very tenuous lunar atmosphere is postulated. Elsmore derives a density for the lunar atmosphere at the surface that is about 2×10^{-13} the density of the earth's sea-level atmosphere. Because of the relatively high probable error of the measurement on which this conclusion is based, the actual atmospheric density may well differ by a factor of 10 from that quoted. Attempts at determining the density of the lunar atmosphere from observations employing optical wavelengths yield systematically very low values. The sensitivity of these methods, however, is much less than that of Elsmore's method.

REFERENCES

1. U. S. DEPARTMENT OF COMMERCE, COAST AND GEODETIC SURVEY, *Manual of Geodetic Triangulation.* Washington: U. S. Government Printing Office, 1959.

2. R. A. HIRVONEN, "The Size and Shape of the Earth," *Advances in Geophysics*, vol. 5, p. 93, H. E. Landsberg and J. van Mieghem, editors. New York: Academic Press Inc., 1956.

3. A. A. MICHELSON, "Measurement of the Velocity of Light Between Mount Wilson and Mount San Antonio," *Astrophys. J.*, **65**, 1, 1927.

4. See p. 101 of reference 2.

5. HAROLD JEFFREYS, *The Earth*, 4th ed. Cambridge: Cambridge University Press, 1959.

6. A. H. COOK, "Developments in Dynamical Geodesy," *Geophys. J. Roy. Astron. Soc.*, **2**, 238, 1959.

7. W. A. HEISKANEN, "The Columbus Geoid," *Transactions of the American Geophysical Union*, **38**, 841, 1957.

8. J. A. O'KEEFE, A. ECKELS, and R. K. SQUIRES, "Vanguard Measurements Give Pear-Shaped Component of Earth's Figure," *Science*, **129**, 565, 1959.

9. NAUTICAL ALMANAC OFFICE OF THE UNITED STATES NAVAL OBSERVATORY, *The American Ephemeris and Nautical Almanac.* Washington: U. S. Government Printing Office, published yearly.

10. G. P. KUIPER, ed., *The Earth as a Planet*, Chapters 7 through 13. Chicago: Chicago University Press, 1954.

11. L. SPITZER (Chapter 7) and F. L. WHIPPLE (Chapter 5), *The Atmospheres of the Earth and Planets*, G. P. Kuiper, ed. Chicago: Chicago University Press, 1952.

12. THOMAS F. MALONE, ed., *Compendium of Meteorology*, sections starting on pp. 3–12, 245–366. Boston: American Meteorological Society, 1951.

13. D. R. BATES, "The Earth's Upper Atmosphere," *Monthly Notices Roy. Astron. Soc.*, **109**, 221, 1949.

14. J. KAPLAN, "The Earth's Atmosphere," *American Scientist*, **41**, 49, 1953.

15. S. K. MITRA, *The Upper Atmosphere*, 2nd ed. Calcutta: Asiatic Society, 1952.

16. AMERICAN GEOPHYSICAL UNION, *Geophysical Monographs*, no. 2, pp. 1–142, 1958.

17. E. H. VESTINE, "Noctilucent Clouds," *J. Roy. Astron. Soc. Canada*, **28**, 249, 303, 1934.

18. See pp. 179 and 219–223 of reference 15.

19. G. V. GROVES, "Air Density in the Upper Atmosphere from Satellite Orbit Observations," *Nature Suppl. no. 4*, **184**, 178, 1959.

20. W. A. SMART, *Spherical Astronomy*, 4th ed. Chapter 3. Cambridge: Cambridge University Press, 1949.

21. B. GARFINKEL, "An Investigation in the Theory of Astronomical Refraction," *Astron. J.*, **50**, 169, 1944.

22. E. SCHOENBERG, "Theoretische Photometrie," *Handbuch der Astrophysik*, vol. 2, part 1, p. 268. G. Eberhard, A. Kohlschütter, H. Ludendorff, eds. Berlin: Springer, 1929.

23. JOEL STEBBINS, "On Atmospheric Extinction," *Proc. Nat. Sci. Found.* p. 75, J. B. Irwin, ed. *Astronomical Photoelectric Conference*, 1953.

24. H. C. VAN DE HULST. See Chapter 3 of reference 11.

25. F. A. JENKINS and H. E. WHITE, *Fundamentals of Optics*, 2nd ed., p. 295. New York: McGraw-Hill, 1950.

26. G. KELLER, W. M. PROTHEROE, P. E. BARNHART, and J. GALLI, "Investigations of Stellar Scintillation and the Behavior of Telescopic Images," *Perkins Observatory Reprints*, no. 39, 1956. Chapter 8 contains a bibliography on this general subject.

27. WILLIAM A. BAUM, "Counting Photons—One by One," *Sky and Telescope*, **14**, 265, 1955.

28. J. H. DeWITT, R. H. HARDIE, and C. K. SEYFERT, "A Seeing Compensator Employing Television Techniques," *Sky and Telescope*, **17**, 8, 1957.

29. MARTIN SCHWARZSCHILD, "Photographs of the Solar Granulation Taken from the Stratosphere," *Astrophys. J.*, **130**, 345, 1959.

30. "Project Stratoscope Obtains Finest Pictures of the Sun," *Sky and Telescope*, **19**, 79, 1959.

31. H. N. RUSSELL, R. S. DUGAN, and J. Q. STEWART, *Astronomy*, Vol. I rev. ed. Chapter 6. Boston: Ginn & Co., 1945.

32. H. UREY, *The Planets*, Chapter 2. New Haven: Yale University Press, 1952.

33. COLLOQUES INTERNATIONAUX DU CENTRE NATIONAL DE LA RECHERCHE SCIENTIFIQUE, *Constantes Fondamentales de l'Astronomie*. Paris: C.N.R.S., 1950.

34. ERNEST W. BROWN, *An Introductory Treatise on the Lunar Theory*, p. 126. Cambridge: Cambridge University Press, 1896.

35. W. DE SITTER, "On the System of Astronomical Constants," *Bull. Astron. Inst. Netherlands*, **8**, 224, 1938.

36. See p. 157 of reference 5.

37. R. B. BALDWIN, *The Face of the Moon*. Chicago: Chicago University Press, 1949.

38. MARY A. BLAGG and K. MÜLLER, *Named Lunar Formations.* London: Percy Lund, Humphries & Co. Ltd., 1935.

39. KAREL ANDEL, "Lunar Map," *Sky and Telescope,* **15,** insert between pp. 126 and 127, 1956. The map is sold by Sky Publishing Corporation, Cambridge 38, Mass. By arrangement with Sky Publishing Co., Andel's map is included in the present book.

40. WALTER GOODACRE, *The Moon.* Bournemouth: Pardy & Sons, 1931.

41. T. GOLD, "The Lunar Surface," *Monthly Notices Roy. Astron. Soc.,* **115,** 585, 1956.

42. See p. 131 of reference 37.

43. G. P. KUIPER, "On the Origin of the Lunar Surface Features," *Proc. Natl. Acad. Sci., U. S.,* **40,** 1096, 1954.

44. J. J. GILVARRY, "The Nature of the Lunar Maria," *Astrophys. J.,* **127,** 751, 1958.

45. J. J. GILVARRY and J. E. HILL, "The Impact of Large Meteorites," *Astrophys. J.,* **124,** 610, 1956.

46. EDISON PETTIT and SETH B. NICHOLSON, "Lunar Radiation and Temperatures," *Astrophys. J.,* **71,** 102, 1930.

47. EDISON PETTIT, "Radiation Measurements on the Eclipsed Moon," *Astrophys. J.,* **91,** 408, 1940.

48. A. DOLLFUS, "Étude des Planètes par la Polarisation de leur Lumière," *Suppléments aux Annales d'Astrophysique,* no. 4, p. 68, 1957.

49. B. LYOT, "Recherches sur la Polarisation de la Lumière des Planètes et de Quelques Substances Terrestres," *Annales de l'Observatoire de Meudon,* **8,** 169, 1929.

50. R. N. BRACEWELL, ed., *Paris Symposium on Radio Astronomy,* p. 3. Stanford: Stanford University Press, 1959.

51. G. P. KUIPER, "The Moon," *J. Geophys. Research,* **64,** 1713, 1959.

52. See p. 47 of reference 50.

LIST OF PLATES

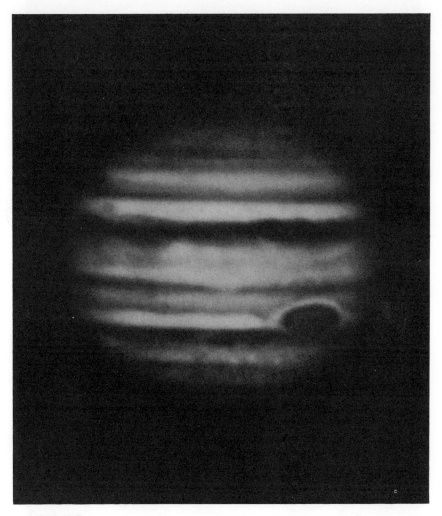

PLATE I

Planet Jupiter in blue light, 200-in. telescope photograph. Cloud bands and the large red spot are seen. Notice the oblateness and darkening toward the limb. (Mt. Wilson and Palomar Observatories photograph)

PLATE II

Planet Saturn. Cassini's division separates ring A (outermost) from ring B. The crape ring is too faint to appear in the photograph. (Photograph by E. C. Slipher, Lowell Observatory)

(a)

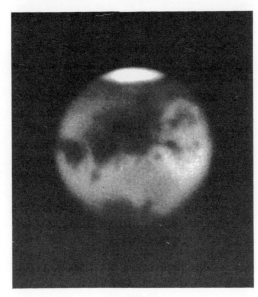

(b)

PLATE III

Mars, July 27, 1939. The south polar cap is at the top. (Photograph by
E. C. Slipher with 27-in. Lamont-Hussey telescope, Bloemfontein, South Africa)

PLATE IV

The moon in full phase, with south pole at the top. Crater Tycho, near the top, is the center of an extended ray system. For identification of other features, see lunar map in the appendix. (Lick Observatory photograph)

PLATE V

Lunar detail. The large crater to the right of center and up is Alphonsus (diameter 130 km), which shows a system of rills. The extensive parallel grooves oriented diagonally from top left were apparently plowed by high-speed masses following grazing trajectories. This is one of the very finest lunar photographs available. (Photograph by Dr. F. G. Pease, Mt. Wilson Observatory)

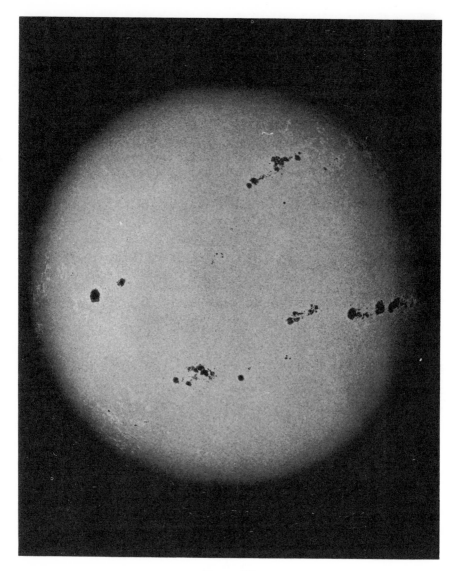

PLATE VI

The sun at sunspot maximum, December 21, 1957. The darkening toward the limb and faculae near the limb are clearly visible. (Mt. Wilson and Palomar Observatories photograph)

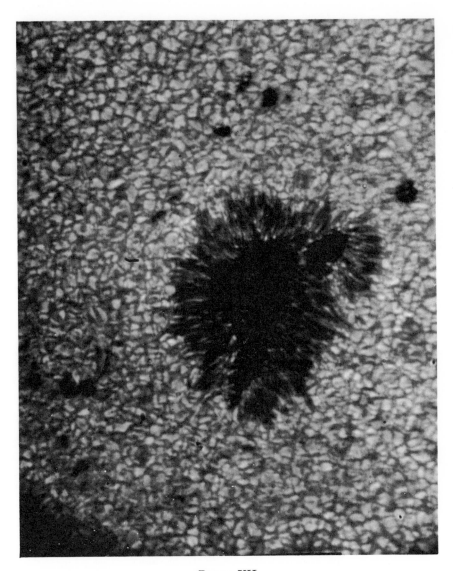

PLATE VII

Solar granulation and a sunspot photographed from a high-altitude balloon, August 17, 1959. (Photograph by Project Stratoscope of Princeton University, sponsored by the Office of Naval Research and the National Science Foundation)

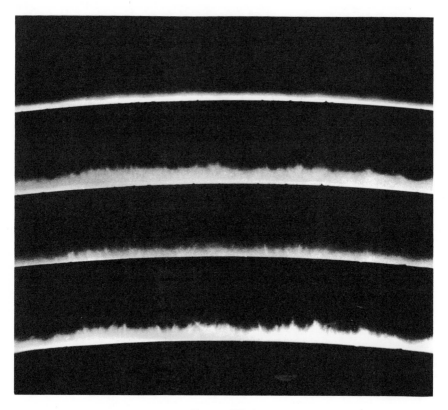

PLATE VIII

Solar spicules observed with a coronograph. Four photographs of the solar limb are shown, taken with different intensities of sunlight. The lower two exposures were obtained with an H_α-filter having a band width of 3 A. The upper two exposures were made with a filter of similar band width but with central transmission at 3 A from H_α. (Sacramento Peak Observatory, Geophysics Research Directorate, AFCRC)

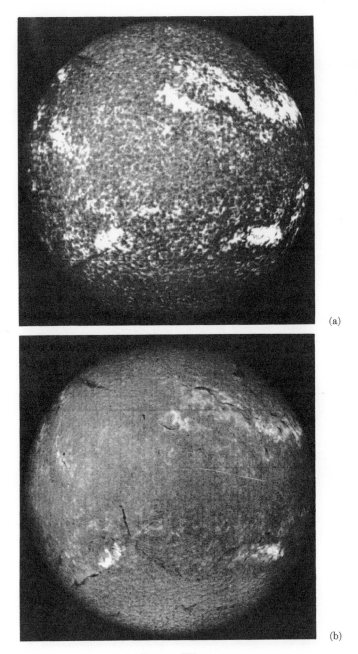

(a)

(b)

PLATE IX

Spectroheliograms of the sun: (a) in the light of the ionized calcium K-line;
(b) in the light of the hydrogen H_α-line. Both photographs were taken on July 11,
1955. (By courtesy of Osservatorio Astrofisico di Arcetri, Italy)

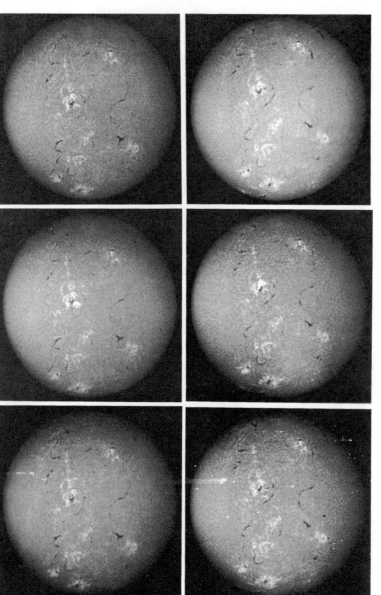

PLATE X

Flare activity in the sun photographed in the light of Hα. In each photograph, north is at the top and east is to the left. The photographs in the top row were taken on July 14, 1959, at 22:00, 22:29, and 23:00 U.T., respectively. The photographs in the lower row were taken the next day at 12:30, 13:14, and 14:50 U.T. Note flares near center of the disk, which appear at maximum in the middle photograph of each row. The flare on the top row was of importance 1 and that on the bottom row of importance 2. (Photographs by McMath-Hulburt Observatory of the University of Michigan)

PLATE XI

Comet Arend-Roland showing a "spike" oriented approximately toward the sun. The explanation of the spike is presented in Fig. 6–9. (Photograph by H. L. Giclas, Lowell Observatory, provided by *Sky and Telescope*)

CHAPTER 4

CELESTIAL DYNAMICS I: THE TWO-BODY PROBLEM

The motion of a planet around the sun, the motion of a satellite around a planet, the motion of a star around the center of the galaxy, all are governed by *central forces*. A central force is one which acts along the line joining the two bodies. The inverse-square law of force, enunciated by Newton, is perhaps the classical example of such a force. Newton's law of gravitation states that *any material body in the universe attracts any other body with a force which varies directly as the product of their masses and inversely as the square of the distance between them, and which acts along the line joining the bodies.* If m_1 and m_2 are the masses and d is the distance between them, then

$$F = - \frac{k^2 m_1 m_2}{d^2},$$ (4-1)

where the constant of proportionality, k^2, has the value 6.67×10^{-8} when the masses are in grams and the distance in centimeters. The negative sign denotes an attractive force. For astronomical purposes we adopt the solar mass as the unit of mass and the astronomical unit as the unit for distance (see Section 2–2). The resulting value of the gravitational constant is

$$k = 0.01720209895.$$ (4-2)

In this chapter we shall discuss the motion of one body under the gravitational influence of a second body. The two will be considered to be isolated in the universe. This constitutes the "two-body" problem in celestial dynamics. The central force involved may be either one of attraction or of repulsion.

4–1 Preliminary definitions. Dynamical concepts are most readily and compactly described by the use of vector notation. With respect to an origin O (Fig. 4–1) fixed relative to the average position of the stars in the solar neighborhood, let

$\mathbf{r}$ = the position vector of a mass m,

$\mathbf{v}$ = the velocity vector of a mass m, and

$\mathbf{a}$ = the acceleration vector of a mass m.

These quantities are functions of time. In terms of the unit vectors

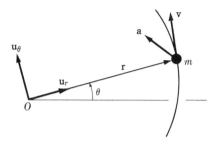

FIG. 4–1. Vector designations for curvilinear motion.

$\mathbf{u}_r$ and $\mathbf{u}_\theta$ (Fig. 4–1) in the polar coordinate system,

$$\mathbf{r} = r\mathbf{u}_r, \tag{4–3}$$

$$\mathbf{v} = \dot{\mathbf{r}} = \dot{r}\mathbf{u}_r + r\dot{\theta}\mathbf{u}_\theta, \tag{4–4}$$

and, since $\dot{\mathbf{u}}_r = \dot{\theta}\mathbf{u}_\theta$ and $\dot{\mathbf{u}}_\theta = -\dot{\theta}\mathbf{u}_r$,

$$\mathbf{a} = \dot{\mathbf{v}} = \ddot{\mathbf{r}} = (\ddot{r} - r\dot{\theta}^2)\mathbf{u}_r + (2\dot{r}\dot{\theta} + r\ddot{\theta})\mathbf{u}_\theta. \tag{4–5}$$

Similarly, we introduce the dynamical concepts:

$$\mathbf{p} = m\mathbf{v} = \text{the linear momentum}, \tag{4–6}$$

$$\mathbf{L} = \mathbf{r} \times m\mathbf{v} = \text{the angular momentum}, \tag{4–7}$$

$$\mathbf{F} = \mathbf{F}(\mathbf{r}, \dot{\mathbf{r}}, t) = \text{the force}, \tag{4–8}$$

$$\mathbf{N} = \mathbf{r} \times \mathbf{F} = \text{the moment, or torque, of } \mathbf{F} \text{ about } O. \tag{4–9}$$

In the following discussion, the force $\mathbf{F}$ will, in most cases, be independent of velocity and, explicitly, of time.

Having defined the dynamical and kinematical concepts, we may state the relationships between them by the *laws of motion*, enunciated by Newton in 1686. Except for special relativity effects due to high velocity and to moving coordinate systems, these laws are adequate to describe mathematically the dynamical behavior of material bodies anywhere in the observable universe. These newtonian laws of motion are:

I. A constant mass remains at rest or moves with constant velocity in a straight line unless acted upon by a force.

II. A mass acted upon by a force moves so that the time rate of change of its linear momentum equals the force.

III. If two masses act on each other, the force exerted by the first on the second is equal in magnitude and opposite in direction to the force exerted by the second on the first.

Law II is expressed mathematically by

$$d(m\mathbf{v})/dt = \mathbf{F}. \tag{4–10}$$

Law III for masses m_1 and m_2 is expressed by

$$\mathbf{F}_1 = -\mathbf{F}_2. \tag{4–11}$$

From Eqs. (4–7), (4–9), and (4–10) we observe that since $\dot{\mathbf{r}} \times m\mathbf{v} = \mathbf{v} \times m\mathbf{v} = 0$,

$$\dot{\mathbf{L}} = \dot{\mathbf{r}} \times m\mathbf{v} + \mathbf{r} \times m\dot{\mathbf{v}} = \mathbf{r} \times \mathbf{F} = \mathbf{N}. \tag{4–12}$$

This is Newton's second law for rotational motion, which states that *the time rate of change of the angular momentum equals the torque (or moment)*.

Consider a constant mass m moving on a straight line with speed v. By Eq. (4–10),

$$m(dv/dt) = F,$$

where F is the force acting on it. Multiplying by v and integrating, we have, since $v\,dt = ds$, an element of distance:

$$\int_{v_0}^{v} mv\,dv = \int_{t_0}^{t} Fv\,dt = \int_{s_0}^{s} F\,ds. \tag{4–13}$$

The integral on the left is $\frac{1}{2}mv^2 - \frac{1}{2}mv_0^2$, which is the change in *kinetic energy* of the mass in the time interval $t - t_0$. The integral on the right is the *work done by F* in the same time interval. Thus we obtain

$$\tfrac{1}{2}m(v^2 - v_0^2) = W(s) - W(s_0), \tag{4–14}$$

where s is the position of m at time t, and s_0 is its position at time t_0.

The work function $W(s)$ is customarily replaced by another scalar function of position, the *potential energy*. This function $V(s)$ is by agreement taken to be $-W(s)$. Thus the difference in potential energy between two positions s_0 and s is the work done by an outside source in moving m from s_0 to s against the force F. Equation (4–14) then states that

$$\tfrac{1}{2}mv^2 + V(s) = \tfrac{1}{2}mv_0^2 + V(s_0) = E. \tag{4–15}$$

This expresses the *conservation of energy* for the motion, and E is called the total energy of the system. Consider, for example, motion under an inverse-square force. Let M be a mass at O (Fig. 4–2) and let m be located at s at time t. The force acting on m is

$$F = -k^2 Mm/s^2.$$

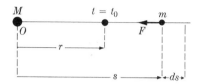

FIG. 4–2. Linear motion of a mass under a force obeying the inverse-square law.

The work done by F in the displacement ds is $-(k^2Mm/s^2)\,ds$. If $s = r$ at time t_0, then by Eq. (4–13),

$$m\int_{v_0}^{v} v\,dv = -k^2Mm\int_{r}^{s} \frac{ds}{s^2}.$$

Hence,

$$\frac{1}{2}\,m(v^2 - v_0^2) = k^2Mm\left[\frac{1}{s} - \frac{1}{r}\right],$$

or

$$\frac{1}{2}\,mv^2 - \frac{k^2Mm}{s} = \frac{1}{2}\,mv_0^2 - \frac{k^2Mm}{r}.$$

If the particle recedes to infinity $(s \to \infty)$, we find

$$\frac{1}{2}\,m(v^2 - v_0^2) + \frac{k^2Mm}{r} = 0.$$

The potential energy at r is, by definition,

$$V(r) = -\frac{k^2Mm}{r}. \tag{4–16}$$

In this case, $V(r)$ is the work done in moving the particle from r to infinity.

When the force depends upon the position in space, we say that a *force field* exists. The potential energy per unit mass at a given point in this field is called the *potential*. For example, if the field is due to an inverse-square force produced by a mass M, then $\mathbf{F} = -(k^2Mm/r^2)\mathbf{u}_r$, and the potential associated with the point at distance r from M is $U = -(k^2M/r)$. We observe that $\mathbf{F}$ and U are related by

$$\mathbf{F} = -m\,\frac{dU}{dr}\,\mathbf{u}_r. \tag{4–17}$$

Note that U is a scalar function of position, whereas $\mathbf{F}$ is a vector function of position.

4–2 Potential due to a sphere. One of the important theorems proved by Newton in his *Principia* (1686) states that a homogeneous sphere attracts an exterior point mass as though the total mass of the sphere were concentrated at its center. We shall present proof, since it is fundamental in any study of gravitational attraction among astronomical objects. Figure 4–3 shows a section of spherical shell of unit thickness and radius a on which an element of surface area is

$$d\sigma = a^2 \sin \varphi \, d\varphi \, d\theta.$$

Let the density of the shell be ρ and be constant over the shell. Let a unit mass be placed, as shown, at $(0, 0, h)$. Then the force of attraction between the surface element $d\sigma$ and the mass at h, directed along the line joining $(0, 0, h)$ and $d\sigma$, is

$$dF = - \frac{k^2 \rho \, d\sigma}{p^2}.$$

By symmetry the components of these vector elements parallel to the xy-plane balance one another as the surface elements $d\sigma$ are summed over the surface. The z-components, however, are additive, and hence for the total force at $(0, 0, h)$ due to the shell, we have

$$F_z = - \int_S \frac{k^2 \rho \, d\sigma \cos \alpha}{p^2}, \tag{4–18}$$

where α is the angle between the z-axis and the direction of the force vector for a given surface element.

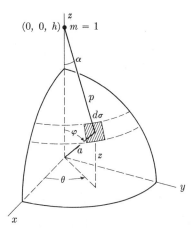

FIG. 4–3. Attraction of a spherical shell on an external mass.

In carrying out the integration of Eq. (4–18), it is convenient to use p as the variable. We have

$$p^2 = a^2 + h^2 - 2ah \cos \varphi,$$

$$\cos \alpha = \frac{h - z}{p} = \frac{h - a \cos \varphi}{p} = \frac{h^2 - a^2 + p^2}{2ph},$$

$$p \, dp = ah \sin \varphi \, d\varphi,$$

$$d\sigma = a^2 \sin \varphi \, d\varphi \, d\theta = \frac{ap \, dp \, d\theta}{h}.$$

Therefore,

$$F_z = - \frac{k^2 \rho a}{2h^2} \int_{\theta=0}^{2\pi} \int_{p=h-a}^{h+a} \frac{h^2 + p^2 - a^2}{p^2} \, dp \, d\theta.$$

This is easily integrated and yields

$$F_z = - \frac{4\pi k^2 \rho a^2}{h^2}. \tag{4–19}$$

But $4\pi a^2 \rho$ is the total mass of the shell. *Hence F_z is the same as if this total mass were concentrated at the origin, which is the center of the shell.*

A solid sphere whose density is a function only of the distance from its center can be considered to be composed of concentric shells of uniform density. For the solid sphere of radius R, the total force on an external point at distance h from its center is

$$F = - \frac{4\pi k^2}{h^2} \int_{a=0}^{R} \rho a^2 \, da.$$

But the total mass of the solid sphere is

$$M = 4\pi \int_{a=0}^{R} \rho a^2 \, da;$$

hence the force due to the solid sphere is

$$F = - \frac{k^2 M}{h^2}, \tag{4–20}$$

and we see that a solid sphere with a spherically symmetric density distribution attracts an exterior mass as if the entire mass of the sphere were concentrated at its center. In all succeeding discussion *we can treat such spherical bodies as if they were point masses.*

The potential due to a spherical body (homogeneous in concentric layers) associated with a point outside the sphere is therefore

$$U = -\frac{k^2 M}{r},$$ (4-21)

where r is the distance from the center of the sphere to the point concerned.

4-3 Two-body equations of motion. Consider two masses m_1 and m_2 situated at points $\mathbf{r}_1$ and $\mathbf{r}_2$ in a fixed coordinate system in which Newton's laws of motion hold (Fig. 4-4). Let $\mathbf{u}_r$ be a unit vector in the direction from m_1 to m_2. Then by the law of gravitation the force due to m_2 which acts on m_1 is $(k^2 m_1 m_2 / r^2)\mathbf{u}_r$, where r is the distance between the masses. Similarly, the force on m_2 due to m_1 is $(-k^2 m_1 m_2 / r^2)\mathbf{u}_r$. Since $\mathbf{u}_r = \mathbf{r}/r$, we have therefore by Newton's second law,

$$m_1 \ddot{\mathbf{r}}_1 = +\frac{k^2 m_1 m_2 \mathbf{r}}{r^3}$$ (4-22)

and

$$m_2 \ddot{\mathbf{r}}_2 = -\frac{k^2 m_1 m_2 \mathbf{r}}{r^3}.$$ (4-23)

We add Eqs. (4-22) and (4-23) to obtain $m_1 \ddot{\mathbf{r}}_1 + m_2 \ddot{\mathbf{r}}_2 = 0$, which may be integrated twice to yield

$$m_1 \mathbf{r}_1 + m_2 \mathbf{r}_2 = \mathbf{c}_1 t + \mathbf{c}_2,$$

where $\mathbf{c}_1$ and $\mathbf{c}_2$ are vector constants. But the center of mass of the system is defined to be

$$\mathbf{R} = \frac{m_1 \mathbf{r}_1 + m_2 \mathbf{r}_2}{M},$$ (4-24)

where $M = m_1 + m_2$. Hence

$$\mathbf{R} = \frac{\mathbf{c}_1}{M} t + \frac{\mathbf{c}_2}{M}.$$ (4-25)

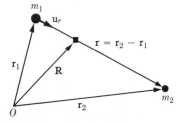

FIG. 4-4. Center of mass for the two-body problem.

Equation (4–25) states that *the center of mass moves uniformly on a straight line in space.*

Equations (4–22) and (4–23) describe the motion of the two masses in space relative to the origin O. To find the motion relative to the center of mass, we write

$$\mathbf{r}_1 = \mathbf{R} + \mathbf{r}_1',$$

$$\mathbf{r}_2 = \mathbf{R} + \mathbf{r}_2',$$

where $\mathbf{r}_1'$ and $\mathbf{r}_2'$ denote the position vectors of m_1 and m_2, respectively, relative to the center of mass. Then $\mathbf{r} = \mathbf{r}_2' - \mathbf{r}_1'$ (Fig. 4–4) and, since $\ddot{\mathbf{R}} = 0$, we have $m_1\ddot{\mathbf{r}}_1 = m_1\ddot{\mathbf{r}}_1'$ and $m_2\ddot{\mathbf{r}}_2 = m_2\ddot{\mathbf{r}}_2'$; hence,

$$m_1\ddot{\mathbf{r}}_1' = + \frac{k^2 m_1 m_2(\mathbf{r}_2' - \mathbf{r}_1')}{r^3} \tag{4–26}$$

and

$$m_2\ddot{\mathbf{r}}_2' = - \frac{k^2 m_1 m_2(\mathbf{r}_2' - \mathbf{r}_1')}{r^3}. \tag{4–27}$$

But $m_1\mathbf{r}_1' + m_2\mathbf{r}_2' = 0$, and therefore $\mathbf{r}_2'$ can be eliminated from Eq. (4–26), while $\mathbf{r}_1'$ can be eliminated from Eq. (4–27), so that

$$m_1\ddot{\mathbf{r}}_1' = -k^2 m_1 m_2\left[1 + \frac{m_1}{m_2}\right]\frac{\mathbf{r}_1'}{r^3}$$

and

$$m_2\ddot{\mathbf{r}}_2' = -k^2 m_1 m_2\left[1 + \frac{m_2}{m_1}\right]\frac{\mathbf{r}_2'}{r^3}.$$

In terms of the total mass M, these equations become

$$\ddot{\mathbf{r}}_1' = - \frac{k^2 M}{r^3}\,\mathbf{r}_1' \tag{4–28}$$

and

$$\ddot{\mathbf{r}}_2' = - \frac{k^2 M}{r^3}\,\mathbf{r}_2'. \tag{4–29}$$

Equations (4–28) and (4–29) give the accelerations of the two masses *relative to the center of mass.* It is apparent that they are of the same form as Eqs. (4–22) and (4–23), with M replacing m_2 or m_1, respectively. If we knew the constants $\mathbf{c}_1$ and $\mathbf{c}_2$ in Eq. (4–25), we could theoretically determine the positions of m_1 and m_2 in space by solving Eqs. (4–28) and (4–29). Since these constants are not known, we shall be content with the solution for the relative motion of one body with respect to the other. If m_1 is considered to be at the origin of the two-body system, we find

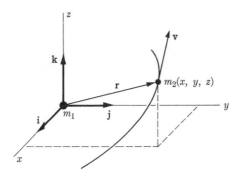

FIG. 4–5. Relative motion of m_2 around m_1.

from Eqs. (4–28) and (4–29),

$$\ddot{\mathbf{r}}_2' - \ddot{\mathbf{r}}_1' = -\frac{k^2 M}{r^3}(\mathbf{r}_2' - \mathbf{r}_1')$$

or

$$\ddot{\mathbf{r}} = -\frac{k^2 M}{r^3}\mathbf{r}, \qquad (4\text{–}30)$$

where $\mathbf{r}$ is the relative radius vector. This equation expresses the acceleration of the mass m_2 in its orbit about m_1. In the planetary system, for example, m_1 would be the sun and m_2 one of the planets. For an earth satellite, m_1 would be the mass of the earth and m_2 that of the satellite.

Equation (4–30) can easily be converted into cartesian form for computational purposes (Fig. 4–5). In terms of the unit vectors $\mathbf{i}$, $\mathbf{j}$, $\mathbf{k}$ shown in Fig. 4–5, $\mathbf{r} = x\mathbf{i} + y\mathbf{j} + z\mathbf{k}$; hence

$$\begin{aligned}
\ddot{x} &= -k^2 M x (x^2 + y^2 + z^2)^{-3/2}, \\
\ddot{y} &= -k^2 M y (x^2 + y^2 + z^2)^{-3/2}, \\
\ddot{z} &= -k^2 M z (x^2 + y^2 + z^2)^{-3/2}.
\end{aligned} \qquad (4\text{–}31)$$

The x-axis is usually taken to be the direction toward the vernal equinox, and the z-axis that toward the north pole of the ecliptic plane. This heliocentric system (Section 1–8) is used for analyses of planetary motions in which m_1 is the mass of the sun. For describing the motion of an earth satellite, one uses the equatorial system (Section 1–4), with the x-axis again pointing to the vernal equinox. Here the xy-plane is the plane of the earth's equator.

Equations (4–31) are second-order differential equations; the solution of each introduces two constants of integration. Hence to specify adequately the position of m_2 relative to m_1 at any time, six initial conditions must be available. If we know the position (three coordinates) and velocity (three components) of m_2 at any *one* instant, its position at any later time can be determined. In astronomical applications, however, these data are usually not available. Therefore one has to rely on the astronomical coordinates (α, δ) for three instants of time to evaluate the constants of integration required in any particular application of Eqs. (4–31). The central problem in orbit theory is to calculate from the observed positions of an object in the sky the elements of the orbit of the object.

4–4 Integrals of area, angular momentum, and of energy. Several fundamental properties of central-force motion can be exhibited by means of Eq. (4–30). These are embodied in the integrals of area, angular momentum, and energy. Although we shall use here for illustrative purposes the inverse-square law of force, it should be emphasized that the *fundamental properties are independent of the exact functional form of the force law.*

In all that follows, the masses m_1 and m_2 will be considered to be spherical bodies, homogeneous in concentric layers, so that they attract each other as if the total mass were concentrated at the centers of the respective spheres. They are in effect "point masses." This assumption has indeed been implicit in the foregoing analysis. We consider the motion of m_2 relative to m_1. From Eq. (4–30) $m_2\ddot{\mathbf{r}} = -k^2 M m_2 \mathbf{r}/r^3$, and, taking the vector product by $\mathbf{r}$, we find

$$\mathbf{r} \times m_2\ddot{\mathbf{r}} = \mathbf{r} \times m_2\dot{\mathbf{v}} = \frac{d}{dt}(\mathbf{r} \times m_2\mathbf{v}) = 0.$$

Thus $\mathbf{r} \times m_2\mathbf{v}$ is a constant of the motion which, by Eq. (4–7), is the angular momentum $\mathbf{L}$. From this we draw two conclusions:

(a) The angular momentum under the inverse-square law of force remains constant with time during the motion of m_2 about m_1.

(b) The vector $\mathbf{L}$ always points in the same direction during the motion, and hence *the orbit lies in a plane containing $\mathbf{r}$ and $\mathbf{v}$.*

From Eqs. (4–3) and (4–4) we find (see Fig. 4–1)

$$\mathbf{r} \times m_2\mathbf{v} = r\mathbf{u}_r \times m_2\{\dot{r}\mathbf{u}_r + r\dot{\theta}\mathbf{u}_\theta\} = m_2 r^2 \dot{\theta}\mathbf{u}_L,$$

where (r, θ) are polar coordinates in the plane of the orbit and $\mathbf{u}_L$ is a unit vector perpendicular to $\mathbf{u}_r$ and $\mathbf{u}_\theta$, that is, perpendicular to the plane of the orbit. We let

$$r^2\dot{\theta} = h, \tag{4–32}$$

where h is a constant. Then the magnitude of the constant angular momentum is $m_2 h$. Furthermore, we note that $\frac{1}{2}r^2\dot{\theta} = \frac{1}{2}h$ is the *rate at which $\mathbf{r}$ sweeps out the area of the orbit.* This is the analytical form of Kepler's second law, which was introduced in Section 2–2. As mentioned above, these conclusions follow for *any* central force whether it obeys the inverse-square law or not.

Again from the equation of motion, Eq. (4–30), $m_2\dot{\mathbf{v}} = -k^2 M m_2 \mathbf{r}/r^3$. Taking the scalar product by $\mathbf{v}$ and rewriting, we find

$$m_2 \mathbf{v} \cdot \dot{\mathbf{v}} = -\frac{k^2 M m_2}{r^3} (\mathbf{v} \cdot \mathbf{r}),$$

$$\frac{d}{dt}\left(\frac{1}{2} m_2 v^2\right) = -\frac{k^2 M m_2}{r^3} [\dot{r}\mathbf{u}_r + r\dot{\theta}\mathbf{u}_\theta] \cdot [r\mathbf{u}_r],$$

$$\frac{d}{dt}\left(\frac{1}{2} m_2 v^2\right) = +k^2 M m_2 \frac{d}{dt}\left(\frac{1}{r}\right).$$

Hence

$$\frac{1}{2} m_2 v^2 - \frac{k^2 M m_2}{r} = E, \tag{4–33}$$

where E is a constant of integration. The first term is the kinetic energy of motion, the second term is the potential energy, and E is the total energy of the system. Equation (4–33) expresses the *conservation of energy* for the two-body motion.

The results obtained from Eqs. (4–32) and (4–33) are called the *areal* and *energy* integrals, respectively.

4–5 Polar equation of the orbit. In terms of radial and transverse components [see Eq. (4–5)], we may write Eq. (4–30) as

$$(\ddot{r} - r\dot{\theta}^2)\mathbf{u}_r + (2\dot{r}\dot{\theta} + r\ddot{\theta})\mathbf{u}_\theta = -\frac{k^2 M}{r^2} \mathbf{u}_r.$$

Hence, equating components, we find

$$\ddot{r} - r\dot{\theta}^2 = -\frac{k^2 M}{r^2} \tag{4–34}$$

and

$$2\dot{r}\dot{\theta} + r\ddot{\theta} = 0.$$

Upon integration the latter yields immediately

$$r^2\dot{\theta} = h, \tag{4–35}$$

the integral of areal velocity. Eliminating $\dot{\theta}$ from Eq. (4–34) by means of Eq. (4–35), we obtain

$$\ddot{r} - \frac{h^2}{r^3} + \frac{k^2 M}{r^2} = 0. \tag{4–36}$$

For convenience, let $k^2M = \mu$ and $r = 1/u$. Then $\dot{r} = -h(du/d\theta)$, and $\ddot{r} = -h^2u^2(d^2u/d\theta^2)$. Substituting in Eq. (4–36), we find

$$\frac{d^2u}{d\theta^2} + u = \frac{\mu}{h^2}, \tag{4–37}$$

which is the differential equation representing the orbit. Its solution is

$$u = A\cos(\theta - \omega) + \frac{\mu}{h^2}, \tag{4–38}$$

where A and ω are constants of integration. Hence,

$$r = \frac{h^2/\mu}{1 + (Ah^2/\mu)\cos(\theta - \omega)}. \tag{4–39}$$

This is identical with the standard polar equation of a conic section, namely

$$r = \frac{p}{1 + e\cos(\theta - \omega)},$$

if we set

$$p = \frac{h^2}{\mu}, \qquad e = \frac{Ah^2}{\mu}.$$

The eccentricity e and the parameter p determine the type and shape of the conic.

It is clear that when $\theta = \omega$, r attains its minimum value, $r_p = p/(1 + e)$, and when $\theta - \omega = 180°$, r attains its maximum value, $r_a = p/(1 - e)$. Furthermore, we observe that if $e = 1$, r_a becomes infinite and $r_p = \frac{1}{2}p$. The type of conic depends upon e. If $e = 1$, the orbit is a parabola; if $e < 1$, it is an ellipse; if $e > 1$, it is an hyperbola; if $e = 0$, it is a circle.

Figure 4–6 shows an orbit for $e < 1$, that is, for an ellipse. If a denotes the semimajor axis, then

$$2a = r_a + r_p.$$

Hence,

$$2a = p\left[\frac{1}{1 - e} + \frac{1}{1 + e}\right]$$

or

$$p = a(1 - e^2).$$

Thus the equation for the ellipse is

$$r = \frac{a(1 - e^2)}{1 + e\cos(\theta - \omega)}, \tag{4–40}$$

which has already been introduced as Eq. (2–3).

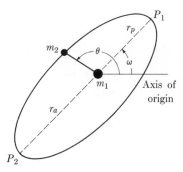

FIG. 4–6. Relative elliptic orbit in the two-body problem.

The constant of integration, A, in Eq. (4–39) is related to the energy of the system. At points P_1 and P_2 in Fig. 4–6, the velocity is entirely transverse, that is, $v = r\dot{\theta} = h/r = hu$. Hence the energy equation, Eq. (4–33), may be written

$$\tfrac{1}{2}m_2 h^2 u^2 - \mu m_2 u - E = 0.$$

Solving for u, we find

$$u = \frac{\mu}{h^2}\left[1 \pm \sqrt{1 + \frac{2Eh^2}{m_2\mu^2}}\right].$$

Equating the maximum value of u to that obtained from Eq. (4–38), we have

$$u_{\text{max}} = \frac{\mu}{h^2} + \frac{\mu}{h^2}\sqrt{1 + \frac{2Eh^2}{m_2\mu^2}} = A + \frac{\mu}{h^2},$$

and hence,

$$A = \frac{\mu}{h^2}\sqrt{1 + \frac{2Eh^2}{m_2\mu^2}}.$$

But $A = \mu e/h^2$, and therefore

$$e = \sqrt{1 + \frac{2Eh^2}{m_2\mu^2}}. \qquad (4\text{–}41)$$

It is clear that if

(a) $E = 0$, then $e = 1$, and the orbit is a parabola;
(b) $E < 0$, then $e < 1$, and the orbit is an ellipse;
(c) $E > 0$, then $e > 1$, and the orbit is an hyperbola.

For an elliptical orbit the areal-velocity constant, h, can be readily determined in terms of the orbital elements. We have

$$h^2 = \mu p = \mu a(1 - e^2) = k^2 M a(1 - e^2).$$

Hence

$$r^2\dot{\theta} = h = k\sqrt{Ma(1 - e^2)}. \tag{4-42}$$

Furthermore, by Eqs. (4–41) and (4–42), $E = -(k^2Mm_2/2a)$, so that Eq. (4–33) may be solved for v^2 to yield

$$v^2 = k^2M\left[\frac{2}{r} - \frac{1}{a}\right], \tag{4-43}$$

where v is the speed of m_2 in its elliptical orbit.

We are now in a position to obtain Kepler's third law (see Section 2–2). Let K denote the area swept out by the radius vector as m_2 moves around the orbit. Then by Eq. (4–42), for an elliptic orbit,

$$\frac{dK}{dt} = \frac{1}{2}k\sqrt{Ma(1 - e^2)},$$

and, upon integrating, we have

$$K = \tfrac{1}{2}k\sqrt{Ma(1 - e^2)}\,t + K_0,$$

where K_0 is a constant. In one period, P, of the motion, the area of the ellipse swept out is

$$\pi ab = \pi a^2(1 - e^2)^{1/2} = \tfrac{1}{2}k\sqrt{Ma(1 - e^2)}\,P,$$

where b is the semiminor axis. Therefore

$$P = \frac{2\pi a^{3/2}}{k\sqrt{M}}. \tag{4-44}$$

This is Kepler's third law. If m_1 is the solar mass and m_2 that of a planet, $M = m_1 + m_2$ is nearly the same for all planets, because the mass of the sun is very large compared with the mass of any planet. If P_1 and P_2 denote the periods of two planets, and a_1 and a_2 their mean distances from the sun, then

$$\left(\frac{P_1}{P_2}\right)^2 = \left(\frac{a_1}{a_2}\right)^3, \quad \text{approximately,}$$

as Kepler's third law asserts.

If the orbit of a celestial object is circular, it follows from Eq. (4–43) that the orbital speed is

$$v_c = k\sqrt{\frac{M}{r}}, \tag{4-45}$$

where r is the radius of the circle. Or, if the object is moving along a parabolic path, it follows from Eq. (4–43), with a indefinitely large, that

$$v_e = k\sqrt{\frac{2M}{r}} \qquad (4\text{–}46)$$

at any point whose distance is r from the center of force. This parabolic velocity, v_e, is called the *velocity of escape*. If the object has this speed of recession from the center of force, it will never return. Some comets, for example, move in approximately parabolic paths. At any given distance r from the sun, their speed is $\sqrt{2}$ times the speed they would have if they moved in a circular path of radius r about the sun.

Equation (4–44) is useful in determining the astronomical value of the gravitational constant k. The International Astronomical Union has adopted the value*

$$k = 0.01720209895. \qquad (4\text{–}47)$$

This value, called the *gaussian gravitational constant* (see Section 2–2), is based on Eq. (4–44), with $P = 365.2563835$ mean solar days, $a = 1$ A.U., and the mass of the sun as unit of mass. The value of k in Eq. (4–47) follows if m_2, the mass of the earth-moon system, is taken as $1/354710$ solar masses. These are the values used by Gauss in his evaluation of k.

The gaussian gravitational constant obviously depends upon the values of a and m_2 assigned to the earth-moon system. These values have been improved since Gauss's time, and it has therefore become customary to keep k fixed and to choose the unit of distance so that Eq. (4–44) will always be satisfied. Note the accuracy with which k has been determined. This far exceeds the accuracy with which the constant of gravitation has been determined in the laboratory, say in cgs units. Hence relative values of a, computed from Eq. (4–44) by means of the gaussian constant, exceed in accuracy the values in absolute units, say kilometers, if k is taken in cgs units and the time measured in seconds.

4–6 Position in the orbit. In this section we shall assume that the elements a, e, i, Ω, ω, T (defined in Section 2–1) of the orbit are known. The period P is also assumed to be known, since we shall restrict the discussion to elliptical orbits. The problem is to express the polar coordinates of a celestial object in its orbit as a function of time. From these one can compute the position of the object in the sky, that is, its right ascension and declination.

---

* See *Trans. Int. Astr. Union*, Vol. VI, p. 20, 1938.

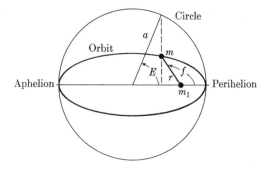

FIG. 4–7. Relationship of the eccentric circle to an elliptic orbit.

Figure 4–7 shows an elliptical orbit together with a circle, the *eccentric circle*, whose diameter equals the major axis of the ellipse. Let the polar coordinates of the planetary mass m at any time t be (r, f), where f is measured from perihelion, that is, the point in the ellipse nearest the sun, m_1.

We define the following quantities:

(a) The *true anomaly*, f, is the angle measured in the direction of motion from perihelion to the radius vector joining sun and planet.

(b) The *eccentric anomaly*, E, is the angle at the center of the circle measured from perihelion in the same direction as f, as shown in the figure.

(c) The *mean anomaly*, M, is defined by the relation

$$M = \frac{2\pi}{P}(t - T) = n(t - T), \qquad (4\text{–}48)$$

where, by Kepler's third law,

$$n = k(1 + m)^{1/2}a^{-3/2}. \qquad (4\text{–}49)$$

In Eq. (4–48) T is the time of perihelion passage. Gaussian units will be used throughout so that k has the value given by Eq. (4–47) and the mass of the sun is unity. The quantity n is in radians per unit time.

With the quantities n and M computed from the above relations for a given time t, the radius vector r and true anomaly f can be determined by means of a relation known as *Kepler's equation*,[*] $M = E - e \sin E$. When e is small, say $e < 0.2$, as for the orbits of planets and the majority of satellites, this equation is solvable for E by a series of successive ap-

---

[*] For a full discussion of this and the following equations see reference 4, pp. 164 ff.

proximations. The computations for r and f are as follows:

(a) Determine a first approximation to E, namely E_0, by the relation

$$E_0 = M + e \sin M + \frac{e^2}{2} \sin 2M. \tag{4-50}$$

(b) Calculate

$$M_0 = E_0 - e \sin E_0 \tag{4-51}$$

and

$$\Delta E_0 = \frac{M - M_0}{1 - e \cos E_0}. \tag{4-52}$$

(c) Then a second approximation for E is

$$E_1 = E_0 + \Delta E_0. \tag{4-53}$$

(d) From this revised value E_1 calculate a new value for M, say M_1, and, if necessary, repeat the approximation process until the value of E satisfies Kepler's equation with as high an accuracy as desired.

(e) Use the final value of E determined by the above process to calculate the values of r and f from the relations

$$r = a(1 - e \cos E) = \frac{a(1 - e^2)}{1 + e \cos f}, \tag{4-54}$$

$$\tan \frac{f}{2} = \sqrt{\frac{1 + e}{1 - e}} \tan \frac{E}{2}. \tag{4-55}$$

When e is near unity, the method sketched above for the solution of Kepler's equation will not converge fast enough to be useful. To obtain a first approximation to E, a graphical method may be used. Let Kepler's equation be rewritten in the form

$$\frac{1}{e}(E - M) = \sin E.$$

Let $y_1 = (1/e)(E - M)$ and $y_2 = \sin E$. Then a plot of y_1 as ordinate and E as abscissa is a straight line of slope $1/e$ and y-intercept equal to $-M/e$. Similarly the curve $y_2 = \sin E$ may be plotted on the same coordinate axes. The abscissa, E_0, of the intersection of straight line and sine curve, where $y_1 = y_2$, is a first approximation to the eccentric anomaly. Successive corrections to E_0 may then be made by Eqs. (4–52) and (4–53).

Analytical methods of solving for E when $e \sim 1$ are outlined in reference 4. More than 100 methods of solving Kepler's equation have been proposed.

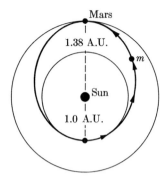

Fig. 4–8. Transfer orbit earth-Mars discussed in the example.

EXAMPLE. As an illustration, suppose that a small mass is ejected from the earth in such a way that it moves only under the influence of the sun until it reaches the orbit of Mars (see Fig. 4–8). When the mass m reaches Mars, suppose that the latter is at its perihelion point, where its distance from the sun is 1.38 A.U. Mars reaches perihelion late in August. We shall assume that the ejection from the earth's orbit takes place at 1.0 A.U. from the sun, and that all orbits are in the plane of the ecliptic.

The orbit of the small mass will have a major axis

$$2a = 2.38 \text{ A.U.},$$

and hence

$$a = 1.19 \text{ A.U.}$$

By Eq. (4–44), the period will be

$$P = \frac{2\pi(1.19)^{3/2}}{0.0172\ldots} = 475 \text{ days.}$$

Therefore

$$n = \frac{2\pi}{P} = 0.01322 \text{ radians/day.}$$

For the orbit of this object, we may write

$$\text{perihelion distance } r_p = a(1 - e) = 1,$$
$$\text{aphelion distance } r_a = a(1 + e) = 1.38.$$

From these, by subtraction, we obtain $e = 0.16$, and the path of the object is given by

$$r = \frac{1.16}{1 + 0.16 \cos f}.$$

TABLE 4–1

Date	$t - T$, days	M, radians	E_0	M_0	Tan $(f/2)$	f, degrees	r, A.U.
Feb. 1	35	0.463	0.5446	0.4618	0.328	36.4	1.03
Mar. 1	64	0.847	0.9796	0.8468	0.632	64.6	1.09
Apr. 1	95	1.255	1.4176	1.2596	1.008	90.4	1.16
May 1	125	1.655	1.8122	1.6569	1.500	112.6	1.24
June 1	156	2.062	2.1923	2.0623	2.285	132.6	1.30
July 1	186	2.460	2.5485	2.4591	3.840	150.8	1.35
Aug. 1	217	2.872	2.9088	2.8717	10.020	168.6	1.37

Calculations for (r, f) by Eqs. (4–48) through (4–55) for various dates are listed in Table 4–1. The ejection is assumed to have taken place late in December so that in one half-period (237.5 days) the object reaches Mars late in August, say August 20.

Column 2 lists the number of days elapsed after the ejection of the object from the earth's orbit. Column 3 contains the mean anomalies for these times, calculated by Eq. (4–48). Column 4 is computed from Eq. (4–50), and lists *first approximations to the eccentric anomaly*. From these values of E_0 the values of M_0 are calculated according to Eq. (4–51). If the latter had differed significantly from the values of M, a correction ΔE_0 would have been computed by Eq. (4–52). To the accuracy adopted here, however, the values of M_0 agree well enough with those of M so that a second approximation to E_0 is unnecessary. If this had been

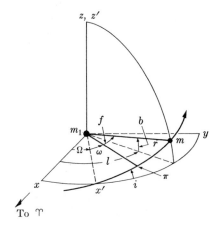

FIG. 4–9. The orbit in space.

required, we would have used Eq. (4–53) for it. Columns 7 and 8 contain the required polar coordinates of the object in its elliptical orbit for the times specified; these values are obtained by means of Eqs. (4–54) and (4–55) with $E = E_0$.

There remains the problem of determining the positions of the object in the sky. These can be found from the values of (r, f) and the orbital elements. In Fig. 4–9, the xy-plane is the plane of the ecliptic. Now let π denote the perihelion point of the orbit; let (r, f) denote the position of m in its orbit about m_1 at time t; let (l, b) be the heliocentric coordinates of m (see Section 1–8). Then

$$x = r \cos b \cos l,$$
$$y = r \cos b \sin l,$$
$$z = r \sin b,$$

or, referred to a new system in which the x'-axis lies along the line of nodes,

$$x' = r \cos b \cos (l - \Omega) = r \cos (f + \omega),$$
$$y' = r \cos b \sin (l - \Omega) = r \sin (f + \omega) \cos i,$$
$$z' = r \sin b = r \sin (f + \omega) \sin i.$$

Hence

$$\tan (l - \Omega) = \tan (f + \omega) \cos i, \qquad (4\text{–}56)$$
$$\tan b = \sin (l - \Omega) \tan i. \qquad (4\text{–}57)$$

With f, ω, Ω, i known, these equations yield (l, b), the heliocentric coordinates of the object. Then Eqs. (1–14), (1–15), and (1–16) with $\lambda = l$ and $\beta = b$ yield the equatorial coordinates (α_1, δ_1) of the body *as seen from the sun*. To find the position (α, δ) *as seen from the earth*, we need the geocentric rectangular equatorial coordinates of the sun (X, Y, Z), given in the *American Ephemeris* for the time of observation. In Fig. 4–10

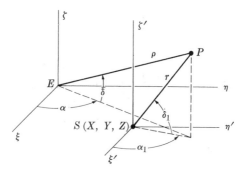

FIG. 4–10. Conversion from heliocentric to geocentric equatorial positions.

let E denote the earth, S the sun, and let the $\xi\eta$-plane be the plane of the earth's equator. Referred to the geocentric origin, the sun has coordinates X, Y, Z. Then for P relative to S, we calculate

$$\xi' = r \cos \delta_1 \cos \alpha_1,$$
$$\eta' = r \cos \delta_1 \sin \alpha_1,$$
$$\zeta' = r \sin \delta_1,$$

and for P relative to E, we find

$$\xi = \xi' + X = \rho \cos \delta \cos \alpha,$$
$$\eta = \eta' + Y = \rho \cos \delta \sin \alpha,$$
$$\zeta = \zeta' + Z = \rho \sin \delta.$$

Hence

$$\tan \alpha = \frac{\eta}{\xi},$$

$$\sin \delta = \zeta(\xi^2 + \eta^2 + \zeta^2)^{-1/2}.$$

4–7 Determination of orbital elements from observed position in the sky. The determination of the elements of an orbit, that is, a, e, i, Ω, ω, and T, from observations of a celestial object in the sky, is in general a time-consuming task. Since there are six constants to be determined, one needs at least three observations of right ascension and declination well spaced in time.

Laplace, Gauss, Olbers, and other noted astronomers have devised methods for solving the fundamental differential equations [Eqs. (4–31)] to obtain the orbital elements.* We shall not dwell in detail on this problem. Instead we shall restrict the discussion to the determination of a circular orbit. The method of attack on this problem is representative of methods used in other instances.

For simplicity, we confine our attention to a planetary object moving around the sun. In this case we need only two complete observations because only four elements have to be determined, namely, a (the radius of the circle), i, Ω, and ω. The eccentricity is zero, and the argument of perihelion is meaningless. The angle ω will be taken to be the *angle in the plane of the orbit from the ascending node to the location of the planet at the time of the first observation.* Had the orbit been an ellipse, ω would be the argument of perihelion, that is, the angle in the plane of the orbit from the line of nodes to the perihelion point.

---

* See references 1, 3, 5, 8 for details of the computations.

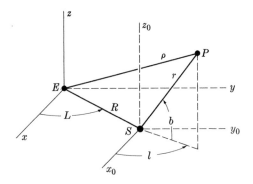

Fig. 4–11. Conversion from geocentric to heliocentric latitude and longitude.

Let the observed data be (α_1, δ_1) at time t_1 and (α_2, δ_2) at time t_2. These are *geocentric coordinates*. Let the celestial longitude and latitude of the planet be (λ, β) and its distance from the earth be ρ. Then by Eqs. (1–11), (1–12), and (1–13) we have at each time t_1 or t_2

$$\sin \beta = \sin \delta \cos \epsilon - \cos \delta \sin \epsilon \sin \alpha,$$
$$\cos \beta \sin \lambda = \sin \delta \sin \epsilon + \cos \delta \cos \epsilon \sin \alpha, \qquad (4\text{–}58)$$
$$\cos \beta \cos \lambda = \cos \delta \cos \alpha,$$

where ϵ is approximately $23°27'.1$.* From these relations, we obtain

$$(\lambda_1, \beta_1) \text{ at time } t_1 \qquad \text{and} \qquad (\lambda_2, \beta_2) \text{ at time } t_2.$$

Let ρ_1 and ρ_2 be the distances of P from the earth at times t_1 and t_2, respectively. Then the rectangular geocentric ecliptic coordinates of P at times t_1 and t_2 are

$$x_1 = \rho_1 \cos \beta_1 \cos \lambda_1, \qquad x_2 = \rho_2 \cos \beta_2 \cos \lambda_2,$$
$$y_1 = \rho_1 \cos \beta_1 \sin \lambda_1, \qquad y_2 = \rho_2 \cos \beta_2 \sin \lambda_2, \qquad (4\text{–}59)$$
$$z_1 = \rho_1 \sin \beta_1, \qquad z_2 = \rho_2 \sin \beta_2.$$

Assuming the same times, let the geocentric coordinates† of the sun be

$$X_1 = R_1 \cos L_1, \qquad X_2 = R_2 \cos L_2,$$
$$Y_1 = R_1 \sin L_1, \qquad Y_2 = R_2 \sin L_2, \qquad (4\text{–}60)$$
$$Z_1 = 0, \qquad Z_2 = 0,$$

---

* A precise value of ϵ for a given time may be found in the *American Ephemeris*.

† These coordinates are tabulated in the *American Ephemeris*.

where R is the distance from the sun to the earth and L is the sun's geocentric longitude. Since the sun is in the ecliptic plane, its latitude is zero.

In Fig. 4–11 let S denote the sun, P the planet, and E the earth. Through S we pass a set of coordinate axes parallel to those through E. We denote the distance SP by r; the longitude of P in this new system is l, and the latitude is b. These are the heliocentric coordinates of P. Then, if x_0, y_0, z_0 denote the cartesian coordinates of P in the heliocentric system, we have

$$x_0 = x - X; \qquad y_0 = y - Y; \qquad z_0 = z - Z.$$

If we express these in terms of the angles at times t_1 and t_2, we find

$$r_1 \cos b_1 \cos l_1 = \rho_1 \cos \beta_1 \cos \lambda_1 - R_1 \cos L_1,$$
$$r_1 \cos b_1 \sin l_1 = \rho_1 \cos \beta_1 \sin \lambda_1 - R_1 \sin L_1, \qquad (4\text{–}61)$$
$$r_1 \sin b_1 = \rho_1 \sin \beta_1;$$

$$r_2 \cos b_2 \cos l_2 = \rho_2 \cos \beta_2 \cos \lambda_2 - R_2 \cos L_2,$$
$$r_2 \cos b_2 \sin l_2 = \rho_2 \cos \beta_2 \sin \lambda_2 - R_2 \sin L_2, \qquad (4\text{–}62)$$
$$r_2 \sin b_2 = \rho_2 \sin \beta_2.$$

In the above equations we know R_1, R_2, L_1, L_2, (or X_1, Y_1, X_2, Y_2) from the information tabulated in the *American Ephemeris* for the dates t_1 and t_2. Furthermore, we know the values of λ and β for the two dates. We have assumed a circular orbit, so $r_1 = r_2 = a$, an unknown yet to be determined. The other unknowns are the values of ρ, l, and b for the two dates.

By squaring and adding Eqs. (4–62), we obtain

$$r_2^2 = \rho_2^2 + R_2^2 - 2\rho_2 R_2 \cos \beta_2 \cos (\lambda_2 - L_2). \qquad (4\text{–}63)$$

But $r_2 = r_1$ and hence, if r_1 is known, ρ_2 may be calculated. Then Eqs. (4–62) yield l_2 and b_2. It is obvious that *an estimate of ρ_1 will yield r_1, l_1, and b_1 by means of Eqs. (4–61) and, indirectly, values of r_2, l_2, b_2 by means of Eqs. (4–63) and (4–62).*

Now we must see how these quantities are related to the orbital elements. In Fig. 4–12, let P_1 and P_2 denote the positions of the planet at times t_1 and t_2, respectively. By the cosine law for the spherical triangle $P_1 N P_2$, we have

$$\cos A = \sin b_1 \sin b_2 + \cos b_1 \cos b_2 \cos (l_2 - l_1), \qquad (4\text{–}64)$$

where A is the arc of a great circle traversed by P in time $t_2 - t_1$. If the coordinates (l_1, b_1) and (l_2, b_2) are known, A can be computed.

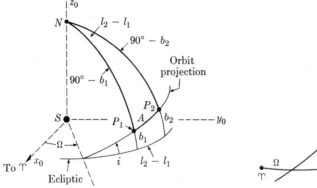

FIG. 4–12. Spherical triangle for circular-orbit calculation.

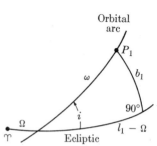

FIG. 4–13. Relationship of i, ω, Ω to circular orbit.

Again, Kepler's third law holds for the planet, namely

$$P = \frac{2\pi a^{3/2}}{k\sqrt{1 + m}} \text{ days.}$$

Therefore, if we neglect the mass m of the planet, its mean daily motion in radians is $2\pi/\text{period} = k/a^{3/2}$. Since the arc A will be traversed in time $t_2 - t_1$ days, we have

$$A = \frac{k}{a^{3/2}} (t_2 - t_1). \tag{4–65}$$

This *expression for A is entirely independent of that given by Eq. (4–64).* The values of A must agree if we have calculated (l_1, b_1) and (l_2, b_2) correctly. The suggested procedure consists of the following steps:

(a) Make an astute guess, or estimate, of ρ_1.
(b) By means of Eq. (4–61) calculate r_1, l_1, b_1.
(c) By means of Eq. (4–63) calculate ρ_2, using $r_2 = r_1$.
(d) Using Eqs. (4–62) and the value of ρ_2 found in step (c), calculate l_2 and b_2.
(e) Use these calculated coordinates in Eq. (4–64) to compute A. Use the value of the radius, $a = r_1 = r_2$, to calculate A from Eq. (4–65). If ρ_1 has been guessed correctly, these must agree. In general, they will not agree.
(f) Repeat the process after finding by trial the changes which must be made to improve the agreement in A calculated in the two ways.

We assume now that this successive approximation process has been carried to the accuracy desired so that the distance a has been found.

This means also that the heliocentric coordinates (l_1, b_1) and (l_2, b_2) are known. We still have to find i, Ω, and ω. These are related by the right spherical triangle shown in Fig. 4–13. By Napier's rules, we have at time t_1

$$\sin (l_1 - \Omega) = \tan (90° - i) \tan b_1, \tag{4-66}$$

and similarly at time t_2

$$\sin (l_2 - \Omega) = \tan (90° - i) \tan b_2.$$

Let $l_2 - \Omega = (l_1 - \Omega) + (l_2 - l_1)$. Then the second of these equations can be written

$$\tan i \{\sin (l_1 - \Omega) \cos (l_2 - l_1) + \cos (l_1 - \Omega) \sin (l_2 - l_1)\} = \tan b_2,$$

and substituting from the first, we have

$$\tan b_1 \cos (l_2 - l_1) + \tan i \cos (l_1 - \Omega) \sin (l_2 - l_1) = \tan b_2.$$

Combining this with Eq. (4–66), we obtain finally

$$\tan (l_1 - \Omega) = \frac{\tan b_1 \sin (l_2 - l_1)}{\tan b_2 - \tan b_1 \cos (l_2 - l_1)}. \tag{4-67}$$

This yields Ω, and i follows immediately from Eq. (4–66). Also, from Fig. 4–13, we observe that

$$\tan \omega = \frac{\tan (l_1 - \Omega)}{\cos i}, \tag{4-68}$$

and that ω and $(l_1 - \Omega)$ are always in the same quadrant.

By this successive approximation process, we have found, in theory, the elements a, i, Ω, and ω which define the orbit completely. When a is known, the period P is given by Kepler's third law, Eq. (4–44).

The successive approximations required in this calculation of the orbital elements can be programmed for a high-speed digital computer.* Although several ambiguous cases may arise in circumstances where the lines of sight from earth to planet are nearly parallel, or happen to be nearly tangent to the unknown orbit, the machine calculation is nonetheless valuable in providing a quick approximate solution for a moving planetoid. This solution may serve as a starting point for calculating future positions and a more refined elliptical orbit.

---

* This program, for an IBM 650, is available in the reports of the Case Computing Center, Case Institute of Technology, Cleveland, Ohio.

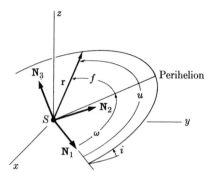

FIG. 4–14. Unit vectors for orbit calculation.

One other case will be considered here. If the radius vector of an object is known *in addition to* its position in the sky at three different times, it is fairly easy to deduce the elements of its orbit. For simplicity, we shall assume that the celestial object is moving around the sun in an elliptical orbit with period P. Let its positions in a heliocentric ecliptic coordinate system be given by the vectors

$$\mathbf{r}_1 = x_1\mathbf{i} + y_1\mathbf{j} + z_1\mathbf{k} \quad \text{at time } t_1,$$
$$\mathbf{r}_2 = x_2\mathbf{i} + y_2\mathbf{j} + z_2\mathbf{k} \quad \text{at time } t_2, \qquad (4\text{–}69)$$
$$\mathbf{r}_3 = x_3\mathbf{i} + y_3\mathbf{j} + z_3\mathbf{k} \quad \text{at time } t_3.$$

Since we assume that the polar coordinates l, b, r are known for each of the three times, the cartesian coordinates defining $\mathbf{r}_1, \mathbf{r}_2$, and $\mathbf{r}_3$ are readily found.

Let $\mathbf{N}_1, \mathbf{N}_2, \mathbf{N}_3$ be unit vectors as shown in Fig. 4–14. The first is directed along the line of nodes toward the ascending node; the second is perpendicular to $\mathbf{N}_1$ in the orbital plane; the third is perpendicular to the orbital plane. The triad forms a right-handed system in which $\mathbf{N}_3 = \mathbf{N}_1 \times \mathbf{N}_2$. Then the reader may verify that

$$\mathbf{N}_1 = (\cos \Omega)\mathbf{i} + (\sin \Omega)\mathbf{j},$$
$$\mathbf{N}_2 = (-\sin \Omega \cos i)\mathbf{i} + (\cos \Omega \cos i)\mathbf{j} + (\sin i)\mathbf{k}, \qquad (4\text{–}70)$$
$$\mathbf{N}_3 = (\sin \Omega \sin i)\mathbf{i} - (\cos \Omega \sin i)\mathbf{j} + (\cos i)\mathbf{k}.$$

From the known values of $\mathbf{r}_1$ and $\mathbf{r}_2$ we obtain a unit vector

$$\frac{\mathbf{r}_1 \times \mathbf{r}_2}{|\mathbf{r}_1 \times \mathbf{r}_2|} \equiv A_1\mathbf{i} + A_2\mathbf{j} + A_3\mathbf{k}$$

which is perpendicular to the orbit and identical with $\mathbf{N}_3$. Therefore,

$$\sin \Omega \sin i = A_1,$$
$$-\cos \Omega \sin i = A_2, \tag{4–71}$$
$$\cos i = A_3,$$

from which we find

$$i = \cos^{-1}(A_3), \tag{4–72}$$

$$\Omega = \tan^{-1}\left(-\frac{A_1}{A_2}\right). \tag{4–73}$$

One can, of course, choose any pair of the vectors $\mathbf{r}_1$, $\mathbf{r}_2$, $\mathbf{r}_3$ for this calculation. The only stipulation is that they be noncollinear.

The equation of the orbit may be written [Eq. (4–40)]

$$r = \frac{a(1 - e^2)}{1 + e \cos f} = \frac{a(1 - e^2)}{1 + e \cos(u - \omega)}, \tag{4–74}$$

where f is the true anomaly and u is the angle in the orbital plane from the ascending node to the radius vector [see Fig. 4–14]. Substituting the three radius vectors, rearranging terms and subtracting, we have

$$e \cos \omega (r_1 \cos u_1 - r_2 \cos u_2) + e \sin \omega (r_1 \sin u_1 - r_2 \sin u_2) = r_2 - r_1, \tag{4–75}$$

$$e \cos \omega (r_1 \cos u_1 - r_3 \cos u_3) + e \sin \omega (r_1 \sin u_1 - r_3 \sin u_3) = r_3 - r_1.$$

But

$$r \cos u = \mathbf{r} \cdot \mathbf{N}_1$$

and

$$r \sin u = \mathbf{r} \cdot \mathbf{N}_2. \tag{4–76}$$

Therefore Eqs. (4–75) can be written

$$e \cos \omega [(\mathbf{r}_1 - \mathbf{r}_2) \cdot \mathbf{N}_1] + e \sin \omega [(\mathbf{r}_1 - \mathbf{r}_2) \cdot \mathbf{N}_2] = r_2 - r_1, \tag{4–77}$$

$$e \cos \omega [(\mathbf{r}_1 - \mathbf{r}_3) \cdot \mathbf{N}_1] + e \sin \omega [(\mathbf{r}_1 - \mathbf{r}_3) \cdot \mathbf{N}_2] = r_3 - r_1.$$

Since the dot products can be computed from Eqs. (4–69) and (4–70) because Ω and i are known, Eqs. (4–77) constitute a pair of linear equations to be solved for $e \cos \omega$ and $e \sin \omega$. If the three vectors are noncollinear, as specified earlier, the determinant of the coefficients in Eqs. (4–77) will not vanish, and there will be a unique solution for $e \cos \omega$ and $e \sin \omega$.

Let the solutions of Eqs. (4–77) be $e \cos \omega = \alpha_1$ and $e \sin \omega = \alpha_2$. Then

$$e = (\alpha_1^2 + \alpha_2^2)^{1/2}, \qquad (4\text{–}78)$$

and

$$\omega = \tan^{-1}\left(\frac{\alpha_2}{\alpha_1}\right). \qquad (4\text{–}79)$$

Since $e > 0$, the quadrant of ω will be determined by the algebraic signs of $e \cos \omega$ and $e \sin \omega$.

With e and ω given, the semimajor axis, a, follows from Eq. (4–74). For this purpose we write

$$a = \frac{r + \mathbf{r} \cdot \mathbf{N}_1 e \cos \omega + \mathbf{r} \cdot \mathbf{N}_2 e \sin \omega}{1 - e^2}. \qquad (4\text{–}80)$$

Any one of the three known values of r may be used in this equation to obtain a.

Since we have assumed the period P to be known, we can confirm the value of a by using Kepler's third law [Eq. (4–44)]. If the period is not known, Eq. (4–44), together with the value of a obtained from Eq. (4–80), can be used to determine P.

The time of perihelion passage is found from Eq. (4–54), $r = a(1 - e \cos E)$, and Kepler's equation, $M = E - e \sin E$. With M given for one time of observation, the value of T can be obtained from

$$M = \frac{2\pi}{P}(t - T). \qquad (4\text{–}81)$$

The example which follows will make this clear.

Thus the elements a, e, i, Ω, ω, and T of an elliptic orbit may be deduced from the known polar space coordinates of the moving object. Actually only two such positions are required. We could have used the constancy of the areal velocity, together with one of Eqs. (4–77), as a condition to obtain a, e, ω. The analysis is more complex, however.

EXAMPLE. The heliocentric ecliptic coordinates of a moving object are given in Table 4–2. The mean daily motion is $2\pi/P = n = 4°.0923$. Find the elements of its orbit.

TABLE 4–2

Date	l	b	r, A.U.
June 1.000	142°45′40″	+6°58′43″	0.34200
June 6.000	166°37′56″	+6°08′47″	0.37022
June 11.000	186°58′43″	+4°35′51″	0.39867

For illustrative purposes the calculations will be carried to five-place accuracy only. In practice, a considerably greater precision is required for such computations.

TABLE 4–2(a)

Date	r	$x = r \cos b \cos l$	$y = r \cos b \sin l$	$z = r \sin b$
June 1.000	0.34200	−0.27025	+0.20542	+0.04155
June 6.000	0.37022	−0.35812	+0.08510	+0.03964
June 11.000	0.39867	−0.39444	−0.04828	+0.03195

$$\mathbf{r}_1 = -0.27025\mathbf{i} + 0.20542\mathbf{j} + 0.04155\mathbf{k},$$
$$\mathbf{r}_2 = -0.35812\mathbf{i} + 0.08510\mathbf{j} + 0.03964\mathbf{k}, \tag{4–82}$$
$$\mathbf{r}_3 = -0.39444\mathbf{i} - 0.04828\mathbf{j} + 0.03195\mathbf{k};$$

$$\mathbf{r}_1 \times \mathbf{r}_2 = \begin{vmatrix} \mathbf{i} & \mathbf{j} & \mathbf{k} \\ -0.27025 & 0.20542 & 0.04155 \\ -0.35812 & 0.08510 & 0.03964 \end{vmatrix},$$

$$\frac{\mathbf{r}_1 \times \mathbf{r}_2}{|\mathbf{r}_1 \times \mathbf{r}_2|} = 0.09023\mathbf{i} - 0.08184\mathbf{j} + 0.99254\mathbf{k}; \tag{4–83}$$

$$i = \cos^{-1}(0.99254) = 7°0'; \tag{4–84}$$

$$\Omega = \tan^{-1}\left(\frac{0.09023}{0.08184}\right) = 47°47'.6. \tag{4–85}$$

Therefore,
$$\mathbf{N}_1 = 0.67181\mathbf{i} + 0.74072\mathbf{j},$$
$$\mathbf{N}_2 = -0.73519\mathbf{i} + 0.66680\mathbf{j} + 0.12187\mathbf{k}, \tag{4–86}$$
$$\mathbf{N}_3 = 0.09023\mathbf{i} - 0.08184\mathbf{j} + 0.99254\mathbf{k}.$$

It is obvious that $\mathbf{N}_3$ and the vector given in Eq. (4–83) are identical, as they should be. Thus

$$\mathbf{r}_1 - \mathbf{r}_2 = 0.08787\mathbf{i} + 0.12032\mathbf{j} + 0.00191\mathbf{k},$$
$$\mathbf{r}_1 - \mathbf{r}_3 = 0.12419\mathbf{i} + 0.25370\mathbf{j} + 0.00960\mathbf{k}. \tag{4–87}$$

From Eqs. (4–87) and (4–86),

$$(\mathbf{r}_1 - \mathbf{r}_2) \cdot \mathbf{N}_1 = 0.14815, \qquad (\mathbf{r}_1 - \mathbf{r}_2) \cdot \mathbf{N}_2 = 0.01586,$$
$$(\mathbf{r}_1 - \mathbf{r}_3) \cdot \mathbf{N}_1 = 0.27135, \qquad (\mathbf{r}_1 - \mathbf{r}_3) \cdot \mathbf{N}_2 = 0.07904.$$

Therefore, by Eqs. (4–77),

$$0.14815e \cos \omega + 0.01586e \sin \omega = 0.02822,$$
$$0.27135e \cos \omega + 0.07904e \sin \omega = 0.05667.$$

Solving these, we obtain $e \sin \omega = 0.09966$ and $e \cos \omega = 0.17981$, so that

$$e = 0.2056, \tag{4–88}$$
$$\omega = 28°59'.8. \tag{4–89}$$

Hence, using $\mathbf{r}_1$ and $\mathbf{N}_1$ and $\mathbf{N}_2$ with the values of e, $e \sin \omega$, and $e \cos \omega$ determined above, we find by Eq. (4–80)

$$a = \frac{0.34200 - (0.02940)(0.17981) + (0.34072)(0.09966)}{1 - 0.04226}$$

or

$$a = 0.38702. \tag{4–90}$$

With this value of a and the value of e given in Eq. (4–88) we obtain at time $t_1 =$ June 1.000,

$$r_1 = a(1 - e \cos E_1) = 0.38702(1 - 0.2056 \cos E_1),$$

which yields

$$\cos E_1 = 0.56576 \quad \text{and} \quad E_1 = 55°32'.7.$$

From Kepler's equation (E_1 must be converted to radians),

$$M_1 = E_1 - e \sin E_1 = 0.96944 - (0.2056)(0.82457)$$
$$= 0.79991 \text{ radians}.$$

Therefore Eq. (4–81) yields ($n = 4°.092339 = 0.07142$ radians)

$$t_1 - T = \frac{M_1}{n} = \frac{0.79991}{0.07142} = 11.200,$$

and since $t_1 =$ June 1.000 $=$ May 32.000, we have

$$T = \text{May } 20.800. \tag{4–91}$$

In summary therefore the elements of the orbit are

$$a = 0.38702 \text{ A.U.} \qquad \Omega = 47°47'.6,$$
$$e = 0.2056, \qquad \omega = 28°59'.8,$$
$$i = 7°0', \qquad T = \text{May } 20.8.$$

As a check on the value of a we use the period $P = 2\pi/n = 87.975$ days. By Kepler's third law [Eq. (4–44)],

$$P = \frac{2\pi a^{3/2}}{k\sqrt{1 + m}},$$

and, since $m \ll 1$, we obtain $a^{3/2} = k/n$. Hence,

$$a = \left(\frac{0.017202}{0.07142}\right)^{2/3} = 0.24086^{2/3}$$

or

$$a = 0.38712,$$

which agrees sufficiently well with the value of a listed above.

4–8 Parabolic and hyperbolic orbits. When a newly discovered comet is moving near the sun, its orbit may be a very elongated ellipse, a parabola, or a hyperbola. These curves may also be followed by space vehicles moving in the gravitational field of a planet. It is difficult to distinguish between these curves when only a small arc is covered by the moving object. Because the elements of a parabolic orbit are more easily calculated than those of other conics, it is customary to start with this approximation when one analyzes the motion of a newly discovered comet. As the object is observed over a greater and greater arc, a satisfactory representation of the motion will usually require the introduction of an elliptic orbit.

In Fig. 4–15 let q denote the distance from the origin O to the vertex V of the parabolic orbit. Then, since $e = 1$, the equation of the orbit becomes

$$r = \frac{2q}{1 + \cos f} = q \sec^2\left(\frac{f}{2}\right), \qquad (4\text{–}92)$$

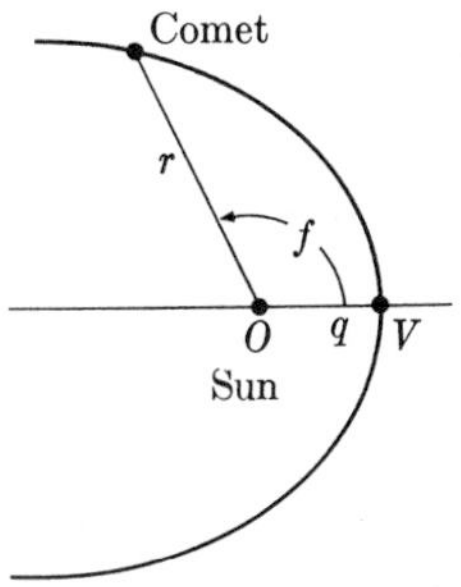

Fig. 4–15. Parameters for a parabolic orbit.

where the true anomaly f is measured from perihelion, the sun being at O. At $r = q$ we find by Eqs. (4–32) and (4–33), with $E = 0$, that $h = k\sqrt{2Mq}$ when the velocity vector is perpendicular to the radius vector. With $\dot{f}$ substituted for $\dot{\theta}$, Eqs. (4–32) and (4–92) yield

$$q^2 \sec^4 \left(\frac{f}{2}\right) \dot{f} = h.$$

Therefore,

$$\sec^4 \left(\frac{f}{2}\right) df = \frac{k\sqrt{2M}}{q^{3/2}} dt,$$

and by integration

$$\tan\left(\frac{f}{2}\right) + \frac{1}{3} \tan^3 \left(\frac{f}{2}\right) = k\sqrt{\frac{M}{2q^3}} (t - T), \qquad (4\text{–}93)$$

where T is the time of perihelion passage and M is the mass of the sun. The comet's mass is negligible compared to that of the sun.

The cubic equation, Eq. (4–93), can be solved for f as a function of t with a digital computer or by graphical or tabular methods. Moulton [4, p. 156] gives the following relations for the direct solution of Eq. (4–93):

$$\cot s = \frac{3k(t - T)}{(2q)^{3/2}}, \qquad (4\text{–}94)$$

$$\cot w = \sqrt[3]{\cot \frac{s}{2}}, \qquad (4\text{–}95)$$

$$\tan\left(\frac{f}{2}\right) = 2 \cot 2w, \qquad (4\text{–}96)$$

where s and w are auxiliary variables. Having obtained f from these equations, one can find r from Eq. (4–92). A tabular solution of Eq. (4–93), known as "Barker's table,"* may also be used.

In general, calculating a preliminary orbit of a comet requires only three observations. These are usually restricted to the time during which the comet is near the sun and is bright enough to be observed with small telescopes. Since the arc covered by the first observations is short, one assumes for purposes of prediction that the orbit is a parabola. Later when the comet has been observed over as much of its path about the sun as possible, a so-called "definitive orbit" is calculated. This orbit may

---

* See Table VI in J. C. Watson, *Theoretical Astronomy*, J. B. Lippincott and Co., Philadelphia, 1892, which gives f as a function of $\log M$, where $M \equiv 75(k/\sqrt{2})q^{-3/2}(t - T)$.

be an ellipse of high eccentricity or it may be very nearly circular. The details of orbit calculation are found in the standard texts by Moulton, Crawford, Williams, and Herget [1, 3, 4, 7].

Hyperbolic orbits are followed by meteors as they fall toward the earth. Whatever its path in outer space may have been, the path of a visible meteor is determined by the dominant attraction of the earth. The influence of the sun is, then, negligible and will be ignored in the following discussion.

The velocity of the meteor as it streaks through the earth's atmosphere can be found by observation. When appropriate corrections for the velocity of the earth are made, the velocity of the comet in its orbit around the sun is known. And since its position is that of the earth, we have the fundamental data necessary for finding the orbital elements of the meteor about the sun.

Figure 4–16 shows the path of a meteor. If the earth had not been present, the meteor would have passed the point O undeflected. Actually as it nears the earth the meteor path is curved, and the object strikes the earth at P. Let

R = the radius of the earth,

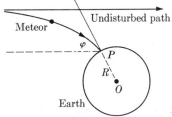

v_1 = the speed relative to the earth which the meteor would have if it remained undeflected, and

v_2 = the observed speed.

FIG. 4–16. Path of a meteor.

Then from a relation for the hyperbola similar to Eq. (4–43) for the ellipse, the speed is given by $v^2 = k^2 M[2/r + 1/a]$. But near the earth, $k^2 M = gR^2$, where g is the gravitational constant on the earth. Therefore,

$$v_1^2 = gR^2 \left(\frac{1}{a}\right), \tag{4–97}$$

$$v_2^2 = gR^2 \left[\frac{2}{R} + \frac{1}{a}\right]. \tag{4–98}$$

From these,

$$v_1^2 = v_2^2 - 2Rg, \tag{4–99}$$

so that v_1 can be found when v_2 is measured

To determine the direction of the velocity vector $\mathbf{v}_2$ we recall that by Kepler's second law,

$$|\mathbf{R} \times \mathbf{v}_2| = v_2 R \sin z = R^2 \dot{\theta} = h,$$

where z is the apparent zenith distance shown in Fig. 4–17. But by the geometry of the hyperbolic orbit,

$$h = k\sqrt{Ma(e^2 - 1)},$$

where M is the mass of the earth (the mass of the meteor being negligible). Equating the two values of h, we obtain

$$a(e^2 - 1) = \frac{v_2^2 R^2 \sin^2 z}{k^2 M} = \frac{v_2^2 \sin^2 z}{g}. \tag{4–100}$$

Thus when v_2 is measured, we can find v_1 from Eq. (4–99) and a from Eq. (4–97). Then Eq. (4–100) yields e.

The original velocity vector, $\mathbf{v}_1$, has the direction of an asymptote to the hyperbolic orbit. This is a concept of geometry from which it follows [8] that

$$\tan\left(\frac{\varphi}{2}\right) = \frac{v_2 - v_1}{v_2 + v_1} \tan\left(\frac{z}{2}\right). \tag{4–101}$$

Suppose that the meteor *appears* at a given azimuth and a given zenith distance. Then the observed zenith distance corrected by the angle φ yields $z_0 = z + \varphi$, the true direction in the sky from which the meteor approached the earth. The right ascension and declination of this point, *the radiant*, can be computed from the zenith distance, azimuth, and the time of observation for a given observing station.

The computation of the velocity components in the heliocentric system for the purpose of determining the orbit of the meteor around the sun requires the velocity components of the earth. Tables for correcting a space velocity for the motion of the earth have been published by Herrick [9]. With the coordinates and velocity components, one can com-

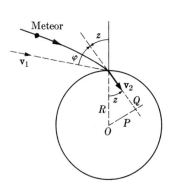

FIG. 4–17. Zenith attraction and velocity relationships for a meteor path.

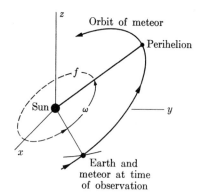

FIG. 4–18. Meteor orbit in space.

pute the orbital elements by using standard formulas [4, pp. 145 ff.]. Note that when the meteor is observed, it must be at one of the nodes of its orbit (see Fig. 4–18). For the date of observation, the longitude of the sun as seen from the earth is given in the *American Ephemeris*. Hence the heliocentric longitude of the earth is obtained at once. If the heliocentric velocity component of the meteor perpendicular to the ecliptic plane is positive, i.e., directed toward the north pole of the ecliptic, then the meteor must have entered the earth's atmosphere at the ascending node of the meteor orbit. The element Ω, therefore, is known. Thus we can give the following summary of the formulas required for the computation of the orbital elements:

$\Omega = $ longitude λ of the sun or $\lambda + 180°$, depending on the observed component of velocity perpendicular to the ecliptic ($\dot{z}$),

$$i = \tan^{-1}\left\{\frac{y\dot{z} - z\dot{y}}{(x\dot{y} - y\dot{x})\sin\Omega}\right\} \quad \text{or} \quad \tan^{-1}\left\{-\frac{z\dot{x} - x\dot{z}}{(x\dot{y} - y\dot{x})\cos\Omega}\right\}, \quad (4\text{–}102)$$

$$a = \left[\frac{2}{r_e} - \frac{\dot{x}^2 + \dot{y}^2 + \dot{z}^2}{k^2 m_1}\right]^{-1} \quad \begin{array}{l}\text{(where } r_e \text{ is the distance} \\ \text{earth-sun and } m_1 \text{ is the} \\ \text{solar mass)},\end{array} \quad (4\text{–}103)$$

$$P = \frac{2\pi a^{3/2}}{k\sqrt{m_1}}, \quad (4\text{–}104)$$

$$e = \left[1 - \frac{(x\dot{y} - y\dot{x})^2}{k^2 m_1 a \cos^2 i}\right]^{1/2}, \quad (4\text{–}105)$$

$$\omega = \tan^{-1}\left[-\frac{1}{k}\sqrt{\frac{a(1 - e^2)}{m_1}}\frac{(x\dot{x} + y\dot{y} + z\dot{z})}{a(1 - e^2) - r_e}\right]. \quad (4\text{–}106)$$

In these formulae, x, y, z and $\dot{x}$, $\dot{y}$, $\dot{z}$ are the *heliocentric cartesian coordinates and velocity components*, respectively, *of the meteor at the time of observation*. If f is the true anomaly of the meteor measured from perihelion (Fig. 4–18) and ω is the argument of perihelion, then $f + \omega = 2\pi$. If the meteor is receding from the sun at the time of observation, the quantity $x\dot{x} + y\dot{y} + z\dot{z} > 0$, and $0° < f < 180°$. If the meteor is approaching the sun, $x\dot{x} + y\dot{y} + z\dot{z} < 0$, and $180° < f < 360°$. This is the situation pictured in Fig. 4–18. Hence the proper quadrant can be deduced for $\omega = 2\pi - f$ by observing the algebraic sign of $x\dot{x} + y\dot{y} + z\dot{z}$.

EXAMPLE. As an illustration of the calculation described above, consider a meteor which was observed on August 1, 1949, at $8^h53^m18^s$ GCT. The observed zenith distance was $z = 42°46'$, and the observed velocity relative to the center of the earth was 20 km/sec. We also have the following data:

$$\text{radius of earth} = R = 6.371 \times 10^8 \text{ cm},$$
$$gR^2 = k^2M = 3.99 \times 10^{20} \text{ cm}^3 \cdot \text{sec}^{-2}.$$

Therefore, by Eq. (4–99),

$$v_1 = [(20 \times 10^5)^2 - 1.252 \times 10^{12}]^{1/2} = 16.6 \text{ km/sec}.$$

This is the speed relative to the earth that the meteor would have had if the earth had not influenced its motion.

By Eq. (4–101) we find the correction for *zenith attraction*, $\varphi = 4°10'$. Hence the true zenith distance is $z_0 = 46°56'$. This value, together with the time and place of observation, enables one to determine the radiant. For brevity we omit this calculation and simply assume that the radiant is at

$$\alpha = 305°,$$

$$\delta = -8°.4,$$

or, in celestial coordinates [see Eqs. (1–11), (1–12), (1–13)],

$$\lambda = 305°.3, \qquad \beta = +10°.7.$$

From the *American Ephemeris* we find, for the date of observation,

$$\lambda \text{ (sun)} = 128°.8,$$
$$r_e \text{ (earth)} = 1.0149 \text{ A.U.}$$

Therefore the heliocentric coordinates of the earth are

$$l = 308°.8,$$
$$b = 0°,$$

and, in cartesian form,

$$x = r_e \cos l = 0.63593 \text{ A.U.},$$
$$y = r_e \sin l = -0.79094 \text{ A.U.,} \qquad (4–107)$$
$$z = 0.$$

The tables in *Lick Bulletin*, **17**, 85, 1935, yield for the heliocentric velocity components of the earth at the time of observation,

$$\dot{x}_e = +22.6, \qquad \dot{y}_e = +18.6, \qquad \dot{z}_e = 0 \quad \text{(km/sec)}.$$

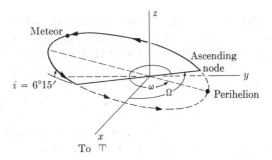

FIG. 4–19. Meteor orbit illustrating the example.

For the meteor relative to the earth we find the velocity components

$$\dot{x}_m = -v_1 \cos \beta \cos \lambda = -16.6 \times 0.56743 = -9.4 \quad (\text{km/sec}),$$
$$\dot{y}_m = -v_1 \cos \beta \sin \lambda = -16.6 \times -0.80157 = +13.3,$$
$$\dot{z}_m = -v_1 \sin \beta = -16.6 \times 0.18846 = -3.1,$$

where λ and β are the celestial coordinates of the radiant. Therefore the velocity components of the meteor relative to the sun are

$$\dot{x} = -9.4 - 22.6 = -32.0 \quad (\text{km/sec}),$$
$$\dot{y} = +13.3 - 18.6 = -5.3, \tag{4–108}$$
$$\dot{z} = -3.1 - 0 = -3.1.$$

Since $\dot{z} < 0$, the meteor entered the atmosphere at the time when the earth was at the descending node of the meteor's orbit. Therefore the longitude of the ascending node is $\Omega = 128°.8$.

With the data from Eqs. (4–107) and (4–108) substituted in Eqs. (4–102) and (4–103), we find

$$i = 6°15' \quad \text{and} \quad a = 1.236 \text{ A.U.}$$

Then Eq. (4–104) yields $P = 2.98$ years. Finally, from Eqs. (4–105) and (4–106) and the values of a and i already computed, we obtain

$$e = 0.492 \quad \text{and} \quad \omega = 308°34'.$$

The element ω is assigned to the fourth quadrant because $x\dot{x} + y\dot{y} + z\dot{z} < 0$ in this example. Hence the meteor was approaching the sun, and the true anomaly was between $180°$ and $360°$. Thus $\omega = 2\pi - f$ must be between $180°$ and $360°$. Since $\tan \omega$ is negative, the assignment to the fourth quadrant follows. The resulting orbit of the meteor around the sun is shown in Fig. 4–19.

4–9 Transfer between orbits. The preceding sections have been devoted to a discussion of the relative motion of one body around another under the force of gravitation. We have found that the motion takes place in a conic section. In particular for a planet moving around the sun or for a satellite moving around a planet, the conic is an ellipse with the center of force at the focus. We now turn our attention to the dynamical aspects of the transfer of a mass particle from one orbit to another. This question is of increasing importance because of the space probes which will traverse interplanetary space in the near future.* In this problem the space vehicle is initially placed in a hyperbolic orbit as viewed from the planet. Then this orbit is matched to an elliptical transfer orbit. In our discussion we shall not consider the local planetary case.

The *impulse* produced by a force $\mathbf{F}$ is defined to be

$$\mathbf{I} = \int_{t_0}^{t_1} \mathbf{F} \, dt, \tag{4–109}$$

where (t_0, t_1) is a finite, though generally small, time interval. By substituting from Newton's second law of motion [Eq. (4–10)] into Eq. (4–109), we find

$$\mathbf{I} = m\mathbf{v}(t_1) - m\mathbf{v}(t_0). \tag{4–110}$$

Thus the impulse is the change in linear momentum in the time interval (t_0, t_1). For our purposes we shall consider the time interval to be short and $\mathbf{F}$ relatively large and of such a form that

$$\lim_{t_1 \to t_0} \int_{t_0}^{t_1} \mathbf{F} \, dt$$

exists. For example, $\mathbf{F}$ may be the suddenly applied thrust of a rocket motor. Application of such an *impulsive force* changes the linear momentum of the mass m, but does not appreciably change its position during the time interval (t_0, t_1).

Any impulse $\mathbf{I}$ applied to a mass m not only changes its velocity but also produces a change in the kinetic energy. Because other nonimpulsive forces do no work during the brief interval (t_0, t_1), the entire change in energy is that in kinetic energy. We have therefore

$$\Delta E = \tfrac{1}{2}m(v_1^2 - v_0^2) = \tfrac{1}{2}m(\mathbf{v}_1 - \mathbf{v}_0) \cdot (\mathbf{v}_1 + \mathbf{v}_0),$$

which may be written

$$\Delta E = \tfrac{1}{2}m\mathbf{I} \cdot (2\mathbf{v}_0 + \mathbf{I}) = \tfrac{1}{2}mI^2 + m\mathbf{I} \cdot \mathbf{v}_0. \tag{4–111}$$

---

* For a comprehensive discussion of these problems see reference 10.

The change in energy will be greatest when I and v_0 have the same direction.

An impulse I will, in general, also cause a change in the angular momentum of the mass m. By definition [Eq. (4–7)], $L = r \times mv$. Hence we may write

$$\Delta L = L_1 - L_0 = r \times (mv_1 - mv_0) = r \times I. \qquad (4\text{–}112)$$

If I is applied in the plane of the original orbit, that is, in the plane defined by r and v_0, then only the magnitude of L changes. The new orbit is coplanar with the first. On the other hand, if I makes an angle with the plane defined by r and v_0, then the new orbit has a different inclination than the original one.

Suppose I is applied in the plane of the orbit. Then it is apparent from Eq. (4–112) that the numerical value of ΔL will be a maximum when I is perpendicular to r, and zero when I is collinear with r. In the latter case the orbit will be swung in its plane as shown by the change of the velocity vector in Fig. 4–20. At the same time the energy changes by an amount ΔE given by Eq. (4–111). But according to Eq. (4–43), the total energy for an elliptical orbit is $E = -(k^2 Mm/2a)$. Hence a

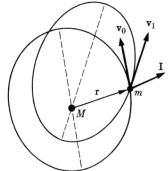

FIG. 4–20. Orbit change by velocity impulse.

change ΔE causes a change in the semimajor axis, amounting to $\Delta a = (2a^2/k^2 Mm) \Delta E$. It is left for the reader to show that there will be an accompanying change in the sidereal period of the motion and in the eccentricity of the orbit.

This brief discussion indicates the direction in which an analysis of the perturbative effects of an impulse must go. In general, the effect of an impulse I may be studied most conveniently by resolving it into mutually orthogonal components:

   S, perpendicular to the orbital plane and directed toward the north pole of the ecliptic,

   T, tangent to the orbit and directed along the velocity vector v,

   N, normal to T in the plane of the orbit and directed toward the concave side.

Moulton [4, pp. 322–334] has given a complete description of the effects of these components on the orbital elements. We shall summarize the

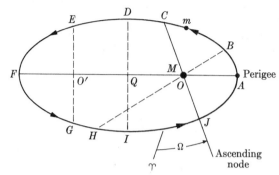

FIG. 4–21. Diagram for velocity-impulse effects listed in Table 4–3.

results here in the form of Table 4–3 and Fig. 4–21. The letters at points on the ellipse define arc boundaries within which an applied impulse will produce the effects shown in the table. By an advance in the line of apsides we mean that the line AF rotates in the direction of motion of m along the ellipse. An advance in the longitude of the node (Ω) is similarly defined by a rotation of line OJ. In the figure the point O' is the focus conjugate to O, which is the center of force, and Q is the geometrical center of the ellipse.

TABLE 4–3

EFFECT OF AN IMPULSE ON ORBITAL ELEMENTS

Element	S	T	N
Longitude of node (Ω)	Advance in BCH, regress in HJB	No effect	No effect
Inclination (i)	Increase in JBC, decrease in CHJ	No effect	No effect
Semimajor axis (a)	No effect	Increases	No effect
Line of apsides (AF)	No effect if ω is reckoned from a fixed point in orbit	Advance in ADF, regress in FIA	Advance in GAE, regress in EFG
Eccentricity (e)	No effect	Increase in IAD, decrease in DFI	Increase in FIA, decrease in ADF

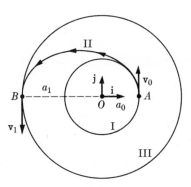

FIG. 4–22. Transfer between two circular orbits.

To illustrate the effects of impulses on the orbital motion of a mass m we shall discuss two transfer problems.

EXAMPLE 1. Consider first the relatively simple problem of the transfer of a unit mass from a circular orbit of radius a_0 to a larger circular orbit of radius a_1. The circles are coplanar, and the transfer orbit is to be an ellipse (sometimes called the Hohmann transfer ellipse [11]). Figure 4–22 illustrates the problem. Let $\mathbf{i}$ and $\mathbf{j}$ be unit vectors as shown. Then

$$\mathbf{v}_0 = \sqrt{\frac{\mu}{a_0}}\,\mathbf{j} = v_0\mathbf{j},$$

$$\mathbf{v}_1 = -\sqrt{\frac{\mu}{a_1}}\,\mathbf{j} = -v_1\mathbf{j},$$

where $\mu = k^2 M$ is the gravitational product at O, and where the circular velocities are given by Eq. (4–45).

By definition, $\mathbf{I}_A = \mathbf{v}_0' - \mathbf{v}_0$ and $\mathbf{I}_B = \mathbf{v}_1 - \mathbf{v}_1'$ are the required impulses at A and B, respectively.

For the elliptical orbit (II) we find from Eq. (4–43) that

$$v_0' = \sqrt{\mu\left(\frac{2}{a_0} - \frac{1}{a}\right)} = \sqrt{\frac{2a_1}{a_0 + a_1}}\,v_0,$$

$$v_1' = \sqrt{\mu\left(\frac{2}{a_1} - \frac{1}{a}\right)} = \sqrt{\frac{2a_0}{a_0 + a_1}}\,v_1,$$

where $2a = a_0 + a_1$. Thus in vector form we have

$$\mathbf{v}_0' = \sqrt{\frac{2a_1}{a_0 + a_1}}\,v_0\mathbf{j} \quad\text{and}\quad \mathbf{v}_1' = -\sqrt{\frac{2a_0}{a_0 + a_1}}\,v_1\mathbf{j}.$$

Therefore,

$$\mathbf{I}_A = \left\{ \left(\frac{2a_1}{a_0 + a_1} \right)^{1/2} - 1 \right\} v_0 \mathbf{j} \quad \text{and} \quad \mathbf{I}_B = -\left\{ 1 - \left(\frac{2a_0}{a_0 + a_1} \right)^{1/2} \right\} v_1 \mathbf{j}.$$

The magnitude of the total impulse required is

$$I = \left[\left(\frac{2a_1}{a_0 + a_1} \right)^{1/2} - 1 \right] v_0 + \left[1 - \left(\frac{2a_0}{a_0 + a_1} \right)^{1/2} \right] v_1.$$

The numerical values of the energy exchanges at A and at B can be computed from Eq. (4–111). We find

$$\Delta E_A = \frac{1}{2} v_0^2 \left[\frac{a_1 - a_0}{a_1 + a_0} \right] \quad \text{and} \quad \Delta E_B = \frac{1}{2} v_1^2 \left[\frac{a_1 - a_0}{a_1 + a_0} \right],$$

so that the total energy change is

$$\Delta E = \frac{1}{2} \left[\frac{a_1 - a_0}{a_1 + a_0} \right] (v_0^2 + v_1^2). \tag{4–113}$$

EXAMPLE 2. Suppose that a satellite moves in a circular orbit of radius a_0 with speed v_0. It is desired to change the inclination of the orbit through an angle θ. We shall consider two ways of doing this.

(a) A direct impulse can be applied to the velocity vector as indicated in Fig. 4–23(a). This must be done in such a way that the magnitude of $\mathbf{v}_0$ is unchanged. Hence we find for the magnitude of the required impulse,

$$I_0 = 2v_0 \sin \left(\frac{\theta}{2} \right). \tag{4–114}$$

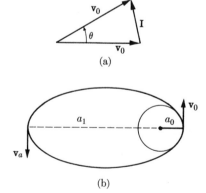

FIG. 4–23. Velocity-impulse diagram for orbit change.

(b) Since in (a) the value of I_0 is quite large when θ is large, we wish to proceed by a method which is more economical of energy. We shall, therefore, change the circular orbit into one of high eccentricity with major axis $2a = a_0 + a_1$, tilt the elliptical orbit about its major axis when the satellite is at apogee, and then return the object to a circular orbit of radius a_0 when the satellite passes through perigee. The impulse budget for this procedure is as follows:

I. The transfer from the circular orbit to the ellipse at perigee requires $I_1 = v_0\{\sqrt{2a_1/(a_0 + a_1)} - 1\}$.

II. Swinging the orbit through an angle θ when the object is at apogee requires $I_2 = 2v_0(a_0/a_1)\sqrt{2a_1/(a_0 + a_1)} \sin(\theta/2)$. This does not change the speed at apogee.

III. Returning the object to a circular orbit of radius a_0 at perigee passage requires a retro-impulse $I_3 = v_0\{\sqrt{2a_1/(a_0 + a_1)} - 1\}$.

Hence the total impulse budget is the sum of these quantities, namely,

$$I = 2v_0\left\{\sqrt{\frac{2a_1}{a_0 + a_1}}\left[\frac{a_0}{a_1} \sin\left(\frac{\theta}{2}\right) + 1\right] - 1\right\}. \qquad (4\text{–}115)$$

This may be written

$$I = 2v_0\left\{\sqrt{1 + e}\left[\frac{1 - e}{1 + e} \sin\left(\frac{\theta}{2}\right) + 1\right] - 1\right\} \quad (0 < e < 1), \qquad (4\text{–}116)$$

where e is the eccentricity of the transfer ellipse.

Table 4–4 indicates the order of magnitude of $I/2v_0$ for representative values of e and θ. In the last line the values of $I_0/2v_0$ for single transfer, given by Eq. (4–114), are listed for comparison. It is clear that for $\theta < 50°$, the double transfer holds no advantage over the single impulse transfer. But for $\theta > 50°$, the double impulse transfer is more economical of impulse energy.

TABLE 4–4

IMPULSE REQUIRED FOR DOUBLE TRANSFER

Example 2

e \ θ	15°	30°	45°	60°	90°
0.2	0.192	0.286	0.376	0.461	0.614
0.5	0.278	0.331	0.381	0.430	0.515
0.8	0.362	0.382	0.400	0.417	0.448
$I_0/2v_0$	0.131	0.259	0.383	0.500	0.707

A transfer from planet to planet in the solar system is more complex than Example 1 shows. The varied inclinations and orientations of the planetary orbits must be taken into account. Limitations of the energy available for producing the required velocity increments also complicate the problem. We have sketched in the discussion above (Example 1) the transfer by a ballistic missile moving only under the gravitational force of the sun. Thus the energy requirement is minimized. We shall not consider the more complex problem here. It has been treated extensively in the current literature [10].

PROBLEMS

Section 4-4:

4-1. Deduce from Eqs. (4-31) the areal velocity integrals $y\dot{z} - z\dot{y} = c_1$, $z\dot{x} - x\dot{z} = c_2$, $x\dot{y} - y\dot{x} = c_3$. How are the constants c_1, c_2, c_3 related to the areal velocity constant h of Eq. (4-32)?

Section 4-5:

4-2. Assume that two planets move around the sun in circular orbits of radii a_1 and a_2 with periods P_1 and P_2. From the fact that the acceleration of a mass moving with constant speed in a circular path of radius a is v^2/a, and from Kepler's third law of motion, show that the accelerations of the two are inversely proportional to the squares of their distances from the sun. What can you say about the forces attracting them to the sun?

4-3. Show that the total energy for a particle of mass m moving in a hyperbolic path of semitransverse axis a in the field of a mass M is $k^2 Mm/2a$. Hence show that the speed at a distance r from the central mass is given by

$$v^2 = k^2 M \left[\frac{2}{r} + \frac{1}{a} \right].$$

4-4. A particle of unit mass moves in a central force field given by $F = -\alpha r$, where $\alpha > 0$. From Eqs. (4-34) and (4-35) show that the differential equation for its orbit is

$$\frac{d^2 u}{d\theta^2} + u = \alpha h^{-2} u^{-3}$$

[compare with Eq. (4-37)]. Show that this is satisfied by an equation of the form $u^2 = A + B \cos 2\theta$, which represents an ellipse with center at the origin.

4-5. A comet moves in an elliptical orbit of eccentricity 0.5. Calculate the ratio of the speed of the comet at perihelion to that at aphelion.

4-6. An artificial satellite is observed to be 200 mi above the earth's surface at perigee and 800 mi above the surface at apogee. Assume that the earth is a sphere 4000 mi in radius. Calculate:

(a) the semimajor axis of the satellite orbit,
(b) the eccentricity of its orbit,

(c) the linear speed of the satellite at perigee,

(d) the period of the satellite in its orbit.

Can you obtain the mass of the satellite from data on its motion? Give reasons for your answer.

4–7. A space probe is sent toward Venus. After some time it is observed to be circling Venus at a mean distance of 400,000 km from the planet's center, with a period of 32.2 days. What is the mass of Venus? The mass of the satellite is negligible compared with the mass of Venus.

Sections 4–5 and 4–6:

4–8. The American artificial planet Pioneer IV reached its closest approach to the sun, 91.7 million mi, on March 17, 1959. At aphelion it will be 106.1 million mi from the sun. Find:

(a) the semimajor axis of its orbit,

(b) the eccentricity of its orbit,

(c) its sidereal period,

(d) the speed with which it passed perihelion.

On March 3, 1959, when the probe was launched, the heliocentric longitude of the earth was $161°32'14''$. On March 17, 1959, when the probe was at perihelion, its estimated heliocentric coordinates were $l = 174°17'14''$, $b = +2°10'$. Find the elements Ω, ω, and i for the orbit.

TABLE 4-5

EPHEMERIS DATA FOR THE SUN AND VENUS

| Date | Sun | | Venus | | |
	Semi-diameter	Geocentric longitude λ	Log radius vector	l	b
Jan. 1	$16'17''.84$	$280°08'43''$	9.8590283	$215°19'59''$	$+2°13'00''$
Feb. 1	$16'15''.70$	$311°41'57''$	9.8613700	$264°43'26''$	$-0°30'44''$
Mar. 1	$16'10''.22$	$339°58'04''$	9.8622799	$309°00'43''$	$-2°42'29''$
Apr. 1	$16'02''.00$	$10°49'08''$	9.8613271	$358°08'56''$	$-3°19'08''$
May 1	$15'54''.08$	$40°09'57''$	9.8590544	$46°02'07''$	$-1°42'03''$
June 1	$15'48''.06$	$70°02'26''$	9.8569019	$95°57'32''$	$+1°09'16''$
July 1	$15'45''.69$	$98°41'32''$	9.8564419	$144°39'03''$	$+3°09'30''$
Aug. 1	$15'47''.39$	$128°16'34''$	9.8580210	$194°54'34''$	$+2°58'27''$
Sept. 1	$15'52''.89$	$158°05'22''$	9.8605037	$244°34'12''$	$+0°40'41''$
Oct. 1	$16'00''.59$	$187°21'09''$	9.8621217	$292°06'15''$	$-1°59'47''$
Nov. 1	$16'08''.96$	$218°06'46''$	9.8618715	$341°09'53''$	$-3°22'53''$
Dec. 1	$16'15''.24$	$248°20'44''$	9.8599268	$28°55'17''$	$-2°29'25''$

4-9. A proposal is made to launch a space probe bearing instruments for surveying Venus at such a time that it arrives in the neighborhood of the planet at perihelion. This occurs on June 22. Data for Venus and the sun are given in Table 4-5. In this problem assume that the orbit of the earth is circular and ignore the attraction of the earth on the probe. Assume the orbit of the probe to be in the ecliptic plane, with its aphelion at 1 A.U. from the sun.

(a) Find the semimajor axis, a, and the eccentricity, e, for the orbit of the probe.

(b) Find the period of its motion, and hence the time required to reach Venus.

(c) On what date should the probe be launched so that it will reach Venus on June 22? Will the earth be in the proper position so that the probe can be launched on the date required?

(d) On the assumption that launching is possible on the required date, calculate the position of the probe on March 1, April 1, May 1, and June 1.

(e) Show on a diagram the relative positions of Venus, the earth, and the probe on the above dates.

(f) By how much would the probe have to be speeded up or slowed down when it reaches Venus for its speed to match that of Venus?

(g) If the orbit of the probe lies in the plane of the ecliptic, how many kilometers will separate it from Venus when it arrives near Venus' perihelion?

(h) What is the angular diameter of the sun as seen from the probe when it is near Venus?

4-10. The orbital elements of the artificial satellite Lunik III, referred to the equator, are

$$P = 15.761 \text{ days}, \qquad T = \text{Nov. } 19.223, 1959, \text{ GCT},$$
$$i = 80°.54, \qquad \omega = 182°.3,$$
$$e = 0.824, \qquad \Omega = 250°.9.$$
$$a = 2.66 \times 10^5 \text{ km},$$

Calculate the position of Lunik III in its orbit, and hence calculate its right ascension and declination, on January 1, 1960, at 0^h GCT. [*Hint:* In this example, the eccentricity is large. Hence, solve Kepler's equation graphically. Plot $y = \sin E$ on rectangular coordinate paper; plot $y = (1/e)(E - M)$ also, and find their intersection.] How far is the satellite from the earth on that date? What are its perigee and apogee distances?

Section 4-8:

4-11. The parabolic orbital elements of Comet Alcock (1959e) are

$$T = \text{August } 17.612 \text{ GCT}, \qquad i = 48°.2,$$
$$\omega = 124°.7, \qquad q = 1.1502 \text{ A.U.}$$
$$\Omega = 159°.2,$$

Calculate the heliocentric coordinates of the comet on November 17, 1959.

4-12. On August 1, 1959, a meteor was observed at longitude 7^h24^m W, latitude $32°15'$ N, coming directly toward the observer. Its apparent zenith distance was $45°30'$, its azimuth $102°$, and the time 3^h12^m a.m. LCT. The observed velocity relative to the center of the earth was 22.5 km/sec. Relevant data from the *American Ephemeris* are:

$$\text{sidereal time, } 0^h \text{ GCT } = 20^h35^m26^s,$$
$$\text{celestial longitude of sun } (\lambda) = 128°.0,$$
$$\text{radius vector of earth } = 1.015 \text{ A.U.}$$

(a) Calculate the values of the velocity at infinity and the zenith attraction.
(b) Find the right ascension and declination of the radiant.
(c) Given the heliocentric velocity components for the earth,

$$\dot{x}_e = +22.5, \qquad \dot{y}_e = +18.5, \qquad \dot{z}_e = 0 \quad \text{(km/sec)},$$

find the heliocentric velocity components of the meteor.
(d) Obtain the elements of the meteor orbit around the sun.

4-13. A space probe was launched in the earth's equatorial plane, toward the east, at local midnight on March 21, 1958, when the sun was precisely at the vernal equinox. Its velocity at launching was 12 km/sec at a zenith distance of $30°$. The radius vector of the earth at launching was 0.9962267 A.U. Find the elements of the orbit of the probe around the sun. [Ignore the earth's rotational velocity and its radius in comparison with its distance from the sun.]

Section 4–9:

4-14. Compute directly the energies in orbits I, II, III of Fig. 4–22 and show that the total energy exchange agrees with that given in Eq. (4–113).

4-15. A satellite of mass m moves in a circular orbit over the north and south poles of the earth. Its period of revolution is P hours. What impulse, and consequently what velocity increment, must be given to the satellite at each passage over the north pole to ensure that it pass over the same point on the equator at each revolution? What energy must be imparted to the satellite to produce the required precession?

Sections 4–8 and 4–9:

4-16. Calculate the total of the velocity increments required to transfer a unit mass from a satellite orbit 1000 km above the earth's surface to Venus. Assume that the orbits of the earth, the satellite, and Venus are circular and lie in the plane of the ecliptic. Choose the most appropriate point in the satellite orbit to eject the unit mass. What is the angular separation between the earth and Venus at the time of launching?

References

1. R. T. Crawford, *Determination of Orbits of Comets and Asteroids*, New York: McGraw-Hill Book Company, Inc., 1930.

2. E. Finlay-Freundlich, *Celestial Mechanics*, New York: Pergamon Press, 1958.

3. Paul Herget, *The Computation of Orbits*, published privately by the author, 1948.

4. F. R. Moulton, *An Introduction to Celestial Mechanics*, 2nd rev. ed., New York: The Macmillan Company, 1958.

5. W. M. Smart, *Celestial Mechanics*, New York: Longmans, Green and Company, 1953.

6. J. C. Watson, *Theoretical Astronomy*, Philadelphia: J. B. Lippincott and Company, 1892.

7. K. P. Williams, *The Calculation of the Orbits of Asteroids and Comets*, Bloomington, Indiana: The Principia Press, 1934.

8. A. C. B. Lovell, *Meteor Astronomy*, pp. 91 ff., Oxford: Clarendon Press, 1954.

9. S. Herrick, Jr., "Tables for the Reduction of Radial Velocities to the Sun," Lick Obs. Bull. **17,** 85, 1935.

10. H. Seifert *et al*, *Space Technology*, New York: John Wiley & Sons, Inc., 1959.

11. W. Hohmann, *Die Erreichbarkeit der Himmelskörper*, Munich: Oldenburg Publishing Co., 1925.

CHAPTER 5

CELESTIAL DYNAMICS II:
THREE- AND n-BODY PROBLEMS

When three or more bodies attract one another according to Newton's law of gravitation, the complexity of the resulting motion increases enormously. As we have seen in Chapter 4, it is possible to formulate the two-body problem mathematically in such a way that, given the initial circumstances of motion of one body about the other, we can predict precisely thereafter its position and velocity in space at any one instant. That is, we have found a closed mathematical solution to the problem. Except under special circumstances, it is impossible to formulate such a solution when three or more bodies are involved.

Consider for a moment the planets of the solar system. Each planet attracts all other planets with a force directly proportional to the product of the respective masses and inversely proportional to the squares of the respective distances between them. The latter change continuously as the planets move in their orbits around the sun. Hence at any instant the forces acting on a given planet, say the earth, are different. Fortunately, however, the predominant force causing the motion of a planet is that of the sun. To a first approximation, the planet moves as though it and the sun constituted a two-body system. This suggests, therefore, that we first study the motion of a planet as a part of a two-body system. Then we determine the deviations from a purely two-body motion that will result from the presence of other disturbing bodies. Such deviations are called *perturbations*.

The principal forces acting on a small satellite in the vicinity of the earth are those due to the earth, the moon, and the sun. Although the earth produces the dominant attraction, the problem is nonetheless a four-body problem.

Other examples of n-body motion might be cited. In this chapter we shall outline the mathematical framework which describes three-body motion. The extension to more than three bodies can be made readily.

5-1 Motion of the center of mass. Consider three particle masses m_1, m_2, m_3 located in a primary inertial system as shown in Fig. 5-1. Their positions are defined by the vectors $\mathbf{r}_1$, $\mathbf{r}_2$, and $\mathbf{r}_3$. We assume that they move under their mutual gravitational attractions. Then the equa-

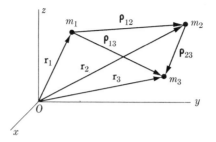

FIG. 5–1. Three point masses in a primary inertial system.

tions of motion are

$$m_1\ddot{\mathbf{r}}_1 = k^2 \left\{ \frac{m_1 m_2}{\rho_{12}^3} \boldsymbol{\rho}_{12} + \frac{m_1 m_3}{\rho_{13}^3} \boldsymbol{\rho}_{13} \right\}, \qquad (5\text{–}1)$$

$$m_2\ddot{\mathbf{r}}_2 = k^2 \left\{ \frac{m_2 m_3}{\rho_{23}^3} \boldsymbol{\rho}_{23} - \frac{m_2 m_1}{\rho_{12}^3} \boldsymbol{\rho}_{12} \right\}, \qquad (5\text{–}2)$$

$$m_3\ddot{\mathbf{r}}_3 = k^2 \left\{ - \frac{m_3 m_2}{\rho_{23}^3} \boldsymbol{\rho}_{23} - \frac{m_3 m_1}{\rho_{13}^3} \boldsymbol{\rho}_{13} \right\}, \qquad (5\text{–}3)$$

where k^2 is the constant of gravitation and the vectors $\boldsymbol{\rho}_{ij}$ are displayed in Fig. 5–1.

Add Eqs. (5–1), (5–2), and (5–3) to obtain

$$m_1\ddot{\mathbf{r}}_1 + m_2\ddot{\mathbf{r}}_2 + m_3\ddot{\mathbf{r}}_3 = 0$$

and integrate twice. We find

$$m_1\mathbf{r}_1 + m_2\mathbf{r}_2 + m_3\mathbf{r}_3 = \mathbf{c}_1 t + \mathbf{c}_2,$$

where $\mathbf{c}_1$ and $\mathbf{c}_2$ are vector constants of integration. But the center of mass is defined by

$$M\mathbf{R} = m_1\mathbf{r}_1 + m_2\mathbf{r}_2 + m_3\mathbf{r}_3,$$

where $M = m_1 + m_2 + m_3$. Hence

$$\mathbf{R} = \frac{\mathbf{c}_1}{M} t + \frac{\mathbf{c}_2}{M}. \qquad (5\text{–}4)$$

The *center of mass either remains at rest or moves uniformly in space on a straight line.* This constitutes a first integral of the equations of motion and is quite analogous to the result found for two-body motion [Eq. (4–25)].

As will be explained later, this result is of no practical significance because it is impossible to establish a reference frame in which the vector constants $\mathbf{c}_1$ and $\mathbf{c}_2$ can be determined from observations.

5–2 Integrals of energy and angular momentum. Let $\mathbf{u}_{12}$, $\mathbf{u}_{23}$, $\mathbf{u}_{13}$ denote unit vectors in the directions $\boldsymbol{\rho}_{12}$, $\boldsymbol{\rho}_{23}$, $\boldsymbol{\rho}_{13}$, respectively (Fig. 5–2). Form the scalar product of Eq. (5–1) by $\dot{\mathbf{r}}_1$, Eq. (5–2) by $\dot{\mathbf{r}}_2$, Eq. (5–3) by $\dot{\mathbf{r}}_3$, and add the resulting equations to obtain

$$\sum_{i=1}^{3} m_i \dot{\mathbf{r}}_i \cdot \ddot{\mathbf{r}}_i$$

$$= k^2 \left\{ \frac{m_1 m_2}{\rho_{12}^2} \mathbf{u}_{12} \cdot (\dot{\mathbf{r}}_1 - \dot{\mathbf{r}}_2) + \frac{m_2 m_3}{\rho_{23}^2} \mathbf{u}_{23} \cdot (\dot{\mathbf{r}}_2 - \dot{\mathbf{r}}_3) + \frac{m_3 m_1}{\rho_{13}^2} \mathbf{u}_{13} \cdot (\dot{\mathbf{r}}_1 - \dot{\mathbf{r}}_3) \right\} .$$

$$(5\text{–}5)$$

But $\dot{\mathbf{r}}_1 - \dot{\mathbf{r}}_2 = -\dot{\boldsymbol{\rho}}_{12}$, $\dot{\mathbf{r}}_2 - \dot{\mathbf{r}}_3 = -\dot{\boldsymbol{\rho}}_{23}$, and $\dot{\mathbf{r}}_1 - \dot{\mathbf{r}}_3 = -\dot{\boldsymbol{\rho}}_{13}$. Furthermore,

$$\dot{\boldsymbol{\rho}}_{ij} = \frac{d}{dt} (\rho_{ij} \mathbf{u}_{ij}) = \dot{\rho}_{ij} \mathbf{u}_{ij} + \rho_{ij} \dot{\mathbf{u}}_{ij} \qquad (i, j = 1, 2, 3; \quad i \neq j),$$

and $\dot{\mathbf{u}}_{ij}$ is perpendicular to $\mathbf{u}_{ij}$ because the latter is a unit vector. Hence $\mathbf{u}_{ij} \cdot \dot{\mathbf{u}}_{ij} = 0$, and $\mathbf{u}_{ij} \cdot \mathbf{u}_{ij} = 1$. Using this information, we find from Eq. (5–5)

$$\sum_{i=1}^{3} m_i \dot{\mathbf{r}}_i \cdot \ddot{\mathbf{r}}_i = -k^2 \left\{ \frac{m_1 m_2}{\rho_{12}^2} \dot{\rho}_{12} + \frac{m_2 m_3}{\rho_{23}^2} \dot{\rho}_{23} + \frac{m_3 m_1}{\rho_{13}^2} \dot{\rho}_{13} \right\} .$$

This may be written

$$\frac{d}{dt} \left\{ \frac{1}{2} \sum_{i=1}^{3} m_i \dot{\mathbf{r}}_i^2 \right\} = \frac{d}{dt} \left\{ \frac{k^2 m_1 m_2}{\rho_{12}} + \frac{k^2 m_2 m_3}{\rho_{23}} + \frac{k^2 m_3 m_1}{\rho_{13}} \right\} . \qquad (5\text{–}6)$$

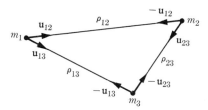

FIG. 5–2. Unit vectors for describing relative motion in a three-mass system.

Now the kinetic energy is $T = \frac{1}{2}(m_1\dot{r}_1^2 + m_2\dot{r}_2^2 + m_3\dot{r}_3^2)$, and we define the potential energy by

$$V = -k^2 \left\{ \frac{m_1 m_2}{\rho_{12}} + \frac{m_2 m_3}{\rho_{23}} + \frac{m_3 m_1}{\rho_{13}} \right\}. \tag{5-7}$$

Then integration of Eq. (5–6) yields

$$T + V = E \qquad \text{(a constant).} \tag{5-8}$$

This expresses the conservation of energy for the system of three masses and constitutes a second integral of the equations of motion.

Again we write $\boldsymbol{\rho}_{12} = \rho_{12}\mathbf{u}_{12}, \boldsymbol{\rho}_{13} = \rho_{13}\mathbf{u}_{13}, \boldsymbol{\rho}_{23} = \rho_{23}\mathbf{u}_{23}$ in Eqs. (5–1), (5–2), and (5–3), form the vector products of these by $\mathbf{r}_1, \mathbf{r}_2$, and $\mathbf{r}_3$, respectively, and add to obtain

$$\sum_{i=1}^{3} \mathbf{r}_i \times m_i\dot{\mathbf{v}}_i = k^2$$

$$\times \left\{ \frac{m_1 m_2}{\rho_{12}^2} (\mathbf{r}_1 - \mathbf{r}_2) \times \mathbf{u}_{12} + \frac{m_2 m_3}{\rho_{23}^2} (\mathbf{r}_2 - \mathbf{r}_3) \times \mathbf{u}_{23} + \frac{m_3 m_1}{\rho_{13}^2} (\mathbf{r}_1 - \mathbf{r}_3) \times \mathbf{u}_{13} \right\}.$$

$$\tag{5-9}$$

The vector differences $\mathbf{r}_i - \mathbf{r}_j$ on the right-hand side of Eq. (5–9) are collinear with the respective unit vectors $\mathbf{u}_{ij}$. Hence the vector products vanish. The equation then may be written

$$\frac{d}{dt} \{\mathbf{r}_1 \times m_1\mathbf{v}_1 + \mathbf{r}_2 \times m_2\mathbf{v}_2 + \mathbf{r}_3 \times m_3\mathbf{v}_3\} = 0. \tag{5-10}$$

The quantity in the braces is the *total angular momentum of the system about the origin O*. Therefore, calling it $\mathbf{L}$, we have

$$\mathbf{L} = \text{constant vector.}$$

This is a third integral of the equations of motion. It is sometimes called the "integral of areas" because it expresses areal-velocity relations analogous to Kepler's second law for the two-body motion. Equation (5–10) expresses the *conservation of angular momentum for the system of three masses*.

As the three bodies move, their position and velocity vectors are always so oriented that the individual angular-momentum vectors $\mathbf{r}_i \times m_i\mathbf{v}_i$ add to yield the constant vector $\mathbf{L}$. The direction of $\mathbf{L}$ defines an *invariable line* in space. Associated with this line is an *invariable plane* perpendicular to $\mathbf{L}$. The name "invariable plane" was given by Laplace.

All the integrals derived above for the problem of three-bodies can be determined for the problem of n bodies. In the solar system, for instance, the invariable plane has an inclination of $1°35'$ to the ecliptic, intermediate between the orbital planes of Jupiter and Saturn. The angular momentum about a line perpendicular to this plane is a maximum; that about a line in this plane is zero. The invariable plane for the solar system passes through the center of mass of the system.

It should be emphasized that each of Eqs. (5–1), (5–2), and (5–3) can be expressed in terms of three equations in cartesian coordinates, in which case nine second-order differential equations are obtained. Therefore, 18 constants must be available for a complete solution to the three-body problem. Since we have found six constants related to the motion of the center of mass, three related to the angular momentum and one related to the total energy, eight integrals remain to be determined. Bruns [1] has shown that when rectangular coordinates are chosen as the dependent variables in the three-body (or n-body) problem, the ten integrals (constants) described above are the only integrals to be expected. In other words, the complete three-body problem is insoluble in closed form.

5–3 The disturbing function. Because it is impossible to observe the positions and velocities required to determine the absolute motions of a group of celestial objects in space, we now consider their relative motions. In the solar system, for example, the motion of the planets relative to the sun is observable, and hence orbit calculations are made for a coordinate system with the sun as the origin. For descriptive purposes, we shall confine the present discussion to a system of three bodies, say the sun and two planets.

We subtract Eq. (5–1) from Eqs. (5–2) and (5–3) to obtain the equations of motion for masses m_2 and m_3 relative to m_1. After some simplification, we find

$$\text{for } m_2: \quad \ddot{\boldsymbol{\rho}}_{12} = -\frac{k^2(m_1 + m_2)}{\rho_{12}^2}\mathbf{u}_{12} + k^2 m_3 \left\{\frac{1}{\rho_{23}^2}\mathbf{u}_{23} - \frac{1}{\rho_{13}^2}\mathbf{u}_{13}\right\}, \quad (5\text{–}11)$$

$$\text{for } m_3: \quad \ddot{\boldsymbol{\rho}}_{13} = -\frac{k^2(m_1 + m_3)}{\rho_{13}^2}\mathbf{u}_{13} - k^2 m_2 \left\{\frac{1}{\rho_{23}^2}\mathbf{u}_{23} + \frac{1}{\rho_{12}^2}\mathbf{u}_{12}\right\}, \quad (5\text{–}12)$$

where the unit vectors involved are shown in Fig. 5–2. The quantities on the right-hand side of Eq. (5–11) are respectively

(a) the acceleration of m_2 due to the action of m_1;
(b) the acceleration of m_2 due to the mass m_3;
(c) the negative of the acceleration of m_1 due to the presence of m_3.

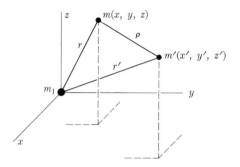

FIG. 5-3. Cartesian coordinates for a three-mass system.

Similar interpretations can be made with regard to the right-hand terms of Eq. (5-12). It is obvious that if $m_3 = 0$, Eq. (5-11) describes the two-body motion of m_2 about m_1. If $m_2 = 0$, Eq. (5-12) describes the motion of m_3 about m_1. These results for the three-body problem can be generalized to the case of n bodies moving relative to any given one.

For convenience in succeeding discussions and for purposes of calculation, we write Eqs. (5-11) and (5-12) in cartesian form. Let m_1 be at the origin and represent the dominant mass. Let m denote the mass whose motion is to be found; let m' denote the mass which disturbs the motion of m around m_1. Designating the position of m by (x, y, z) and that of m' by (x', y', z'), we find from Eq. (5-11) and by reference to Fig. 5-3 that the motion of m is described by

$$\ddot{x} = -\frac{k^2 M x}{r^3} + \frac{k^2 m'(x' - x)}{\rho^3} - \frac{k^2 m' x'}{r'^3}, \qquad (5\text{-}13)$$

$$\ddot{y} = -\frac{k^2 M y}{r^3} + \frac{k^2 m'(y' - y)}{\rho^3} - \frac{k^2 m' y'}{r'^3}, \qquad (5\text{-}14)$$

$$\ddot{z} = -\frac{k^2 M z}{r^3} + \frac{k^2 m'(z' - z)}{\rho^3} - \frac{k^2 m' z'}{r'^3}, \qquad (5\text{-}15)$$

where $M = m_1 + m$.

Since

$$\rho = [(x - x')^2 + (y - y')^2 + (z - z')^2]^{1/2},$$

we may write

$$\frac{\partial \rho^{-1}}{\partial x} = \frac{x' - x}{\rho^3}; \qquad \frac{\partial \rho^{-1}}{\partial y} = \frac{y' - y}{\rho^3}; \qquad \frac{\partial \rho^{-1}}{\partial z} = \frac{z' - z}{\rho^3}.$$

And since x', y', z', r' are independent of x, y, z, we also have

$$\frac{\partial}{\partial x}\left[\frac{xx' + yy' + zz'}{r'^3}\right] = \frac{x'}{r'^3},$$

$$\frac{\partial}{\partial y}\left[\frac{xx' + yy' + zz'}{r'^3}\right] = \frac{y'}{r'^3},$$

$$\frac{\partial}{\partial z}\left[\frac{xx' + yy' + zz'}{r'^3}\right] = \frac{z'}{r'^3}.$$

Therefore Eqs. (5–13), (5–14), and (5–15) can be written in the symmetric forms

$$\ddot{x} = -\frac{k^2 M x}{r^3} + k^2 m' \frac{\partial}{\partial x}\left\{\frac{1}{\rho} - \frac{xx' + yy' + zz'}{r'^3}\right\},$$

$$\ddot{y} = -\frac{k^2 M y}{r^3} + k^2 m' \frac{\partial}{\partial y}\left\{\frac{1}{\rho} - \frac{xx' + yy' + zz'}{r'^3}\right\},$$

$$\ddot{z} = -\frac{k^2 M z}{r^3} + k^2 m' \frac{\partial}{\partial z}\left\{\frac{1}{\rho} - \frac{xx' + yy' + zz'}{r'^3}\right\}.$$

We introduce the function R by the definition

$$R = k^2 m' \left\{\frac{1}{\rho} - \frac{xx' + yy' + zz'}{r'^3}\right\}. \qquad (5\text{--}16)$$

This is the *disturbing*, or perturbative, *function* corresponding to the mass m'. Then we have

$$\ddot{x} = -\frac{k^2 M x}{r^3} + \frac{\partial R}{\partial x}, \qquad (5\text{--}17)$$

$$\ddot{y} = -\frac{k^2 M y}{r^3} + \frac{\partial R}{\partial y}, \qquad (5\text{--}18)$$

$$\ddot{z} = -\frac{k^2 M z}{r^3} + \frac{\partial R}{\partial z}, \qquad (5\text{--}19)$$

which are the equations of motion for m under the influence of m'.

If there are additional masses disturbing the motion of m, the disturbing function is taken to be the sum of the disturbing functions corresponding to the individual masses. If

$$R_i = k^2 m'_i \left\{\frac{1}{\rho_i} - \frac{xx'_i + yy'_i + zz'_i}{r_i'^3}\right\}$$

is the disturbing function for the ith mass, the equations of motion for m take the form

$$\ddot{x} = -\frac{k^2 M x}{r^3} + \sum_{i=1}^{n-2} \frac{\partial R_i}{\partial x}, \tag{5-20}$$

$$\ddot{y} = -\frac{k^2 M y}{r^3} + \sum_{i=1}^{n-2} \frac{\partial R_i}{\partial y}, \tag{5-21}$$

$$\ddot{z} = -\frac{k^2 M z}{r^3} + \sum_{i=1}^{n-2} \frac{\partial R_i}{\partial z}, \tag{5-22}$$

where n is the total number of masses in the system.

In general, these equations have no closed mathematical solution. The meaning of the term *insolvability in closed form* should be made very clear. It is this: given, at a certain instant t_0, the positions and velocities of three or more point masses attracting one another according to the law of gravitation, we cannot predict the future motion for any arbitrary time interval.

Under certain special circumstances and for special configurations of the attracting masses, solutions to the n-body problem can be found. These have been known since the time of Lagrange (1772). We shall examine in later sections some of these special cases.

5–4 Perturbations. Each planet of the solar system moves in its orbit under the dominant force of the sun. To this extent, the sun-planet system constitutes a two-body problem. The other planets, however, exert forces which change with the relative positions of the objects and disturb the two-body motion. The resulting deviations from the orbit are usually small. Given the elements a, e, i, ω, Ω, and T of the two-body orbit at any instant, we wish to calculate the changes, i.e., *perturbations*, in these elements as functions of time. Similar perturbations arise in the motion of a man-made satellite around the earth. We shall defer treatment of these to a later section.

Consider the motion of a mass m around m_1 in the presence of a small mass m' as discussed in Section 5–3. The equations of motion are Eqs. (5–17), (5–18), and (5–19). If m' is neglected so that $R = 0$, these equations can be solved to yield

$$x = x(t, a, e, i, \omega, \Omega, T),$$
$$y = y(t, a, e, i, \omega, \Omega, T),$$
$$z = z(t, a, e, i, \omega, \Omega, T),$$

where, from a knowledge of the initial conditions, it is possible to evaluate the constants of integration in terms of the orbital elements. The solution is a Keplerian ellipse traced by m about m_1, called the *intermediary orbit*. It has a tangent in common with the true orbit and represents the orbit that m would have if all disturbing influences were removed. If, in addition, the velocity in the Keplerian ellipse at any given instant is the same as that in the true orbit, the elements of the ellipse are called *osculating elements*.

We now consider $R \neq 0$ for small m' and determine the time variations in a, e, i, Ω, ω, and T caused by the presence of the disturbing mass. This procedure is the *variation-of-parameters* technique due to Lagrange. We shall not elaborate upon the details here, which can be found in reference 2, pp. 402 ff. The resulting differential equations for the time dependence of the orbital elements are

$$\frac{da}{dt} = \frac{2}{na} \frac{\partial R}{\partial M}, \tag{5–23}$$

$$\frac{di}{dt} = -\frac{1}{na^2\sqrt{1 - e^2} \sin i} \left\{ \frac{\partial R}{\partial \Omega} - \cos i \frac{\partial R}{\partial \omega} \right\}, \tag{5–24}$$

$$\frac{de}{dt} = \frac{\sqrt{1 - e^2}}{na^2 e} \left\{ \sqrt{1 - e^2} \frac{\partial R}{\partial M} - \frac{\partial R}{\partial \omega} \right\}, \tag{5–25}$$

$$\frac{d\Omega}{dt} = \frac{1}{na^2\sqrt{1 - e^2} \sin i} \frac{\partial R}{\partial i}, \tag{5–26}$$

$$\frac{d\omega}{dt} = \frac{\sqrt{1 - e^2}}{na^2 e} \frac{\partial R}{\partial e} - \frac{\cos i}{na^2\sqrt{1 - e^2} \sin i} \frac{\partial R}{\partial i}, \tag{5–27}$$

$$\frac{dM}{dt} = n - \frac{2}{na} \frac{\partial R}{\partial a} - \frac{(1 - e^2)}{na^2 e} \frac{\partial R}{\partial e}, \tag{5–28}$$

where $M = n(t - T)$ and $n = 2\pi/P$, the mean daily motion.

In order to use Eqs. (5–23) through (5–28), R must be expressed in terms of the osculating elements of the orbit at some epoch. An illustrative example will be given in Section 5–12.

In many practical applications it is more convenient to resolve the perturbing acceleration into rectangular components as indicated in Fig. 5–4; more specifically, let

$R =$ the radial component directed positively outward from m_1 along r;

$S =$ the transverse component in the plane of the orbit, perpendicular to R, and making an angle less than 90° with the velocity vector;

$W =$ the component perpendicular to R and S and positive on the north side of the orbital plane.

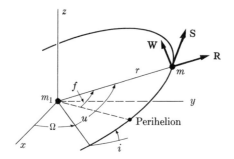

FIG. 5–4. Components of perturbing acceleration in elliptic motion: $\mathbf{R}$ is radial; $\mathbf{S}$ is perpendicular to $\mathbf{R}$ and in the orbital plane; $\mathbf{W}$ is normal to the orbital plane.

It can be shown that the differential equations for the time variations of the osculating elements then become

$$\frac{da}{dt} = \frac{2}{nr\sqrt{1-e^2}} \{(er \sin f)R + a(1-e^2)S\}, \tag{5–29}$$

$$\frac{di}{dt} = \frac{r \cos u}{na^2\sqrt{1-e^2}} W, \tag{5–30}$$

$$\frac{de}{dt} = \frac{\sqrt{1-e^2}}{nae} \left\{(e \sin f)R + \left[\frac{a(1-e^2)}{r} - \frac{r}{a}\right]S\right\}, \tag{5–31}$$

$$\frac{d\Omega}{dt} = \frac{r \sin u}{na^2\sqrt{1-e^2}\,\sin i} W, \tag{5–32}$$

$$\frac{d\omega}{dt} = -\frac{\sqrt{1-e^2}}{nae} \left\{(\cos f)R - \sin f\left[1 + \frac{r}{a(1-e^2)}\right]S + \frac{er \sin u \cot i}{a(1-e^2)} W\right\}, \tag{5–33}$$

$$\frac{dM}{dt} = n + \frac{1}{na} \left\{\left[\frac{(1-e^2)\cos f}{e} - \frac{2r}{a}\right]R - \frac{(1-e^2)\sin f}{nae}\left[1 + \frac{r}{a(1-e^2)}\right]S\right\}. \tag{5–34}$$

Where it is desirable to introduce perturbing accelerations tangent to the orbit and normal to it, Eqs. (5–29) to (5–34) are also useful. Let

$T =$ the component of the perturbing acceleration tangent to the orbit in the direction of motion;

$N =$ the component of the perturbing acceleration normal to the curve, directed positively toward the interior of the orbit.

Then the reader may show that the components T and N can be found from the relations

$$S = \frac{1 + e \cos f}{\sqrt{1 + e^2 + 2e \cos f}} \, T + \frac{e \sin f}{\sqrt{1 + e^2 + 2e \cos f}} \, N, \quad (5\text{--}35)$$

$$R = \frac{e \sin f}{\sqrt{1 + e^2 + 2e \cos f}} \, T - \frac{1 + e \cos f}{\sqrt{1 + e^2 + 2e \cos f}} \, N, \quad (5\text{--}36)$$

where f is the true anomaly. It is apparent from Eqs. (5–35) and (5–36) that for a circular orbit ($e = 0$), we have $S = T$ and $R = -N$. The resolution into tangential and normal components is useful in the discussion of atmospheric drag on a satellite moving close to the earth's surface.

The theory of perturbations as outlined above yields the values of (x, y, z) for the mass m in terms of the elements of the osculating orbit and the time rates of change of the latter. The resulting changes or perturbations in the elements appear in the form of infinite series. Many terms have to be calculated to yield accurate results for any one epoch. Such analytical expressions for the changes in the elements are called *general perturbations*.

Quite frequently a numerical method of calculating the orbit is used. Starting with the set of osculating elements and their time rates of change, one performs a step-by-step integration to obtain the elements at successive time intervals. For each integration the positions of the disturbed and the disturbing bodies are known, and the disturbing function can be calculated. Perturbations calculated in this way by numerical integration of the equations of motion are called *special perturbations*.

If a satellite moves in proximity to an oblate planet, the planet's oblateness causes perturbations in the orbit of the satellite. The resulting changes (see Section 5–10 for details) in the orbital elements are usually large.

5–5 The restricted three-body problem. In 1772 Lagrange discussed a three-body problem which can be solved in closed form. The so-called "restricted problem of three-bodies" is of particular importance in discussions of space probes moving in the gravitational fields of the earth and the moon.

Let two massive bodies move in circular orbits about their center of mass. A third body of negligible mass moves about the two under their combined attraction. By negligible mass is meant a particle so small that it does not disturb the two larger masses. This would be the case, for example, for a space vehicle moving in the gravitational field of the earth-moon system. Here, however, the assumption of circular orbits breaks down. The unit of mass is chosen such that the sum of the large masses is unity. Then let m denote the lesser and $1 - m$ the greater mass.

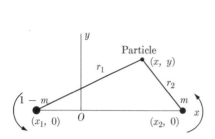

FIG. 5-5. Restricted three-body problem.   FIG. 5-6. Angular motion of axes in the restricted three-body problem.

Consider an xy-coordinate system rotating about O with the masses as shown in Fig. 5-5. The origin is taken to be the center of mass of the system, and $x_2 - x_1 = 1$. Furthermore the unit of time is chosen such that the gravitational constant $k^2 = 1$.

We shall assume here that the particle moves in the plane of motion of $(1 - m)$ and m; that is, in the xy-plane. If the initial position and velocity vectors of the particle lie in this plane, the particle will always remain in the plane because there is no force component perpendicular to the plane. The motion may be generalized to three dimensions without undue complication.*

To obtain the equations of motion we shall use the Lagrangian method in dynamics [3]. Let the mass of the particle be m' and choose a fixed coordinate system, as shown in Fig. 5-6, whose origin coincides with that of the rotating xy-system. Then for m' we have

$$x_0 = x \cos \omega t - y \sin \omega t,$$
$$y_0 = x \sin \omega t + y \cos \omega t.$$

Now by Kepler's third law [Eq. (4-44)],

$$P = \frac{2\pi}{\sqrt{(1 - m) + m}} = 2\pi,$$

because we have taken the distance between the large masses to be unity and $k^2 = 1$. Therefore the angular velocity $\omega = 2\pi/P = 1$, and

$$x_0 = x \cos t - y \sin t, \tag{5-37}$$

$$y_0 = x \sin t + y \cos t. \tag{5-38}$$

---

* See reference 2, pp. 281 ff. for a discussion of the three-dimensional case.

Furthermore $r_1 = \sqrt{(x - x_1)^2 + y^2}$ and $r_2 = \sqrt{(x - x_2)^2 + y^2}$, where x_1 and x_2 are fixed in the rotating system. Hence we obtain for the kinetic and potential energies, respectively,

$$T = \frac{m'}{2}(\dot{x}_0^2 + \dot{y}_0^2) = \frac{m'}{2}\{(\dot{x} - y)^2 + (\dot{y} + x)^2\}, \tag{5–39}$$

$$V = -m'\left\{\frac{1 - m}{[(x - x_1)^2 + y^2]^{1/2}} + \frac{m}{[(x - x_2)^2 + y^2]^{1/2}}\right\}. \tag{5–40}$$

In terms of the Lagrangian function $L = T - V$, the equations of motion are

$$\frac{d}{dt}\left(\frac{\partial L}{\partial \dot{x}}\right) - \frac{\partial L}{\partial x} = 0,$$

$$\frac{d}{dt}\left(\frac{\partial L}{\partial \dot{y}}\right) - \frac{\partial L}{\partial y} = 0.$$

Using Eqs. (5–39) and (5–40), we find for the system under consideration,

$$\ddot{x} - 2\dot{y} = x - \frac{(1 - m)(x - x_1)}{r_1^3} - \frac{m(x - x_2)}{r_2^3}, \tag{5–41}$$

$$\ddot{y} + 2\dot{x} = y - \frac{(1 - m)y}{r_1^3} - \frac{my}{r_2^3}. \tag{5–42}$$

Let $U = \frac{1}{2}(x^2 + y^2) + (1 - m)/r_1 + m/r_2$. Then these equations become

$$\ddot{x} - 2\dot{y} = \frac{\partial U}{\partial x}, \qquad \ddot{y} + 2\dot{x} = \frac{\partial U}{\partial y}.$$

Multiply the first by $2\dot{x}$, the second by $2\dot{y}$, and add to obtain

$$2\dot{x}\ddot{x} + 2\dot{y}\ddot{y} = 2\frac{\partial U}{\partial x}\dot{x} + 2\frac{\partial U}{\partial y}\dot{y},$$

which may be integrated to yield

$$\dot{x}^2 + \dot{y}^2 = 2U - C, \tag{5–43}$$

where C is a constant. This integral is called Jacobi's integral. It resembles the energy integral in the two-body system [Eq. (4–33) of Section 4–4].

By Eq. (5–43) the speed $v = (\dot{x}^2 + \dot{y}^2)^{1/2}$ is seen to be a function of position in the xy-plane. The constant C depends on the initial position

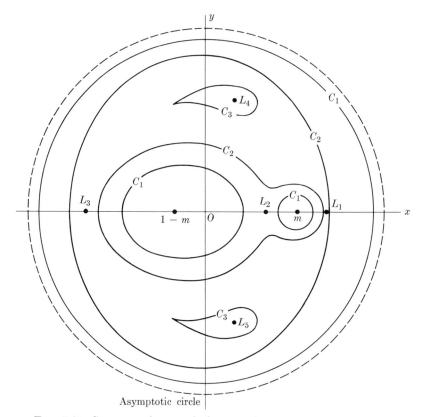

FIG. 5–7. Contours of zero velocity; restricted three-body problem.

and velocity of the particle. There will clearly be curves of zero velocity given by

$$x^2 + y^2 + \frac{2(1 - m)}{\sqrt{(x - x_1)^2 + y^2}} + \frac{2m}{\sqrt{(x - x_2)^2 + y^2}} = C. \quad (5\text{–}44)$$

A few of these contour curves ($C_1 > C_2 > C_3$) are exhibited in Fig. 5–7. It is apparent from Eq. (5–43) that motion of the particle can occur only in those regions of the xy-plane for which $2U - C > 0$. The contour curves, therefore, for a given value of C mark the boundaries of the regions within which motion can take place. These areas of possible orbital motion can be described briefly as follows:

Case I. For very large C, $2U - C$ will be positive either if $x^2 + y^2$ is very large or if r_1 or r_2 is very small. In Fig. 5–7, let C_1 denote such a large value of C. As C increases, the outer oval contour denoted by C_1

moves closer and closer to the asymptotic circle. The motion is restricted to the region outside the nearly circular contour or within one of the closed ovals surrounding $(1 - m)$ or m. No motion can take place if the particle is between these ovals and the outer contour because, in this region, $2U - C_1 < 0$.

If we consider, for example, the mass $(1 - m)$ to be the sun, the mass m to be the earth, the particle to be the moon, and ignore the eccentricity of the earth's orbit and the mass of the moon, then $r_2 \ll r_1$, and the moon will be within the oval surrounding m. It cannot escape from this stable position.

Case II. As C is allowed to decrease, the inner ovals expand, and the outer circular contours move toward the origin. The curves denoted by C_2 in Fig. 5–7 show an intermediate situation in which the ovals have joined to form a single closed contour around the two principal mass points. Stable motion of the particle results if it lies within the region surrounding $(1 - m)$ and m, or outside of the region marked by the oval contour C_2.

Case III. If C is diminished further, the regions of stability become larger, and motion can take place everywhere outside of the closed contours C_3 in Fig. 5–7.

As C is decreased, there will be points where the ovals and the outer quasi-circular contours have common tangents. These are labeled L_1 and L_3 in Fig. 5–7. There also will be a point, L_2, where the ovals surrounding $(1 - m)$ and m merge. Furthermore, when C is decreased sufficiently, the points L_4 and L_5 are obtained. The points $L_1, L_2, \ldots, L_5$ are called the *Lagrangian points* of the plane and are of special significance in the three-body problem.

Let the equation of the contour curves be

$$f(x, y) \equiv x^2 + y^2 + \frac{2(1 - m)}{r_1} + \frac{2m}{r_2} = C. \qquad (5\text{–}45)$$

At a Lagrangian point the coalescing contour curves have double tangents. In analytic geometry it is shown that double points such as L_1, L_2, and L_3 can be determined by setting $\partial f/\partial x = 0$ and $\partial f/\partial y = 0$. Applying these conditions to Eq. (5–45), we find

$$x - \frac{(1 - m)(x - x_1)}{r_1^3} - \frac{m(x - x_2)}{r_2^3} = 0,$$

$$y - \frac{(1 - m)y}{r_1^3} - \frac{my}{r_2^3} = 0.$$

These are the values of $\partial U/\partial x$ and $\partial U/\partial y$ appearing on the right-hand sides of Eqs. (5–41) and (5–42). Furthermore, along a contour curve,

$\dot{x} = \dot{y} = 0$. Hence the equations of motion *at the double points* reduce to $\ddot{x} = \ddot{y} = 0$. No accelerating forces act on the particle at these points. A similar situation can be shown to exist at the points L_4 and L_5. These points and the locations of $(1 - m)$ and m form equilateral triangles. We conclude that if the small-mass particle is initially at rest at one of the Lagrangian points, it will remain there forever unless it is acted upon by forces other than those due to $(1 - m)$ and m.

The distances from the mass m to the collinear Lagrangian points on the x-axis are given by the following relations:

$$\text{for } L_1: \quad d_1 = \left(\frac{m}{3}\right)^{1/3} + \frac{1}{3}\left(\frac{m}{3}\right)^{2/3} - \frac{1}{9}\left(\frac{m}{3}\right)^{3/3} + \cdots,$$

$$\text{for } L_2: \quad d_2 = \left(\frac{m}{3}\right)^{1/3} - \frac{1}{3}\left(\frac{m}{3}\right)^{2/3} - \frac{1}{9}\left(\frac{m}{3}\right)^{3/3} + \cdots,$$

$$\text{for } L_3: \quad d_3 = 2 - \frac{7}{12}m - \frac{1127}{20736}m^3 + \cdots.$$

Because of perturbations, no particle initially at one of the Lagrangian points is likely to remain there indefinitely. If it is disturbed, the question of the stability of its subsequent motion is of importance. We shall not discuss this question in detail. In general, the motion around one of the collinear Lagrangian points is unstable. A disturbed particle will wander away from the vicinity of the point to an indefinitely large distance. Under special initial conditions, however, it is possible for the particle to move around the critical point in an elliptical path.

Stable motion around one of the Lagrangian points L_4 or L_5 can occur if the lesser mass satisfies the inequality $m < 0.0385$. An extensive study of periodic motions around the Lagrangian points has been made by F. R. Moulton [4].

In the foregoing discussion we have restricted the three-body problem in a severe way. As early as 1772, Lagrange showed that *three arbitrary mass points* could move in a configuration which would maintain its geometrical form if certain conditions were fulfilled. These are: (a) the resultant force acting on each mass passes through their common center of mass; (b) this resultant force is directly proportional to the distance of each mass from the common center of mass; (c) the initial velocities are proportional to the respective distances of the point masses from the center of mass, and the direction of these velocity vectors is such that they form equal angles with the lines from the respective masses to the center of mass. The only configurations which will satisfy these conditions are: (1) three masses situated at the vertices of an equilateral triangle; (2) three masses lying on a straight line. The Lagrangian points described above represent these two configurations.

Two examples representative of motion near Lagrangian points may be mentioned. The first of these is the "gegenschein" or counterglow. It has been suggested that the gegenschein is sunlight reflected by a swarm of small particles moving around the sun in an orbit similar to that of the earth. However, the particles are trapped in concentrated form near the point L_1 on the side of the earth opposite the sun. As the earth moves in its orbit, the swarm also moves in an orbit, but always remains in the neighborhood of the Lagrangian point.

The second example is the group of Trojan asteroids. These small planetary bodies are located at the Lagrangian points L_4 and L_5 of the sun-Jupiter system. They orbit the sun with the same period as Jupiter, always remaining near the vertex of an equilateral triangle whose base is the line sun-Jupiter.

5–6 Euler's equations for rigid-body motion. Let a body consisting of rigidly connected masses rotate about an instantaneous axis through O (Fig. 5–8) with an angular velocity $\boldsymbol{\omega}$. We consider O to be either fixed in space or, if moving, to be the center of mass of the system. At any instant, the linear velocity of a mass m_j located at $\mathbf{r}_j$ is $\boldsymbol{\omega} \times \mathbf{r}_j$, and its angular momentum is

$$\mathbf{L} = \mathbf{r}_j \times m_j\mathbf{v}_j = m_j[\mathbf{r}_j \times (\boldsymbol{\omega} \times \mathbf{r}_j)].$$

Expanding the triple vector product, we have

$$\mathbf{L} = m_j[(\mathbf{r}_j \cdot \mathbf{r}_j)\boldsymbol{\omega} - (\mathbf{r}_j \cdot \boldsymbol{\omega})\mathbf{r}_j].$$

Hence for an assemblage of n masses,

$$\mathbf{L} = \boldsymbol{\omega} \sum_{j=1}^{n} m_j r_j^2 - \sum_{j=1}^{n} m_j(\mathbf{r}_j \cdot \boldsymbol{\omega})\mathbf{r}_j. \tag{5–46}$$

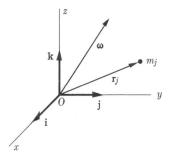

FIG. 5–8. Reference frame for rigid-body motion.

Let

$$\mathbf{r} = x\mathbf{i} + y\mathbf{j} + z\mathbf{k}, \qquad \boldsymbol{\omega} = \omega_x\mathbf{i} + \omega_y\mathbf{j} + \omega_z\mathbf{k}, \qquad \mathbf{L} = L_x\mathbf{i} + L_y\mathbf{j} + L_z\mathbf{k}$$

be the vectors of position, angular velocity, and angular momentum expressed in a fixed cartesian coordinate system. Then Eq. (5–46) can be separated into the components

$$L_x = \omega_x \sum_{j=1}^{n} m_j(y_j^2 + z_j^2) - \omega_y \sum_{j=1}^{n} m_j x_j y_j - \omega_z \sum_{j=1}^{n} m_j x_j z_j,$$

$$L_y = -\omega_x \sum_{j=1}^{n} m_j y_j x_j + \omega_y \sum_{j=1}^{n} m_j(x_j^2 + z_j^2) - \omega_z \sum_{j=1}^{n} m_j y_j z_j,$$

$$L_z = -\omega_x \sum_{j=1}^{n} m_j z_j x_j - \omega_y \sum_{j=1}^{n} m_j z_j y_j + \omega_z \sum_{j=1}^{n} m_j(x_j^2 + y_j^2).$$

The quantities $\sum m(y_j^2 + z_j^2)$, $\sum m(z_j^2 + x_j^2)$, and $\sum m(x_j^2 + y_j^2)$ are moments of inertia, which we shall denote by I_{xx}, I_{yy}, I_{zz}. The quantities $\sum m_j x_j y_j$, $\sum m_j y_j z_j$, $\sum m_j z_j x_j$ are products of inertia which will be denoted by I_{xy}, I_{yz}, I_{zx}. These are the fundamental inertial parameters of the system. For a continuous medium, the sums are to be replaced by integrals over the given volume.

Newton's second law for rotational motion is

$$\frac{d\mathbf{L}}{dt} = \mathbf{N}, \tag{5–47}$$

where $\mathbf{N}$ is the torque about O. If $\mathbf{L}$ as defined in the fixed cartesian system is used in this equation, the moments and products of inertia, as well as $\boldsymbol{\omega}$, change with time. The resulting equations of motion are complicated. If the coordinate system is properly chosen, however, the equations of motion can be written in terms of inertial parameters which are constants characteristic of any given mass distribution.

Let λ, μ, ν (Fig. 5–9) be the direction cosines of a line in space. By definition, the moment of inertia of a system of n discrete masses about l is

$$I_l = \sum_{j=1}^{n} m_j d_j^2$$

or, for a continuous mass distribution,

$$I_l = \int_V (d^2)\, \delta\, dV,$$

where V is the volume over which the integration is to be taken and δ is the density.

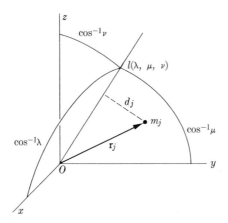

FIG. 5-9. Moment of inertia about a line $l(\lambda, \mu, \nu)$.

If we express I_l in terms of moments and products of inertia in the spatially fixed cartesian system, we obtain

$$I_l = \lambda^2 I_{xx} + \mu^2 I_{yy} + \nu^2 I_{zz} - 2\mu\lambda I_{xy} - 2\lambda\nu I_{xz} - 2\mu\nu I_{yz}. \quad (5\text{--}48)$$

This yields the moment of inertia about any line through O when the parameters I_{xx}, I_{yy}, I_{zz}, etc. are known. These can be calculated most easily if the x-, y-, and z-axes coincide with the symmetry axes of the body.

Let ξ, η, ζ be the coordinates of a point on any one of the lines through O, at distance ρ from the origin. Then $\lambda = \xi/\rho$, $\mu = \eta/\rho$, $\nu = \zeta/\rho$, and Eq. (5-48) can be written

$$I_{xx}\xi^2 + I_{yy}\eta^2 + I_{zz}\zeta^2 - 2\xi\eta I_{xy} - 2\xi\zeta I_{xz} - 2\eta\zeta I_{yz} = \rho^2 I_l. \quad (5\text{--}49)$$

This represents a quadric surface if we set $\rho^2 I_l = 1$. Associated with O, therefore, there is a quadric surface which can be shown to be an ellipsoid. The distance from O to a point P on this ellipsoid is related to the moment of inertia about the line OP by $\rho = 1/\sqrt{I_l}$. The surface is called the *momental ellipsoid* for the body at O.

If the coordinate axes had been chosen such that no cross-product terms appeared in Eq. (5-49), the quadric would have been of the form

$$I_1\xi^2 + I_2\eta^2 + I_3\zeta^2 = \rho^2 I_l = 1,$$

where I_1, I_2, I_3 are moments of inertia about the axes of symmetry of the ellipsoid. These are called the *principal moments of inertia* at O and may be calculated once and for all for a given mass distribution. The corresponding axes are the *principal axes* of the body.

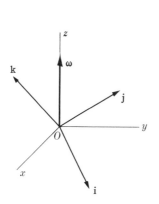

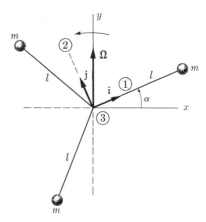

FIG. 5–10. Vectors **i**, **j**, **k** define the directions of the principal axes of a rotating body.

FIG. 5–11. Rotating framework of three masses.

When a body rotates about one of its principal axes, the angular-momentum vector is collinear with the angular-velocity vector. That is,

$$\mathbf{L} = I\boldsymbol{\omega},$$

where I is a principal moment of inertia.

It can be shown [5] that any rigid body has three real principal moments of inertia, and that the principal axes of the body are mutually perpendicular. In what follows, we shall consider the principal moments of inertia to be known. They will be designated by I_1, I_2, I_3.

Let **i**, **j**, **k** denote a triad of unit vectors along the principal axes of a rigid body which is rotating with angular velocity $\boldsymbol{\omega}$. For simplicity we assume that $\boldsymbol{\omega}$ is directed along the z-axis of a cartesian coordinate system fixed in space, as shown in Fig. 5–10. As the body rotates, the triad defines a coordinate system rotating with it. Then, by Newton's second law [Eq. (5–47)],

$$\frac{d\mathbf{L}}{dt} = \dot{L}_1\mathbf{i} + \dot{L}_2\mathbf{j} + \dot{L}_3\mathbf{k} + \boldsymbol{\omega} \times \mathbf{L} = \mathbf{N}. \qquad (5\text{--}50)$$

Here L_1, L_2, L_3 are the components of **L** in the rotating coordinate system, and the angular velocity is $\boldsymbol{\omega} = \omega_1\mathbf{i} + \omega_2\mathbf{j} + \omega_3\mathbf{k}$. Hence Eq. (5–50) may be written

$$\frac{d\mathbf{L}}{dt} = (\dot{L}_1 + \omega_2 L_3 - \omega_3 L_2)\mathbf{i} + (\dot{L}_2 + \omega_3 L_1 - \omega_1 L_3)\mathbf{j}$$

$$+ (\dot{L}_3 + \omega_1 L_2 - \omega_2 L_1)\mathbf{k} = N_1\mathbf{i} + N_2\mathbf{j} + N_3\mathbf{k}. \qquad (5\text{--}51)$$

But $L_1 = I_1\omega_1$, $L_2 = I_2\omega_2$, $L_3 = I_3\omega_3$. Therefore, equating components in Eq. (5–51), we find

$$I_1\dot{\omega}_1 + \omega_2\omega_3[I_3 - I_2] = N_1,$$
$$I_2\dot{\omega}_2 + \omega_3\omega_1[I_1 - I_3] = N_2, \qquad (5\text{–}52)$$
$$I_3\dot{\omega}_3 + \omega_1\omega_2[I_2 - I_1] = N_3.$$

These are *Euler's equations of motion*. Their use is subject to the limitation that the origin of coordinates be fixed in space or, *if in translational motion, that the center of mass of the body be taken as O*.

EXAMPLE 1.   Consider three equal masses mounted on a coplanar framework at the vertices of an equilateral triangle as shown in Fig. 5–11. Let the masses rotate in the xy-plane with constant angular speed $\omega = \dot{\alpha}$ about an axis pointing instantaneously toward the reader. By symmetry we see that the principal axes may be taken as shown, the axis ③ pointing toward the reader. Now let the whole xy-plane be rotated slowly with constant angular velocity Ω about the y-axis so that the x-axis moves into the plane of the paper. This angular velocity is represented by Ω in Fig. 5–11. Then

$$\omega_1 = \Omega \sin \alpha, \qquad \omega_2 = \Omega \cos \alpha, \qquad \omega_3 = \omega = \dot{\alpha},$$

and

$$\dot{\omega}_1 = \Omega\dot{\alpha} \cos \alpha, \qquad \dot{\omega}_2 = -\Omega\dot{\alpha} \sin \alpha, \qquad \dot{\omega}_3 = 0.$$

Also, we find

$$I_1 = \tfrac{3}{2}ml^2, \qquad I_2 = \tfrac{3}{2}ml^2, \qquad I_3 = 3ml^2.$$

Euler's equations, therefore, are

$$3ml^2\Omega\omega \cos \alpha = N_1,$$
$$-3ml^2\Omega\omega \sin \alpha = N_2,$$
$$0 = N_3.$$

Thus the torque components generated by the slow rotation of the xy-plane about the y-axis can be evaluated. We may interpret these further. The torque components can be resolved in the directions of the x-axis, the y-axis, and the z-axis (perpendicular to the other two and pointed toward the reader). We have

$$N_x = N_1 \cos \alpha - N_2 \sin \alpha = 3ml^2\Omega\omega,$$
$$N_y = N_1 \sin \alpha + N_2 \cos \alpha = 0,$$
$$N_z = N_3 = 0.$$

These are the *pitching moment*, the *yawing moment*, and the *rolling moment*, respectively. The z-axis about which these three masses rotate tends to dip downward.

EXAMPLE 2. Consider rotation about a point under no torque. The rotation of the earth to a *first approximation* can be cited as an example, the torques due to the sun and moon being relatively small. Consider that the rigid body has axes of symmetry such that $I_1 = I_2 \neq I_3$. Then, if I denotes either I_1 or I_2, we have from Euler's equations,

$$I\dot{\omega}_1 + \omega_2\omega_3(I_3 - I) = 0,$$
$$I\dot{\omega}_2 + \omega_3\omega_1(I - I_3) = 0,$$
$$I_3\dot{\omega}_3 = 0.$$

From the last equation it is clear that ω_3 is constant in time. The first two may be written

$$\dot{\omega}_1 + \omega_2\left[\left(\frac{I_3}{I} - 1\right)\right]\omega_3 = 0 \qquad (5\text{--}53)$$

and

$$\dot{\omega}_2 - \omega_1\left[\left(\frac{I_3}{I} - 1\right)\right]\omega_3 = 0. \qquad (5\text{--}54)$$

Let $k = [(I_3/I) - 1]\omega_3$, a constant. Differentiate Eq. (5–53) with respect to time and substitute Eq. (5–54) into it to obtain

$$\ddot{\omega}_1 + k^2\omega_1 = 0,$$

which has the solution

$$\omega_1 = a \sin (kt + b).$$

Then from Eq. (5–53)

$$\omega_2 = -a \cos (kt + b).$$

Hence $\omega_1^2 + \omega_2^2 = a^2$. The projection of the tip of the $\boldsymbol{\omega}$-vector on the ①②-plane moves on a circle of radius a, as shown in Fig. 5–12. The period of the motion is $2\pi/k$, or

$$P = \frac{2\pi}{\omega_3}\left[\frac{I}{I_3 - I}\right].$$

The vector $\boldsymbol{\omega}$ rotates about the ③-axis with this period, always maintaining a constant angle with the axis. The magnitude of $\boldsymbol{\omega}$ is

$$|\boldsymbol{\omega}| = \sqrt{\omega_1^2 + \omega_2^2 + \omega_3^2} = \text{constant}.$$

For the earth, $I/(I_3 - I)$ is approximately 300, and the value of ω_3 is 2π radians per day. Hence $P \sim 300$ days. This periodic motion of the

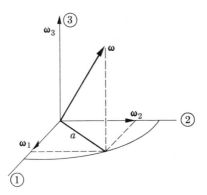

FIG. 5–12. Angular-velocity components for torque-free motion.

vector $\boldsymbol{\omega}$ about the earth's axis of symmetry constitutes the principal part of the phenomenon known observationally as the *variation of latitude*. The observed period of motion is 433 days. The discrepancy may be attributed to the elastic properties of the earth; that is, it is not a perfectly rigid body [6].

5–7 Euler's angles and the equations of motion. The angular speeds $\dot{\omega}_1,\ \dot{\omega}_2,\ \dot{\omega}_3$ *do not measure* the time rate of change of the three angles which describe the orientation of a rigid body in space. Euler's angles are introduced to provide such an orientation. Let a system of cartesian axes fixed in space be $Ox_0,\ Oy_0,\ Oz_0$ (Fig. 5–13). Let the principal axes in the body be ①, ②, ③. Then the angles $\theta,\ \varphi,\ \psi$ indicated in Fig. 5–13 are called Euler's angles, and it is clear that

$$\omega_1 = \dot{\theta}\cos\psi + \dot{\varphi}\sin\theta\sin\psi,$$
$$\omega_2 = -\dot{\theta}\sin\psi + \dot{\varphi}\sin\theta\cos\psi,$$
$$\omega_3 = \dot{\varphi}\cos\theta + \dot{\psi}.$$

By means of these equations, the Eulerian equations of motion, Eq. (5–52), can be transformed into terms of $\theta,\ \varphi,\ \psi$. If there is symmetry about an axis, that is, $I_1 = I_2 = I \neq I_3$, we find the following equations of motion:

for θ: $\dfrac{d}{dt}(I\dot{\theta}) - I\dot{\varphi}^2\sin\theta\cos\theta + I_3\dot{\varphi}(\dot{\psi} + \dot{\varphi}\cos\theta)\sin\theta = N_\theta;$ (5–55)

for φ: $\dfrac{d}{dt}[I\dot{\varphi}\sin^2\theta + I_3(\dot{\psi} + \dot{\varphi}\cos\theta)\cos\theta] = N_\varphi;$ (5–56)

for ψ: $\dfrac{d}{dt}[I_3(\dot{\psi} + \dot{\varphi}\cos\theta)] = N_\psi.$ (5–57)

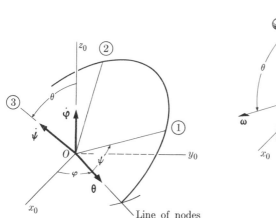

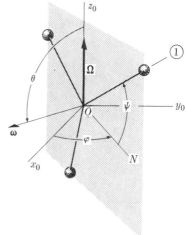

FIG. 5–13. Euler's angles defined.   FIG. 5–14. Euler's angles for the three-mass system of Fig. 5–11.

These are equivalent in every way to the Eulerian equations of motion. When they are solved for θ, φ, ψ as functions of time, the orientation of the body in space at any instant is known.

EXAMPLE 1. Consider the three-mass system shown in Fig. 5–11. We illustrate Euler's angles for this case by reference to Fig. 5–14. The plane of the masses intersects the $x_0 y_0$-plane of the fixed coordinate system in the line of nodes ON. This is the x-axis of Fig. 5–11. The axis of rotation of the masses lies in the $x_0 y_0$-plane. Hence $\theta = \pi/2$. Therefore, $\dot\theta = 0$, $\dot\psi \equiv \omega = $ constant, and $\dot\varphi \equiv \Omega = $ constant. By Eqs. (5–55) through (5–57), $N_\theta = 3ml^2\Omega\omega$, $N_\varphi = 0$, and $N_\psi = 0$, a result which agrees with that obtained from Euler's equations.

EXAMPLE 2. The motion of a spinning top (Fig. 5–15) can be described by means of Euler's angles and Eqs. (5–55) through (5–57). Here the torque due to gravity is
$$N_\theta = mgl \sin\theta,$$

where l is the distance from the fixed point O to the center of gravity of the top. The other torque components, N_φ and N_ψ, are zero. The equations of motion then become

$$I\ddot\theta - I\dot\varphi^2 \sin\theta \cos\theta + I_3\dot\varphi(\dot\psi + \dot\varphi \cos\theta) \sin\theta = mgl \sin\theta, \qquad (5\text{–}58)$$

$$\frac{d}{dt}\left[I\dot\varphi \sin^2\theta + I_3(\dot\psi + \dot\varphi \cos\theta) \cos\theta\right] = 0, \qquad (5\text{–}59)$$

$$\frac{d}{dt}\left[I_3(\dot\psi + \dot\varphi \cos\theta)\right] = 0. \qquad (5\text{–}60)$$

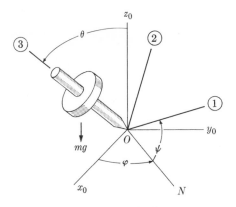

FIG. 5-15. The spinning top.

By Eq. (5-60), $\dot{\psi} + \dot{\varphi} \cos \theta = \omega_3 =$ a constant; and by Eq. (5-59),

$$I\dot{\varphi} \sin^2 \theta + I_3(\dot{\psi} + \dot{\varphi} \cos \theta) \cos \theta = L_\varphi = \text{a constant.}$$

Therefore,

$$\dot{\varphi} = \frac{L_\varphi - I_3\omega_3 \cos \theta}{I \sin^2 \theta}. \tag{5-61}$$

This is called the *precessional velocity*, which is a function of θ only.

A detailed analysis* of Eq. (5-58) shows that, in general, θ is oscillatory in time. This phenomenon is called *nutation*. The precession represented by Eq. (5-61) is also oscillatory. For a critical value of θ, however, $\dot{\varphi}$ becomes constant, and a steady precession results. If $\dot{\psi} \gg \dot{\varphi}$, the spin velocity dominates the motion, and the steady precession is given by

$$\dot{\varphi} \simeq \frac{mgl}{I_3\dot{\psi}}. \tag{5-62}$$

5-8 Potential due to an oblate spheroid. Consider any bounded distribution of matter of total mass M, as shown in Fig. 5-16. Let O be the center of mass and let a unit mass be placed at P, a distance r from O. Then the potential at P due to an element dm is

$$dU = \frac{-k^2 \, dm}{(\rho^2 + r^2 - 2r\rho \cos \varphi)^{1/2}}.$$

* See, for example, reference 3, pp. 135 ff.

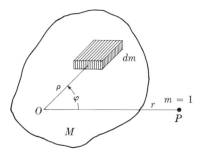

FIG. 5–16. Potential at an external point due to a bounded mass.

Integrating over the entire mass, we have

$$U = -k^2 \int_M \frac{dm}{(\rho^2 + r^2 - 2r\rho \cos \varphi)^{1/2}}. \qquad (5\text{-}63)$$

Now by the binomial expansion,

$$(\rho^2 + r^2 - 2r\rho \cos \varphi)^{-1/2}$$

$$= \frac{1}{r} \left\{ 1 - 2 \left(\frac{\rho}{r}\right) \cos \varphi + \left(\frac{\rho}{r}\right)^2 \right\}^{-1/2}$$

$$= \frac{1}{r} \left\{ 1 + \frac{\rho}{r} \cos \varphi - \frac{1}{2} \left(\frac{\rho}{r}\right)^2 + \frac{3}{2} \left(\frac{\rho}{r}\right)^2 \cos^2 \varphi + \cdots \right\},$$

in which terms of higher powers than two in (ρ/r) have been neglected. Inserting this expression into Eq. (5–63) and integrating term by term, we find

$$U = -\frac{k^2}{r} \int_M dm - \frac{k^2}{r^2} \int_M \rho \cos \varphi \, dm$$

$$+ \frac{k^2}{2r^3} \int_M \rho^2 \, dm - \frac{3k^2}{2r^3} \int_M \rho^2 \cos^2 \varphi \, dm.$$

The first integral is $-(k^2 M/r)$, the potential at P, if the entire mass of the distribution had been concentrated at O. The second integral vanishes because O has been chosen as the center of mass. The third integral is the polar moment of inertia, I_0, of the mass. The fourth integral may be written

$$\int_M (\rho^2 - \rho^2 \sin^2 \varphi) \, dm = I_0 - I,$$

where I is the moment of inertia about the line OP. Combining all these expressions, we have

$$U = -\frac{k^2 M}{r} - \frac{k^2}{2r^3}\{2I_0 - 3I\}.$$

But $2I_0 = I_1 + I_2 + I_3$, where I_1, I_2, I_3 are the principal moments of inertia of M. Hence we finally have

$$U = -\frac{k^2 M}{r} - \frac{k^2}{2r^3}[I_1 + I_2 + I_3 - 3I]. \qquad (5\text{-}64)$$

This has been derived on the assumption that ρ/r is small. Actually, however, it is a very good approximation for bodies such as the earth, where I_1, I_2, and I_3 do not differ appreciably. Note that if these were equal, as for a sphere, U would be $-(k^2 M/r)$, which is the value obtained in Eq. (4–21).

Such an expression for the potential is required in any discussion of satellites moving in proximity to the earth. The earth is an oblate spheroid, so we may set $I_1 = I_2 = I_e$ and write

$$U = -\frac{k^2 M}{r} - \frac{k^2}{2r^3}\{2I_e + I_3 - 3I\}, \qquad (5\text{-}65)$$

where I_e is the moment of inertia of the earth about an axis in the equatorial plane and I_3 is the moment of inertia about the earth's axis of rotation.

5–9 Acceleration due to gravity on the earth's surface. The gravitational acceleration for an object located on the earth's surface has components due to the gravitational attraction of the earth's mass and to the centripetal acceleration resulting from the earth's rotation. At the equator, where the centripetal component is at a maximum, it constitutes only about one-third of one per cent of the gravitational component. Both components vary with latitude because of the ellipsoidal shape of the earth. The total acceleration is 978.039 cm·sec^{-2} at the equator and 983.217 cm·sec^{-2} at the poles.

In Section 3–1 we found that the shape of the earth may be described with accuracy by an ellipsoid of revolution about its minor axis. This figure is produced by the combined effects of gravitational attraction and the earth's rotation, if the earth is assumed to be nonrigid. A knowledge of the shape of the earth, together with the assumption that the mass distribution in the earth's interior has rotational symmetry about the earth's axis and symmetry about the equatorial plane, permits a derivation of an equation yielding the acceleration due to gravity as a function of position. This expression, first developed by Clairaut in 1743, is of

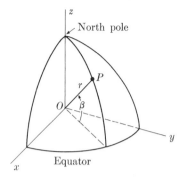

FIG. 5–17. Geocentric latitude and radius vector for a point on the earth.

interest because its accuracy does not depend upon a detailed knowledge of the mass distribution in the earth's interior.

For a point P on the earth's surface the total potential is

$$U = -\frac{k^2 M}{r} - \frac{k^2}{2r^3}\{2I_e + I_3 - 3I\} - \frac{1}{2}\omega^2(x^2 + y^2), \quad (5\text{–}66)$$

which follows from Eq. (5–65). The term $\frac{1}{2}\omega^2(x^2 + y^2)$ is due to the centripetal acceleration, ω being the angular velocity of rotation of the earth.

From Eq. (5–48) and the discussion following it, the moment of inertia I about the line OP (see Fig. 5–17) is given by

$$I = I_1\lambda^2 + I_2\mu^2 + I_3\nu^2 = I_e(\lambda^2 + \mu^2) + I_3\nu^2,$$

where λ, μ, ν are the direction cosines of OP. Now, $\lambda = x/r$, $\mu = y/r$, $\nu = z/r$, and since β is the geocentric latitude,

$$\lambda^2 + \mu^2 = \cos^2\beta, \qquad \nu^2 = \sin^2\beta,$$

and

$$I = I_e + (I_3 - I_e)\sin^2\beta. \qquad (5\text{–}67)$$

Inserting this in Eq. (5–66) and carrying out some algebraic reduction, we find

$$U = -\frac{k^2 M}{r}\left[1 - \frac{\epsilon I_3}{r^2 M}\left(\frac{3}{2}\sin^2\beta - \frac{1}{2}\right) + \frac{1}{2}\frac{\omega^2 r^3 \cos^2\beta}{Mk^2}\right], \quad (5\text{–}68)$$

where

$$\epsilon = \frac{I_3 - I_e}{I_3}$$

is called the *mechanical ellipticity*.

The value of ϵI_3 may be obtained by assuming that the geoid can be closely approximated by an ellipsoid of revolution, as discussed in Section 3–1. On such a surface the potential U has everywhere the constant value U_0. If tidal and some meteorological effects are disregarded, the ocean can be considered to have a level surface on which the potential function is constant. From Eq. (3–7), neglecting the term in f^2 (which is appropriate for the order of approximation carried so far), we have

$$\frac{1}{r} = \frac{1 + f \sin^2 \beta}{A},$$

where A is the equatorial radius and f is the oblateness of the earth. Inserting this in Eq. (5–68) and letting U_0 be the value of U at the ellipsoidal equipotential surface, we find that

$$U_0 = -\left[\left(\frac{k^2 M}{A} + \frac{1}{2}\frac{k^2 \epsilon I_3}{A^3} + \frac{1}{2}\omega^2 A^2\right)\right.$$
$$\left. + \left(\frac{k^2 M}{A}f + \frac{3}{2}f\frac{k^2 \epsilon I_3}{A^3} - \frac{3}{2}\frac{k^2 \epsilon I_3}{A^3} - \omega^2 A^2 f - \frac{1}{2}\omega^2 A^2\right)\sin^2 \beta + \cdots\right].$$

Since by hypothesis U_0 is constant, the coefficient of $\sin^2 \beta$ must be zero. Therefore,

$$\epsilon I_3 = \frac{2A^3}{k^2}\left[\frac{(k^2 Mf/A) - \omega^2 A^2(f + \frac{1}{2})}{3(1 - f)}\right]$$

or, with terms in f^2 neglected,

$$\epsilon I_3 = \frac{2}{3}\frac{A^3}{k^2}\left[\frac{k^2 M}{A}f - \frac{\omega^2 A^2}{2} - \frac{3}{2}\omega^2 A^2 f\right]. \tag{5–69}$$

Let m be the ratio of centripetal force to gravity at the equator; for a spherical earth,[*]

$$m = \frac{\omega^2 A^3}{k^2 M}.$$

For the earth, $m = 0.003461$, and this is very close in value to f. Hence the terms where mf is a factor in Eq. (5–69) may be neglected, and we obtain

$$\epsilon I_3 = \frac{2}{3}MA^2\left(f - \frac{m}{2}\right). \tag{5–70}$$

---

[*] We use the symbol m for this ratio according to the usual notation. The reader must be careful not to confuse m with the symbol for mass used elsewhere in this chapter.

Equation (5–68) may now be written

$$U = -\frac{k^2 M}{r}\left[1 - \frac{A^2}{r^2}\left(f - \frac{m}{2}\right)\left(\sin^2 \beta - \frac{1}{3}\right) + \frac{1}{2}\frac{\omega^2 r^3 \cos^2 \beta}{Mk^2}\right], \quad (5\text{–}71)$$

and the magnitude of the gravitational acceleration along the direction of r is

$$g = \frac{\partial U}{\partial r} = \left[\frac{k^2 M}{r^2} - \frac{3k^2 M A^2}{r^4}\left(f - \frac{m}{2}\right)\left(\sin^2 \beta - \frac{1}{3}\right) - \omega^2 r \cos^2 \beta\right].$$
$$(5\text{–}72)$$

In particular, at the equator, where $r = A$ and $\beta = 0$, we have

$$g_e = \frac{k^2 M}{A^2}\left(1 + f - \frac{3}{2}m\right). \quad (5\text{–}73)$$

The value of g at other latitudes may be derived from Eq. (5–72).

It is convenient to rewrite Eq. (5–72) by means of the following relationships which are correct to the first power of f:

$$r = A(1 - f\sin^2 \beta),$$

$$\frac{1}{r^2} = \frac{1}{A^2}(1 + 2f\sin^2 \beta),$$

$$\frac{1}{r^4} = \frac{1}{A^4}(1 + 4f\sin^2 \beta).$$

After some algebraic manipulation, we obtain from Eq. (5–72)

$$g = \frac{k^2 M}{A^2}\left[\left(1 + f - \frac{3}{2}m\right) + \left(\frac{5}{2}m - f\right)\sin^2 \beta\right]$$

which, to the accuracy maintained so far, can be written as

$$g = g_e[1 + (\tfrac{5}{2}m - f)\sin^2 \beta]. \quad (5\text{–}74)$$

In this equation we may, without sacrificing accuracy, replace β with the geographic latitude, ϕ, [compare Eqs. (3–7) and (3–10)].

The international gravity formula is based on the development of Clairaut's theory to include terms in f^2. The result [7] is

$$g = 978.049(1 + 0.0052884\sin^2 \phi - 0.00000059\sin^2 2\phi). \quad (5\text{–}75)$$

The acceleration due to gravity at the equator given by this formula should be modified according to Jeffreys' results as discussed in Section 2–3.

The modification required in Kepler's third law [Eq. (2–7)] for a satellite moving in a circular orbit in the equatorial plane of an oblate planet can now be made. When the term in ω^2 is omitted, Eq. (5–72) yields, for an object orbiting around an oblate planet, the gravitational acceleration:

$$g = \frac{k^2 M}{r^2}\left[1 + \frac{A^2}{r^2}\left(f - \frac{m}{2}\right)\right].$$

The planet's oblateness causes an increase in the gravitational attraction between planet and satellite. The resulting orbital motion is the same as if

$$k^2\left[1 + \frac{A^2}{r^2}\left(f - \frac{m}{2}\right)\right]$$

were substituted for k^2 in the elementary theory of the two-body problem. Kepler's third law as derived for two point masses, namely

$$\frac{P^2}{a^3} = \frac{4\pi^2}{k^2(m_1 + m_2)},$$

therefore becomes

$$\frac{P^2}{a^3} = \frac{4\pi^2}{k^2(m_1 + m_2)[1 + (A^2/a^2)(f - m/2)]},$$

or, since $(A^2/a^2)(f - m/2) \ll 1$,

$$\frac{P^2}{a^3} = \frac{4\pi^2[1 - (A^2/a^2)(f - m/2)]}{k^2(m_1 + m_2)}.$$

This is Eq. (2–7) presented in Chapter 2.

5–10 Perturbations of artificial satellites. We denote by the term "artificial satellites" those man-made objects that are projected into orbits around the earth. In general, their paths will be within a few thousand miles of the earth's surface. If, in particular, the distance of the satellite from the earth's surface is no more than 1000 miles, its motion is strongly influenced by the oblateness of the earth and by the resistance due to the earth's atmosphere. Although the motion of such a satellite under the influences of the earth, moon, and sun is in effect a four-body problem, the perturbing effects due to the sun and moon have been shown to be small [8, 9].

(a) *Perturbations due to oblateness.* It is apparent from Eq. (5–65) that the potential due to the distributed mass of the oblate earth differs by a factor proportional to $1/r^3$ from that which would prevail if the earth were spherical. It is this factor which causes the perturbations in nearby satellite motions. We shall ignore here the very small contribution to the potential caused by the earth's rotation.

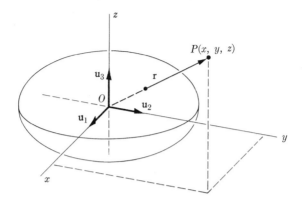

FIG. 5–18. Unit mass in the gravitational field of a spheroid.

In Fig. 5–18 let P denote the position of a unit mass located at (x, y, z) in a coordinate system whose axes are the principal axes of the rigid body. Then the force per unit mass at P is given by

$$-\nabla U = X\mathbf{u}_1 + Y\mathbf{u}_2 + Z\mathbf{u}_3,$$

where U is the potential obtained in Eq. (5–65), and where

$$X = -\frac{\partial U}{\partial x}, \qquad Y = -\frac{\partial U}{\partial y}, \qquad Z = -\frac{\partial U}{\partial z}.$$

Here $\mathbf{u}_1, \mathbf{u}_2, \mathbf{u}_3$ are unit vectors along the principal axes.

By Eq. (5–48) we have $I = I_1\lambda^2 + I_2\mu^2 + I_3\nu^2$, where $\lambda = x/r$, $\mu = y/r$, and $\nu = z/r$ are the direction cosines of the line OP. Hence Eq. (5–64) can be written

$$U = -\frac{k^2 M}{r} - \frac{k^2 I_0}{r^3} + \frac{3k^2}{2r^5}\,(I_1 x^2 + I_2 y^2 + I_3 z^2), \qquad (5\text{–}76)$$

where I_0 is the polar moment of inertia given by $2I_0 = I_1 + I_2 + I_3$. The gradient can then be written

$$\nabla U = \left[\frac{k^2 M}{r^2} + \frac{3k^2 I_0}{r^4} - \frac{15k^2}{2r^6}\,(I_1 x^2 + I_2 y^2 + I_3 z^2)\right]\mathbf{u}_r$$

$$+ \frac{3k^2}{r^5}\,[x I_1 \mathbf{u}_1 + y I_2 \mathbf{u}_2 + z I_3 \mathbf{u}_3], \qquad (5\text{–}77)$$

where $\mathbf{u}_r$ is a unit vector in the direction of $\mathbf{r}$. The force per unit mass, $-\nabla U$, is observed to have a radial component *as well as components which are inversely proportional to r^5, but are nonradial.* Thus the force field is *not* a central force field, and consequently perturbations of the two-body motion of a satellite must occur.

For descriptive purposes we shall introduce the concept of a *torque field* surrounding the mass M. We define the moment, or torque, at any point in the space around M as

$$\mathbf{N} = -\mathbf{r} \times \nabla U = -\frac{3k^2}{r^5} [yz(I_3 - I_2)\mathbf{u}_1 + zx(I_1 - I_3)\mathbf{u}_2 + xy(I_2 - I_1)\mathbf{u}_3].$$
(5–78)

For the earth, $I_1 = I_2 = I_e$, and Eq. (5–78) can be written

$$\mathbf{N} = -\frac{3k^2 z}{r^5} (I_3 - I_e)(y\mathbf{u}_1 - x\mathbf{u}_2).$$
(5–79)

At any given instant, imagine the satellite of unit mass attached to its orbital plane. The torque $\mathbf{N}$ tends to twist the system instantaneously. Consider, in particular, three cases.

Case I. Suppose $z = 0$, that is, the satellite is moving in the plane of the equator. Then $\mathbf{N}$ vanishes and, by Newton's second law for rotational motion, the angular momentum $\mathbf{L}$ is constant in time. The satellite motion simulates that of the simple two-body case, but there are, however, contributions to the force field which do not obey the inverse-square law. Hence certain perturbations result.

Case II. Suppose $y = 0$, so that the satellite moves in the xz-plane. Then $\mathbf{N} = -(3k^2 zx/r^5)(I_3 - I_e)\mathbf{u}_2$, a vector along the y-axis. This situation does not lead to a twisting of the satellite's orbital plane. However, as x, z, r change with time, the orbit turns in its own plane about the y-axis, with the result that the perigee point rotates.

Case III. Consider motion in *any* orbital plane passing through the poles of the earth. According to Eq. (5–79), $\mathbf{N}$ lies in a plane perpendicular to such an orbital plane. Hence no twisting of the orbital plane in space is to be expected, but the orbit may turn in its own plane.

In general, however, the orbital plane of the satellite will behave in this torque field like the plane of symmetry of a gyroscope. The angular velocity of the satellite in its orbit corresponds to the spin of the gyro. The orbital plane will precess, and at the same time the perigee of the orbit will revolve in the plane of the orbit. This means that the orbital elements Ω and ω (see Sections 2–1 and 4–6) change with time.

We shall not exhibit the details of the perturbation calculations required to determine these rates of change. (The reader is referred to the original sources [10, 11].)   The resulting time rates of change are, to the first order in J,

$$\dot{\Omega} = -\frac{Jn \cos i}{a^2(1 - e^2)^2}, \tag{5-80}$$

$$\dot{\omega} = \frac{Jn(2 - \frac{5}{2}\sin^2 i)}{a^2(1 - e^2)^2}, \tag{5-81}$$

where

$$J = \frac{3}{2}\left[\frac{I_3 - I_e}{Mr_e^2}\right] = 1.624 \times 10^{-3},$$

$M =$ the mass of the earth,

$r_e =$ the mean equatorial radius of the earth,

$i =$ the inclination of the satellite orbit to the celestial equator,

$e =$ the eccentricity of the satellite orbit,

$a =$ the semimajor axis of the satellite orbit *in units of the earth's equatorial radius r_e,*

$n =$ the mean angular velocity (rad/sec or deg/day) of the satellite in its orbit. By Kepler's third law $n^2 a^3 = k^2 M$.

It is clear from Eqs. (5–80) and (5–81) that $\dot{\Omega} = 0$ for a polar orbit ($i = 90°$), and hence there is no precession of the line of nodes. There is, however, a regression or advance of the line of apsides. If $\sin^2 i < 0.8$ and the motion is direct, then $\dot{\omega} > 0$, and the line of apsides advances. If $\sin^2 i > 0.8$, the perigee point moves in a direction opposite to the motion of the satellite in its orbit.

Consider as a representative example a satellite whose period is two hours, whose orbital inclination is 45°, and whose orbital eccentricity is 0.2. From Eqs. (5–80) and (5–81) we find

$$\dot{\Omega} = -3.37 \text{ deg/day},$$
$$\dot{\omega} = +3.58 \text{ deg/day}.$$

Figure 5–19 represents the motion. The satellite S moves counterclockwise around the orbit, as shown. The perigee point moves forward in the direction of the satellite's motion. In 101 days the perigee point will have moved through a complete revolution. The ascending node N moves clockwise, with Ω decreasing, and completes a revolution in 107 days. Let the reader calculate the perigee distance for this satellite and see whether the object will strike the earth.

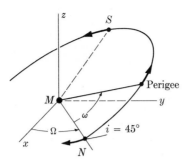

Fig. 5–19. Regression of node and advance of perigee for an earth satellite.

It should be clearly pointed out that the above first-order secular perturbations due to oblateness are not the only ones present. Both short- and long-period perturbations are also found in the elements of the orbit. The secular effects indicated in Eqs. (5–80) and (5–81) are merely the most notable of the perturbations arising from this cause.

(b) *Perturbations due to atmospheric resistance.* The force per unit mass due to air resistance, acting on a spherical body, is given approximately by [12]

$$F = \frac{C_D A \rho V^2}{2m},$$

(5–82)

where

$\rho = \rho(r)$ is the atmospheric density at distance r from the center of the earth,

$A =$ the cross-sectional area of the body,

$m =$ the mass of the body,

$V =$ its speed,

$C_D =$ a drag coefficient.

The coefficient C_D has a value of about 1 for spheres that are large compared with the mean free path of gas molecules. If the sphere is small compared with the mean free path and the molecules either adhere to its surface or are totally reflected, then C_D is of the order of magnitude 2.

We assume here that the earth is spherical, and that the mass of the satellite is negligible compared with that of the earth. The small satellite moves essentially in an elliptical orbit as in the two-body problem discussed in Chapter 4. The elements of its orbit, however, change with time at rates which have been given in Section 5–4. Because the force due to a resisting medium is tangential, we find from Eqs. (5–35) and (5–36) that

$$S = \frac{1 + e \cos f}{\sqrt{1 + e^2 + 2e \cos f}} \, [-b\rho V^2], \tag{5–83}$$

$$R = \frac{e \sin f}{\sqrt{1 + e^2 + 2e \cos f}} \, [-b\rho V^2], \tag{5–84}$$

where $b = C_D A / 2m$. The above equations, together with $W = 0$, substituted into Eqs. (5–29) through (5–34) yield

$$\frac{di}{dt} = 0 \quad \text{and} \quad \frac{d\Omega}{dt} = 0,$$

$$\frac{da}{dt} = \frac{-2b\rho V^2 \sqrt{1 + e^2 + 2e \cos f}}{n\sqrt{1 - e^2}}, \tag{5–85}$$

$$\frac{de}{dt} = \frac{-2b\rho V^2 \sqrt{1 - e^2}}{na} \left\{ \frac{e + \cos f}{\sqrt{1 + e^2 + 2e \cos f}} \right\}, \tag{5–86}$$

$$\frac{d\omega}{dt} = \frac{-2b\rho V^2 \sqrt{1 - e^2}}{nae} \left\{ \frac{\sin f}{\sqrt{1 + e^2 + 2e \cos f}} \right\}, \tag{5–87}$$

$$\frac{dM}{dt} = n + \frac{2b\rho V^2 (1 - e^2) \sin f}{nae(1 + e \cos f)} \left\{ \frac{1 + e^2 + e \cos f}{\sqrt{1 + e^2 + 2e \cos f}} \right\}. \tag{5–88}$$

We observe that, because of the presence of the $(\sin f)$-term in Eqs. (5–87) and (5–88), the rates of change of ω and M are periodic. Therefore we shall not consider them further. Equations (5–85) and (5–86) show, however, that a and e change secularly with time, and these perturbations will be discussed in more detail.

Integration of Eqs. (5–85) and (5–86) is facilitated by introducing the eccentric anomaly E as the variable of integration. By the theory described in Section 4–6, we have

$$r = a(1 - e \cos E) = \frac{a(1 - e^2)}{1 + e \cos f}, \tag{5–89}$$

$$n(t - T) = E - e \sin E, \tag{5–90}$$

$$h^2 = \mu a(1 - e^2). \tag{5–91}$$

Furthermore, the speed $V = (\dot{r}^2 + r^2 \dot{\theta}^2)^{1/2}$ may be written

$$V = \frac{h}{a(1 - e^2)} \, [1 + e^2 + 2e \cos f]^{1/2}. \tag{5–92}$$

Substituting from Eqs. (5–89), (5–90), (5–91), and (5–92) into Eqs. (5–85) and (5–86), we obtain

$$\frac{da}{dE} = -2b\rho a^2 \frac{[1 + e \cos E]^{3/2}}{[1 - e \cos E]^{1/2}},$$ (5–93)

$$\frac{de}{dE} = -2b\rho a \frac{[1 + e \cos E]^{1/2}}{[1 - e \cos E]^{1/2}} (1 - e^2) \cos E.$$ (5–94)

Integration over one cycle in E yields

$$\Delta a = -4ba^2 \int_0^\pi \rho(r) \frac{(1 + e \cos E)^{3/2}}{(1 - e \cos E)^{1/2}} dE,$$ (5–95)

$$\Delta e = -4ba(1 - e^2) \int_0^\pi \rho(r) \left(\frac{1 + e \cos E}{1 - e \cos E}\right)^{1/2} \cos E \, dE.$$ (5–96)

For applications it is convenient to express a and ρ in terms of their values at the surface of the earth. Let these values be a_e and ρ_0. Then Eqs. (5–95) and (5–96) become

$$\Delta\left(\frac{a}{a_e}\right) = -4b\rho_0 a_e \left(\frac{a}{a_e}\right)^2 \int_0^\pi \left(\frac{\rho}{\rho_0}\right) \frac{(1 + e \cos E)^{3/2}}{(1 - e \cos E)^{1/2}} dE,$$ (5–97)

$$\Delta e = -4b\rho_0 a_e \left(\frac{a}{a_e}\right)(1 - e^2) \int_0^\pi \left(\frac{\rho}{\rho_0}\right) \left(\frac{1 + e \cos E}{1 - e \cos E}\right)^{1/2} \cos E \, dE.$$ (5–98)

The integrations in Eqs. (5–97) and (5–98) can be carried out with sufficient accuracy by Simpson's rule. It must be remembered that $\rho = \rho(r)$ and $r = a(1 - e \cos E)$.

Let the radius vector of the satellite at perigee be $r_p = a(1 - e)$, and that at apogee be $r_a = a(1 + e)$. Then the changes in r_p and r_a per orbital revolution of the satellite can be obtained from Eqs. (5–93), (5–94), (5–97), and (5–98). We find

$$\Delta\left(\frac{r_p}{a_e}\right) = -4b\rho_0 a_e (1 - e) \left(\frac{a}{a_e}\right)^2 \int_0^\pi \left(\frac{\rho}{\rho_0}\right) \left[\frac{1 + e \cos E}{1 - e \cos E}\right]^{1/2} (1 - \cos E) \, dE,$$ (5–99)

$$\Delta\left(\frac{r_a}{a_e}\right) = -4b\rho_0 a_e (1 + e) \left(\frac{a}{a_e}\right)^2 \int_0^\pi \left(\frac{\rho}{\rho_0}\right) \left[\frac{1 + e \cos E}{1 - e \cos E}\right]^{1/2} (1 + \cos E) \, dE.$$ (5–100)

These may be converted into time rates of change by the introduction of Kepler's third law [Eq. (4–44)] and the superficial gravity g_0 of the earth. By Kepler's third law, $n^2 a^3 = g_0 a_e^2$. Introducing this quantity into

Eqs. (5–99) and (5–100), we find

$$\frac{d}{dt}\left(\frac{r_p}{a_e}\right) = -\frac{c}{\pi}\left(\frac{g_0}{a_e}\right)^{1/2}\left(\frac{a}{a_e}\right)^{1/2}(1-e)$$

$$\times \int_0^\pi \left(\frac{\rho}{\rho_0}\right)\left[\frac{1+e\cos E}{1-e\cos E}\right]^{1/2}(1-\cos E)\,dE, \qquad (5\text{–}101)$$

$$\frac{d}{dt}\left(\frac{r_a}{a_e}\right) = -\frac{c}{\pi}\left(\frac{g_0}{a_e}\right)^{1/2}\left(\frac{a}{a_e}\right)^{1/2}(1+e)$$

$$\times \int_0^\pi \left(\frac{\rho}{\rho_0}\right)\left[\frac{1+e\cos E}{1-e\cos E}\right]^{1/2}(1+\cos E)\,dE, \qquad (5\text{–}102)$$

where $c = 2b\rho_0 a_e$. If r_p, r_a, and a are expressed in units of earth radii, a_e, these equations yield directly the time rates in the perigee and apogee distances.

By constructing a model atmosphere, calculating the time rates of change in r_p, r_a, a, and e, and comparing the results with observations, one can establish the function $\rho = \rho(r)$ which most closely approximates the physical atmosphere of the earth. If the satellite orbit is eccentric and the atmospheric density diminishes rapidly with increasing r, the range in E in the integrals of Eqs. (5–101) and (5–102) is limited to values near $E = 0$. It is clear that under these circumstances, r_a/a_e decreases more rapidly than r_p/a_e, which in turn has the effect (over a period of time) of decreasing the eccentricity of the orbit and the major axis. When the satellite in its now nearly circular orbit remains for longer and longer periods within the atmospheric resistance layer, it gradually spirals in toward extinction.

5–11 Precession and nutation. In Section 5–10 we associated with an oblate spheroid a torque field

$$\mathbf{N} = -\frac{3k^2 z}{r^5}(I_3 - I_e)(y\mathbf{u}_1 - x\mathbf{u}_2). \qquad (5\text{–}103)$$

This is the *torque about the origin produced on unit mass* by the nonspherical distribution of matter. A mass m_1 placed at (x, y, z) (see Fig. 5–18) will experience a torque $m_1\mathbf{N}$.

Alternatively we may consider m_1 to generate an equal and opposite torque in the nonspherical mass M. Thus the earth, for example, would be subjected to a torque

$$\mathbf{N} = \frac{3k^2 m_1 z}{r^5}(I_3 - I_e)(y\mathbf{u}_1 - x\mathbf{u}_2) \qquad (5\text{–}104)$$

due to the presence of the moon or the sun, either one of which may be taken as the mass m_1. The torques produced by the moon and the sun cause the *luni-solar precession* and *nutation*. We shall analyze this motion by means of the Eulerian angles introduced in Section 5–7.

Reference to Fig. 5–13 shows that the appropriate torque components are

$$N_\theta = N_1 \cos \psi - N_2 \sin \psi,$$
$$N_\varphi = (N_1 \sin \psi + N_2 \cos \psi) \sin \theta + N_3 \cos \theta,$$
$$N_\psi = N_3,$$

where, for this application,

$$N_1 = \frac{3k^2 m_1 zy}{r^5} (I_3 - I_e),$$

$$N_2 = -\frac{3k^2 m_1 zx}{r^5} (I_3 - I_e),$$

$$N_3 = 0.$$

Thus we have

$$N_\theta = \frac{3k^2 m_1 z(I_3 - I_e)}{r^5} [y \cos \psi + x \sin \psi], \qquad (5\text{--}105)$$

$$N_\varphi = \frac{3k^2 m_1 z(I_3 - I_e)}{r^5} [y \sin \psi - x \cos \psi] \sin \theta, \qquad (5\text{--}106)$$

$$N_\psi = 0, \qquad (5\text{--}107)$$

where x, y, z, r clearly are measured in the coordinate system attached to the rotating spheroid (Fig. 5–18). Equations (5–105) through (5–107) give the torque components to be used with Eqs. (5–55), (5–56), and (5–57).

From Eqs. (5–57) and (5–107) we have

$$\frac{d}{dt} [I_3(\dot\psi + \dot\varphi \cos \theta)] = N_\psi = 0. \qquad (5\text{--}108)$$

Hence, since I_3 is constant, we find

$$\dot\psi + \dot\varphi \cos \theta = \text{constant} = \omega_3. \qquad (5\text{--}109)$$

This is the total angular velocity of rotation of the earth about its axis. Under the prevailing torque field it remains constant.

With this value in mind, we write Eqs. (5–55) and (5–56) as

$$I_e \ddot\theta - I_e \dot\varphi^2 \sin \theta \cos \theta + I_3 \omega_3 \dot\varphi \sin \theta = N_\theta, \qquad (5\text{--}110)$$

$$\frac{d}{dt} [I_e \dot\varphi \sin^2 \theta + I_3 \omega_3 \cos \theta] = N_\varphi. \qquad (5\text{--}111)$$

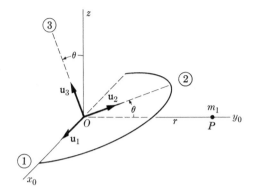

FIG. 5–20. Orientation of frame of reference for the torque produced on the earth by a mass at P.

In the application to the earth which we are considering, let the $x_0 y_0$-plane (Fig. 5–13) be the plane of the ecliptic and let the z_0-axis point toward the northerly side of this plane. The line of nodes is the intersection of the plane of the earth's equator and the ecliptic, and defines the direction toward the vernal equinox.

The general solution of Eqs. (5–110) and (5–111) is quite complex. For illustrative purposes, we shall therefore discuss a simplified situation.

Consider an instant when the ①-axis coincides with the x_0-axis and the perturbing mass m_1 is on the y_0-axis, as shown in Fig. 5–20. The point O is the center of the earth. The point P is the center of the moon or of the sun, depending on which perturbing body we are considering. The distance r is large compared with the radius of the earth. If R denotes the radius of the earth, the values of r for the moon and sun, respectively, are $60R$ and $23{,}225R$.

At the specified instant we have $\varphi = 0$, $\psi = 0$, $x = 0$, $y = r \cos \theta$, and $z = -r \sin \theta$. Hence by Eqs. (5–105) and (5–106) the torque components are

$$N_\theta = -K \sin \theta \cos \theta,$$
$$N_\varphi = 0,$$

where

$$K = \frac{3k^2 m_1 (I_3 - I_e)}{r^3}.$$

Applying these torque components to Eqs. (5–110) and (5–111), we find

$$I_e \ddot{\theta} - I_e \dot{\varphi}^2 \sin \theta \cos \theta + I_3 \omega_3 \dot{\varphi} \sin \theta = -K \sin \theta \cos \theta, \quad (5\text{–}112)$$

$$I_e \dot{\varphi} \sin^2 \theta + I_3 \omega_3 \cos \theta = c_1, \quad (5\text{–}113)$$

where c_1 is a constant of integration.

From Eq. (5–113) we have

$$\dot{\varphi} = \frac{c_1 - I_3\omega_3 \cos\theta}{I_e \sin^2\theta}. \tag{5–114}$$

In general, we would substitute this value of $\dot{\varphi}$ into Eq. (5–112), solve for θ as a function of time, resubstitute this solution into Eq. (5–114), and solve for φ as a function of time. The reader will note, however, that Eqs. (5–112) and (5–114) are very similar to those derived in Section 5–7 for the spinning top, namely Eqs. (5–58) and (5–61). The motion, therefore, will be a *precession and a nutation*.

To obtain the steady precession we set $\theta = \theta_0$ (a constant) and obtain from Eq. (5–112)

$$\dot{\varphi}^2 \cos\theta_0 - \frac{I_3\omega_3\dot{\varphi}}{I_e} = \frac{K}{I_e} \cos\theta_0.$$

Now in the case of the earth the ratio is $\dot{\varphi}/\omega_3 \simeq 10^{-7}$. Hence, if we neglect the first term above in comparison with the second, we obtain

$$\dot{\varphi} \cong -\frac{K \cos\theta_0}{I_3\omega_3}. \tag{5–115}$$

The value of $\dot{\varphi}$ thus found is the approximate precessional velocity. The line of nodes, defining the direction to the vernal equinox, moves in the negative φ-direction at a constant rate. This motion is the *mean precession of the equinox*. The equinox moves westward along the ecliptic about $50''.2$ per year, completing a circuit in about 26,000 years.

In this simplified illustration we have considered only one disturbing body, namely, the moon or the sun. The order of magnitude of the torques due to these can be readily estimated. In Fig. 5–20, we assume that the *obliquity of the ecliptic* is $\theta = 23°.5$, and that the perturbing body is at P. Then the values for the moon and sun are 2.83×10^{29} dyne·cm and 1.31×10^{29} dyne·cm, respectively, for $K \sin\theta \cos\theta$. It is apparent that the moon is the dominant contributor to the luni-solar precession. The proximity of the moon more than compensates for its small mass compared with that of the sun. Both bodies perturb the motion of the earth, and in a full discussion of the problem their influence would have to be considered simultaneously.

As the relative positions of moon and sun with respect to the earth change, the torques affecting the earth vary. In the general solutions of Eqs. (5–110) and (5–111), therefore, we would expect to find periodic variations in θ, that is, *nutation*, superposed on the precessional motion. These motions of the earth cause changes in the tabulated right ascensions and declinations of the stars because the latter are referred to the celestial equator and to the vernal equinox. The right ascensions increase with time because of the mean precessional motion of the equinox. The declinations increase or decrease, depending upon the position of the stars relative to the pole of the ecliptic.

A full discussion of precessional effects would necessarily include those due to the other planets of the solar system. The planetary precession is evidenced in a small motion of the pole of the ecliptic, a motion which is only about one fortieth as fast as the motion of the celestial pole resulting from the luni-solar precession. When the two precessions are combined, the result is called *general precession*. Values for the precessional and nutational constants may be found elsewhere [13].

5–12 Dynamics of the earth-moon system. The theory of the motion of the moon is extremely complex. This is due in large part to its relatively large mass and to its proximity to the earth. The problem can be regarded as a three-body problem, with the earth as the primary mass, m_1, the moon as the perturbed mass, m, and the sun as the perturbing mass, m' (Fig. 5–21). We wish to find the motion of m relative to m_1 under the action of m'. Let the position of the moon be denoted by (x, y, z), and that of the sun by (x', y', z'). Then the disturbing function as defined in Section 5–3, Eq. (5–16), is

$$R = k^2 m' \left\{ \frac{1}{\rho} - \frac{xx' + yy' + zz'}{r'^3} \right\},$$

where ρ and r' have the significance shown in Fig. 5–21. The equations of motion for the moon are, therefore,

$$\ddot{x} + \frac{k^2 M' x}{r^3} = \frac{\partial R}{\partial x}, \tag{5–116}$$

$$\ddot{y} + \frac{k^2 M' y}{r^3} = \frac{\partial R}{\partial y}, \tag{5–117}$$

$$\ddot{z} + \frac{k^2 M' z}{r^3} = \frac{\partial R}{\partial z}, \tag{5–118}$$

where $M' = m_1 + m$.

The orbit of the moon is inclined about 5° to the plane of the ecliptic. Hence, for illustrative purposes, we shall make the approximation that the inclination is negligible, i.e., earth, moon, and sun lie in the same plane. Furthermore we shall neglect the eccentricity of the earth's orbit around the sun. This is 0.06, and the inclusion of so small an eccentricity in the analysis will have only a second-order effect on the results. The problem then reduces to a two-dimensional one. Let ψ be the celestial longitude of the moon (Fig. 5–22) and let ψ' be the celestial longitude of the sun.

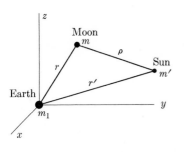

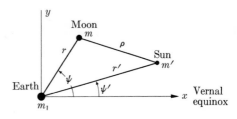

FIG. 5–21. The earth-moon system.

FIG. 5–22. Two-dimensional orientation of earth-moon system.

Then, by the cosine law,

$$\rho^2 = r^2 + r'^2 - 2rr' \cos(\psi - \psi'),$$

or

$$\rho^2 = r'^2 \left\{ 1 + \left(\frac{r}{r'}\right)^2 - 2\left(\frac{r}{r'}\right) \cos(\psi - \psi') \right\},$$

and

$$\frac{1}{\rho} = \frac{1}{r'} \left\{ 1 + \left(\frac{r}{r'}\right)^2 - 2\left(\frac{r}{r'}\right) \cos(\psi - \psi') \right\}^{-1/2}. \qquad (5–119)$$

Expanding Eq. (5–119) and retaining only terms in r/r' of degree two or less, we obtain

$$\frac{1}{\rho} = \frac{1}{r'} \left\{ 1 + \left(\frac{r}{r'}\right) \cos(\psi - \psi') + \frac{1}{4}\left(\frac{r}{r'}\right)^2 + \frac{3}{4}\left(\frac{r}{r'}\right)^2 \cos 2(\psi - \psi') \right\}.$$

Furthermore,

$$\frac{xx' + yy'}{r'^3} = \frac{rr' \cos(\psi - \psi')}{r'^3},$$

and, since the earth's orbit is assumed to be circular, we will denote its radius r' by a'. Then the disturbing function becomes

$$R = \frac{k^2 m'}{a'} \left\{ 1 + \frac{1}{4}\left(\frac{r}{a'}\right)^2 + \frac{3}{4}\left(\frac{r}{a'}\right)^2 \cos 2(\psi - \psi') \right\}. \qquad (5–120)$$

The central problem of lunar theory is the expression of R in terms of the orbital elements of the moon. The formulas necessary for this computation are given in Chapter 4, Section 4–6, Eqs. (4–48) through (4–55). From Eq. (4–54) we write, *for the moon's orbit*,

$$r = a(1 - e \cos E) = a(1 - e \cos[E - M + M])$$
$$= a\{1 - e \cos(E - M) \cos M + e \sin(E - M) \sin M\}.$$

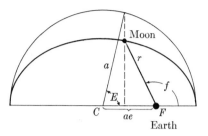

FIG. 5–23. Geometry of the elliptical path of the moon.

But for small e, Kepler's equation yields $M \simeq E$. Hence, $\cos (E - M) \sim 1$ and $\sin (E - M) \simeq E - M = e \sin M + (e^2/2) \sin 2M$. Thus we have

$$r = a\{1 - e \cos M + e^2 \sin^2 M\},$$

and, squaring, we obtain

$$r^2 = a^2 \left\{ 1 + \frac{3e^2}{2} - 2e \cos M - \frac{e^2}{2} \cos 2M \right\}. \tag{5–121}$$

In the above expansions, we have neglected powers of $e > 2$. In an analogous way, by using the geometry of the elliptical path (Fig. 5–23), we have

$$\frac{r}{a} \cos f = \cos E - e$$

and

$$\frac{r}{a} \sin f = \sqrt{1 - e^2} \sin E$$

which can be expressed in terms of e and M, where M is the mean anomaly. After a considerable amount of trigonometric manipulations, we find

$$\frac{r}{a} \cos f = \left(1 - \frac{3}{8} e^2 \right) \cos M + \frac{1}{2} e \cos 2M + \frac{3}{8} e^2 \cos 3M - \frac{3e}{2} \tag{5–122}$$

and

$$\frac{r}{a} \sin f = \left(1 - \frac{5}{8} e^2 \right) \sin M + \frac{1}{2} e \sin 2M + \frac{3}{8} e^2 \sin 3M. \tag{5–123}$$

Now $\psi = \Omega + \omega + f$ because we have assumed that the moon, sun, and earth are in the same plane. Hence the last term in the disturbing function [Eq. (5–120)] may be expressed in terms of the orbital elements. We write

$$\left(\frac{r}{a} \right)^2 \cos 2(\psi - \psi') = \left(\frac{r}{a} \right)^2 \cos \{2f + 2(\Omega + \omega - \psi')\},$$

and expand to obtain

$$\left(\frac{r}{a}\right)^2 \cos 2(\psi - \psi') = 2\left(\frac{r}{a}\right)^2 \cos^2 f \cos 2(\Omega + \omega - \psi')$$

$$- \left(\frac{r}{a}\right)^2 \cos 2(\Omega + \omega - \psi') - 2\left[\frac{r}{a}\sin f\right]\left[\frac{r}{a}\cos f\right]\sin 2(\Omega + \omega - \psi').$$

$$(5\text{-}124)$$

Substituting from Eqs. (5–121), (5–122), and (5–123) into Eq. (5–124) and thence into Eq. (5–120), and neglecting terms with powers of $e > 2$, we obtain for R the value

$$R = \frac{k^2 m'}{a'^3}\, a^2 \left[\frac{3}{8}\, e^2 - \frac{1}{2}\, e \cos M - \frac{1}{8}\, e^2 \cos 2M \right.$$

$$+ \frac{15}{8}\, e^2 \cos \{2(\Omega + \omega - \psi')\} \qquad \text{(A)}$$

$$- \frac{9}{4}\, e \cos \{2(\Omega + \omega - \psi') + M\}$$

$$+ \frac{3}{4} \cos \{2(\Omega + \omega - \psi') + 2M\} \qquad \text{(B)}$$

$$- \frac{15}{8}\, e^2 \cos \{2(\Omega + \omega - \psi') + 2M\}$$

$$+ \frac{3}{4}\, e \cos \{2(\Omega + \omega - \psi') + 3M\}$$

$$\left. + \frac{3}{4}\, e^2 \cos \{2(\Omega + \omega - \psi') + 4M\} \right]. \qquad (5\text{-}125)$$

The complexity of the perturbing function indicates the difficulty of describing the moon's motion with precision. Terms in R such as $3k^2 m' a^2 e^2 / 8a'^3$ indicate secular variations. Those involving $\cos M$ and $\cos 2M$ are *elliptic terms* and are not strictly perturbations, but represent motion in a Keplerian orbit. Of the other terms we mention only two, namely, the *evection* with a period of about one month and the *variation* with a period of about 15 days. These are denoted by (A) and (B), respectively, in Eq. (5–125). Other terms have appropriate names.

To illustrate the effect of the evection on the orbital elements and on the longitude of the moon, we shall sketch the method of analysis. Similar analyses could be made for the other terms in R. The differential equations for the variations in the orbital elements are given in Section 5–4, Eqs. (5–23) through (5–28).

Let

$$A = \frac{15}{8} n'^2 a^2 e^2 \cos 2(\Omega + \omega - \psi')$$

denote the evection, with $n'^2 = k^2 m'/a'^3$. Then, since $\partial R/\partial M = \partial A/\partial M = 0$, we find that, due to the evection,

$$\frac{de}{dt} = \frac{15}{4} \frac{n'^2}{n} e \sin 2(\Omega + \omega - \psi'), \tag{5-126}$$

$$\frac{da}{dt} = 0, \tag{5-127}$$

$$\frac{d\omega}{dt} = \frac{d(\Omega + \omega)}{dt} = \frac{15}{4}\left(1 - \frac{1}{2} e^2\right)\frac{n'^2}{n} \cos 2(\Omega + \omega - \psi'), \tag{5-128}$$

$$\frac{dM}{dt} = n - \frac{15}{4}\frac{n'^2}{n}(1 + e^2) \cos 2(\Omega + \omega - \psi'). \tag{5-129}$$

We have assumed in our discussion that $i = 0$. Hence $di/dt = 0$, and $d\Omega/dt = 0$ because R does not contain i. Furthermore, we note that the evection contributes nothing to the perturbation in the semimajor axis a.

Integration of Eqs. (5–126), (5–128), and (5–129) yields

$$e = e_0 + \frac{15}{8}\frac{n'}{n} e_0 \cos 2(\Omega_0 + \omega_0 - \psi_0'),$$

$$\Omega + \omega = (\Omega + \omega)_0 + \frac{15}{8}\left(1 - \frac{1}{2} e_0^2\right)\frac{n'}{n} \sin 2(\Omega_0 + \omega_0 - \psi_0'),$$

$$M = M_0 + \frac{15}{8}\frac{n'}{n}(1 + e_0^2) \sin 2(\Omega_0 + \omega_0 - \psi_0').$$

The subscripts refer to the osculating orbit of reference. In integrating Eq. (5–126), for instance, we bear in mind that $\psi' = \Omega' + \omega' + n'(t - t_0)$, where n', Ω', and ω' refer to the sun's *apparent* orbital motion around the earth. Furthermore, we have assumed that $e = e_0$ is constant on the right-hand side of Eq. (5–126). Thus we find that the perturbations are

$$\delta e = \frac{15}{8}\frac{n'}{n} e_0 \cos 2(\Omega_0 + \omega_0 - \psi_0'), \tag{5-130}$$

$$\delta(\Omega + \omega) = \frac{15}{8}\frac{n'}{n}\left(1 - \frac{1}{2} e_0^2\right)\sin 2(\Omega_0 + \omega_0 - \psi_0'), \tag{5-131}$$

$$\delta M = \frac{15}{8}\frac{n'}{n}(1 + e_0^2)\sin 2(\Omega_0 + \omega_0 - \psi_0'). \tag{5-132}$$

For comparison with observations, however, we wish to find the perturbation in longitude, $\delta\psi$. Since $\psi = \Omega + \omega + f$ and $f = M + 2e \sin M$, we have

$$\delta\psi \cong \delta(\Omega + \omega) + \delta M + 2e \cos M \; \delta M + 2 \sin M \; \delta e,$$

and substituting from Eqs. (5–130), (5–131), and (5–132), we obtain

$$\delta\psi \cong \frac{15}{4} \frac{n'}{n} e \sin [2(\Omega_0 + \omega_0 - \psi_0') + M]. \tag{5–133}$$

The perturbation in ψ, the longitude of the moon, due to evection, has a period of about one month and an amplitude of about one degree.

In a similar way, the perturbations in the orbital elements due to the other terms in R can be found. If the effects of the eccentricity of the earth's orbit and the inclination of the moon's orbit are taken into account, other perturbations arise, the most important of which is the motion of the node. The inclusion of i in the disturbing function and the use of Eq. (5–26) lead to

$$\frac{d\Omega}{dt} = -\frac{3}{4} \frac{n'^2}{n} \frac{i}{\sin i} \cong -\frac{3}{4} \frac{n'^2}{n},$$

since $i/\sin i \cong 1$. Hence

$$\delta\Omega = -\frac{3}{4} \frac{n'^2}{n} (t - t_0) = -\left(\frac{3}{4} \frac{n'}{n}\right) n' \delta t. \tag{5–134}$$

The node moves along the ecliptic toward the west. The quantity $n' = 2\pi/P_e$, where P_e is the period of the earth's motion around the sun. Also $n = 2\pi/P_m$, where P_m is the moon's sidereal period. Therefore we may write Eq. (5–134) in the form

$$\frac{\delta\Omega}{2\pi} = -\left(\frac{3}{4} \frac{P_m}{P_e}\right) \frac{\delta t}{P_e} = -0.0555 \frac{\delta t}{P_e},$$

and we find that Ω changes by 2π radians when $\delta t = 6580$ days, or approximately 18 years.

The brief discussion above indicates the magnitude of the investigation required to describe the moon's motion. The exhaustive work by E. W. Brown [14] indicates 155 periodic terms in the expression for the moon's longitude which have amplitudes exceeding $0''.1$; and there are 500 smaller ones which must be reckoned with in cases of resonance. One may summarize the principal secular effects, however, by stating that the *line of nodes regresses at an average rate of one revolution in 18.6 years* and the *line of apsides (the major axis of the orbit) advances* $[(d\omega/dt) > 0]$ *at an average rate of one revolution in 8.9 years.*

5-13 Effects of relativity. The general theory of relativity predicts three effects which can be tested by observation. These are: (a) the motion of the perihelion of Mercury; (b) the deflection of light rays near the sun; (c) the displacement of spectral lines toward the red in the spectrum of the light emitted by a massive star. We shall briefly consider the first two of these, since they represent phenomena related to the solar system.

(a) *Motion of Mercury's perihelion.* Observations of the planet Mercury over many years revealed that the perihelion of the planet was advancing more rapidly than could be accounted for by perturbations calculated according to newtonian theory. Attempts to explain the excess of some 43″ of arc per century by the action of a resisting medium were unsuccessful. The general relativity theory of Einstein [15, pp. 88 ff.] does predict, within the errors of measurement, the observed advance of perihelion.

According to Einstein's theory, the law of force can be expressed as

$$F(r) = -\frac{\mu}{r^2}\left[1 + \frac{3h^2}{c^2 r^2}\right], \qquad (5\text{-}135)$$

where $\mu = k^2(m_1 + m_2)$, m_1 is the solar mass, m_2 is the mass of the planet, c is the velocity of light, and h is the areal-velocity constant which was introduced in Chapter 4, Eq. (4-35). The ratio h/r is the transverse velocity of the planet in its orbit. Thus Eq. (5-135) states that, for a nearly circular orbit, the relativity force supplement, $3h^2/c^2 r^2$, is proportional to the ratio of the squares of the speed of the planet to the speed of light.

The first term in Eq. (5-135) is the newtonian force. For the earth, the ratio of the relativity correction to the newtonian force is about 3×10^{-8}.

The perturbative force $R = -(3\mu h^2/c^2 r^4)$ is radial. Therefore we may determine its effect on the longitude of perihelion, ω, by using Eq. (5-33) with $S = W = 0$. We have

$$\dot{\omega} = \frac{3\mu h^2 \sqrt{1 - e^2} \cos f}{naec^2 r^4}, \qquad (5\text{-}136)$$

where f is the true anomaly.

Setting $r = a(1 - e^2)/(1 + e \cos f)$, $h^2 = \mu a(1 - e^2)$, $\mu = n^2 a^3$, and $\dot{\omega} = (d\omega/df)\dot{f} = (h/r^2)(d\omega/df)$, we obtain

$$d\omega = \frac{3n^2 a^2}{ec^2(1 - e^2)} \cos f(1 + e \cos f)^2 \, df.$$

To find the change in ω in one revolution of the planet in its orbit, we

TABLE 5–1

Planet	Theory	Observations
Mercury	$43''.03$	$42''.56 \pm 0''.94$
Venus	$8''.63$	$8''.4 \pm 4''.8$
Earth	$3''.84$	$4''.6 \pm 2''.7$
Mars*	$1''.35$	$1''.5 \pm 0''.04$

* The result for Mars is unpublished. We quote it by kind permission of Dr. Clemence.

integrate from 0 to 2π, obtaining

$$\Delta\omega = \frac{3n^2a^2}{ec^2(1 - e^2)} \int_0^{2\pi} \cos f(1 + e \cos f)^2 \, df$$

$$= \frac{6\pi n^2 a^2}{c^2(1 - e^2)}. \tag{5–137}$$

This is the advance of perihelion *in radians per revolution of the planet.* If we express this quantity in *fractions of a revolution per revolution* and use the planet's period P in place of n, we find for the advance of perihelion,

$$\Delta\omega = \frac{12\pi^2 a^2}{c^2 P^2 (1 - e^2)}. \tag{5–138}$$

Clemence [16] has compared theory with observation and finds the results shown in Table 5–1. Here the advance of perihelion is expressed in *seconds of arc per century.* We observe a satisfactory agreement between theory and observation. As more and better observations of the planets accumulate, the predictions of the theory can be more firmly verified.

According to Eqs. (5–30) and (5–32), the inclination of a planetary orbit and the longitude of the node (Ω) are not affected by the relativistic perturbing force. However, it is clear from Eqs. (5–29), (5–31), and (5–34) that the semimajor axis, a, the eccentricity, e, and the mean anomaly, M, may be influenced. We leave it as a problem for the reader to show that a and e have no secular variations due to the relativistic modification in the law of force.

(b) *Deflection of light.* For a photon traveling with the speed of light, theory shows [15, pp. 90 ff.] that $h \to \infty$ and that the differential equation describing the orbit therefore is

$$\frac{d^2u}{df^2} + u = 3mu^2,$$

where $u = 1/r$ and f is the true anomaly of the path followed by the photon. Here m is a constant of integration in the theory of relativity which corresponds to the power of a central mass to produce a field of acceleration. The reader may compare it with the quantity μ in Eq. (4–37).*

Integration of this equation by successive approximations yields a path (in cartesian coordinates)

$$x = R - \frac{m}{R} \frac{x^2 + 2y^2}{\sqrt{x^2 + y^2}} . \qquad (5\text{–}139)$$

The angle in radians between asymptotes to this curve is

$$\alpha = \frac{4m}{R} ,$$

where R is a constant of integration and represents physically the distance from the center of the perturbing mass to the photon path. Obviously, if R in Eq. (5–139) becomes very large, there results the straight line $x = R$.

For the sun $R = 6.97 \times 10^5$ km, $m = 1.47$ km, and hence $\alpha = 1''.75$. Observations at eclipses of the deflection of starlight by the sun have yielded values of α ranging from $1''.6 \pm 0.3$ to $2''.2$.

(c) *Tests using data from artificial satellites.* The general theory of relativity predicts that the time scale in which atomic phenomena occur is influenced by the gravitational potential in which the atomic phenomenon is located. It has been suggested that a test of this theory could be made by transporting a cesium or thalium "atomic" clock in a satellite and comparing its timekeeping with that of a similar clock at a ground station. According to the theory, the known difference of gravitational potential to which these two atomic clocks are subjected should produce a systematic difference in the rate at which they run.

Another test of the theory is provided by an artificial satellite, since the motion of its perigee is similar to the motion of perihelion described above in part (a). For a satellite whose maximum distance from the earth is 600 mi and whose minimum is 200 mi, McVittie [17] has calculated that the perigee moves at a rate of 1324" of arc per century. This calculation was made on the basis of a spherical earth and ignores the drag due to the atmosphere. The cumulative effect is what counts here. Thus observations of a satellite over a long period should provide substantial evidence in support of the general theory of relativity. It should be pointed out,

---

* For a more complete definition of m the reader may consult reference 15, p. 87.

however, that this effect is enmeshed with the perturbative effects due to the earth's oblateness, to air drag, to the moon, and so forth. Hence at present the practicality of such a test is questionable.

PROBLEMS

Section 5–2:

5–1. Show that the projections of the angular-momentum vector $\mathbf{L}$ [Eq. (5–10)] on the cartesian coordinate planes are

$$\sum_{i=1}^{3} m_i\{y_i\dot{z}_i - z_i\dot{y}_i\} = c_1,$$

$$\sum_{i=1}^{3} m_i\{z_i\dot{x}_i - x_i\dot{z}_i\} = c_2,$$

$$\sum_{i=1}^{3} m_i\{x_i\dot{y}_i - y_i\dot{x}_i\} = c_3,$$

where c_1, c_2, c_3 are the constant cartesian components of $\mathbf{L}$. Interpret these relations in terms of areal velocity. What is the magnitude of $\mathbf{L}$?

5–2. If $\mathbf{L} = c_1\mathbf{i} + c_2\mathbf{j} + c_3\mathbf{k}$ in a cartesian reference frame, show that the invariable plane through the origin for a system of three mass points will have the equation $c_1x + c_2y + c_3z = 0$.

Section 5–3:

5–3. Show that Eqs. (5–17), (5–18), and (5–19) can be written as

$$\ddot{r} - r\dot{\theta}^2 + \frac{k^2 M}{r^2} = \frac{\partial R}{\partial r},$$

$$\frac{d}{dt}[r^2\dot{\theta}] = \frac{\partial R}{\partial \theta}$$

in terms of polar coordinates in the xy-plane. Show also that the disturbing function becomes

$$R = k^2 m' \left\{ \frac{1}{\rho} - \frac{r}{r'^2} \cos(\theta - \theta') \right\}.$$

5–4. From the equations of Problem 5–3 solve the three-body problem when the central mass m_1, the perturbed mass m, and the perturbing mass m' are always situated at the vertices of an equilateral triangle. Compare the period of m around m_1 in this solution with the period which would arise if m' were not present.

5–5. A vehicle initially moves around the earth in a circular path of radius a. Then a small but constant tangential thrust from its motor is applied. What effect will this thrust have on the elements of the vehicle's orbit?

Section 5–5:

5–6. A space platform whose mass is negligible compared with the mass of the moon is to be placed in orbit between earth and moon in such a way that it always remains on the line joining earth to moon. If the mass of the moon is 1/81.3 that of the earth and the moon's orbit is considered to be circular with radius 384,405 km, how far from the moon should the platform be located?

5–7. A small space vehicle finds itself trapped at the Lagrangian point L_1 (Fig. 5–7) of the sun-earth system. If the orbit of the earth is considered to be a circle, will the vehicle ever be in the shadow of the earth? (Data for this problem will be found in Table VI of the appendix.)

Section 5–10:

5–8. The artificial satellite Vanguard II, launched February 17, 1959, was found to have an orbit with $e = 0.166$, $i = 32°.9$, and $P = 126$ min. At what rate and in what direction does the ascending node move? At what rate and in what direction does its perigee move?

5–9. A comet moves in a resisting medium which opposes the motion with a tangential force per unit mass equal to $-(\alpha V/r^2)$, where α is a small constant and V is the speed in the orbit. Calculate the change Δa in the semimajor axis and the change Δe in the eccentricity per orbital revolution, on the assumption that e is small so that powers of $e > 2$ can be neglected.

Section 5–12:

5–10. The disturbing function for the moon, Eq. (5–125), contains *the variation*

$$B = \frac{3k^2 m' a^2}{4a'^3} \cos 2[\Omega + \omega + M - \psi'].$$

Calculate the perturbations in the orbital elements arising from this term. Calculate also the perturbation due to this term which is present in the longitude of the moon, ψ.

5–11. Analyze the effects of the secular term $3k^2 m' a^2 e^2/8a'^3$ in Eq. (5–125) on the orbital elements of the moon.

5–12. Show that the semimajor axis a and the eccentricity e undergo no secular variations due to the relativistic perturbing force shown in Eq. (5–135).

REFERENCES

1. H. BRUNS, "Über die Integrale des Vielkörper-Problems," *Acta Mathematica*, 11, 25, 1887.

2. F. R. MOULTON, *Celestial Mechanics*, 2nd rev. ed. New York: Macmillan & Co., 1914.

3. S. W. McCUSKEY, *An Introduction to Advanced Dynamics*, Chapter 2. Reading, Mass.: Addison-Wesley Publishing Co., Inc., 1959.

4. F. R. MOULTON, *Periodic Orbits*, Chapter V through VII. Washington: Carnegie Institution of Washington, Publ. No. 161, 1920.

5. A. G. WEBSTER, *Dynamics of Particles and of Rigid, Elastic, and Fluid Bodies*, pp. 567 ff. New York: Hafner Publishing Co., Inc., 1949.

6. H. JEFFREYS, *The Earth*, 4th ed., p. 211. Cambridge: Cambridge University Press, 1959.

7. W. A. HEISKANEN and F. A. V. MEINESZ, *The Earth and Its Gravitational Field*, Chapter 3. New York: McGraw-Hill Book Co., Inc., 1958.

8. L. SPITZER, JR., "Perturbations of a Satellite Orbit," *J. Brit. Interplanetary Soc.*, **9**, 131, 1950.

9. Y. KOZAI, "On the Effects of the Sun and the Moon Upon the Motion of a Close Earth Satellite," *Smithsonian Institution Astrophysical Observatory*, Special Report No. 22, 1959.

10. Y. KOZAI, "The Motion of Close Earth Satellite," *Astron. J.* **64**, 367, 1959.

11. J. A. O'KEEFE, A. ECKELS, and R. K. SQUIRES, "The Gravitational Field of the Earth," *Astron. J.* **64**, 245, 1959.

12. T. E. STERNE, "An Atmospheric Model, and Some Remarks on the Inference of Density from the Orbit of a Close Earth Satellite," *Astron. J.* **63**, 81, 1958.

13. C. W. ALLEN, *Astrophysical Quantities*, p. 18. London: The Athlone Press, 1955.

14. E. W. BROWN, *An Introductory Treatise on the Lunar Theory*. Cambridge: Cambridge University Press, 1896. Also *Memoirs of the Roy. Astron. Soc.* **54**, 1900; **57**, 1905; **59**, 1908.

15. A. S. EDDINGTON, *The Mathematical Theory of Relativity*. Cambridge: Cambridge University Press, 1957.

16. G. M. CLEMENCE, *Revs. Modern Phys.* **19**, 361, 1947; see also R. L. DUNCOMBE, *Astron. J.* **61**, 174, 1956.

17. G. C. McVITTIE, *Science* **127**, 505, 1958.

18. E. W. BROWN and C. A. SHOOK, *Planetary Theory*. Cambridge: Cambridge University Press, 1933.

19. H. C. PLUMMER, *An Introductory Treatise on Dynamical Astronomy*. Cambridge: Cambridge University Press, 1918.

20. H. S. SEIFERT, ed., *Space Technology*. New York: John Wiley and Sons, Inc., 1959.

CHAPTER 6

THE SUN AND INTERPLANETARY SPACE

6–1 The sun. The study of the sun may be divided into several broad areas according to the methods of analysis and observation customarily employed. These areas are:

(I)   theory of the internal structure;
(II) theory of the structure of the outer layers;
(III) observational studies of the phenomena detected in the outer layers.

This section follows the above outline and concludes with a summary of the nature of solar radiation.

I. Internal Structure

The aim of investigations of the internal structure of the sun is to determine the density, the temperature, the pressure, and the chemical composition existing at each point in the solar interior. A successful model of the sun's structure must account for the observed mass, radius, and luminosity of the sun. The parameters describing the structure must, at the same time, satisfy a number of equations which relate them. Thus investigations of the structure of the sun (or of any other star) require the construction, by mathematical calculations, of a synthetic gaseous model that satisfies known physical laws, and whose mass, radius, and luminosity are matched to those of the sun. We shall sketch here only the essentials of the analysis. The reader will find in references 1, 2, and 3 detailed discussions of the methods of analysis.

To describe conditions in the interior of the sun, which is assumed to be spherically symmetric, we may use the following basic functions of the distance r from the center:

$M(r) =$ the mass contained inside a sphere of radius r;

$T(r) =$ the temperature at distance r;

$L(r) =$ the net rate at which energy flows out of a sphere of radius r;

$P(r) =$ the pressure at distance r.

In addition to these functions, the relative abundance of the chemical elements as a function of r should be given in a complete description of the solar structure. However, for practical computations, it has been

222

found sufficient to specify the relative abundance per unit mass of hydrogen, X, of helium, Y, and the combined abundance of all the heavier elements, Z. The sum of X, Y, and Z is unity.

The basic functions listed above must satisfy Eqs. (6–1) through (6–8) that follow.

The condition for hydrostatic equilibrium is

$$dP(r) = -\frac{GM(r)}{r^2}\rho(r)\,dr, \qquad (6\text{–}1)$$

where G is the gravitational constant and $\rho(r)$ is the density. This follows from Eq. (2–38), since the acceleration due to gravity at distance r from the center is $GM(r)/r^2$. The effects of solar rotation are neglected here.

The rate at which $M(r)$ changes with r in a spherically symmetric structure is

$$dM(r) = 4\pi r^2\rho(r)\,dr. \qquad (6\text{–}2)$$

This is also called the *conservation-of-mass* condition.

According to the principle of energy conservation, the difference in net energy flows, $dL(r)$, between two concentric spherical surfaces separated by a distance dr, is given for matter in thermal equilibrium by

$$dL(r) = 4\pi r^2\rho(r)\epsilon(r)\,dr, \qquad (6\text{–}3)$$

where $\epsilon(r)$ is the energy released by thermonuclear processes per unit mass, per unit time. The energy released is a function of the temperature, the density, and the chemical composition.

Energy can be transported through the interior of a star by radiation or by convection. Conduction can be shown to play only a negligible role. If the energy is transported by radiation alone, the star has a stable stratification. In this case, the equation

$$dT(r) = -\frac{3\kappa\rho(r)}{4acT^3}\frac{L(r)}{4\pi r^2}\,dr \qquad (6\text{–}4)$$

is valid. This is called the equation of *radiative equilibrium*. Here c is the velocity of light, a is the radiation constant (7.569×10^{-15} erg·cm^{-3}·deg^{-4}), and κ is the mean absorption coefficient per unit mass, which is a function of the chemical composition, the temperature, and the density. If the energy is transported by convective currents, the adiabatic equation is valid, that is,

$$\frac{dP(r)}{P(r)} = \gamma\,\frac{d\rho(r)}{\rho(r)}. \qquad (6\text{–}5)$$

For an ideal monatomic gas, $\gamma = \frac{5}{3}$. In the computations, one must decide at each value of r whether Eq. (6–4) or Eq. (6–5) is to be used.

Convection currents cannot occur so long as the temperature gradient obeys the inequality

$$\frac{dT}{dr} > \left(1 - \frac{1}{\gamma}\right) \frac{T(r)}{P(r)} \frac{dP(r)}{dr}. \tag{6-6}$$

Finally, the equation of state must be obeyed, i.e.,

$$P(r) = \frac{\overline{R}}{\mu} \rho(r) T(r) + \frac{1}{3} aT^4(r). \tag{6-7}$$

The contribution of radiation pressure to the total pressure is given by the last term. In the case of the sun, this contribution is very small. In Eq. (6-7), $\overline{R}$ is the gas constant, and μ is the mean molecular weight. At the high temperatures found in the interior of the sun, hydrogen, helium, and many of the heavier elements are fully ionized. When ionized, hydrogen (with molecular weight 1) yields two particles, and helium (with molecular weight 4) yields three particles. For the heavier elements, the molecular weight is nearly twice the atomic number N, while the number of particles obtained in ionization is $N + 1$. As a result of complete ionization, the mean molecular weight can be expressed approximately as

$$\mu = \frac{1}{2X + \frac{3}{4}Y + \frac{1}{2}Z}. \tag{6-8}$$

The equations listed above are solved simultaneously by numerical methods, subject to the following initial and boundary conditions:
At the center, where $r = 0$,

$$M(0) = 0, \qquad L(0) = 0.$$

At the surface, where $r = R$ (the solar radius), $M(R)$ and $L(R)$ must equal the total mass and the total luminosity of the sun, respectively. Also

$$\rho(R) = 0, \qquad P(R) = 0.$$

The results of such calculations indicate that the chemical composition of the sun by weight is about 80% hydrogen, 18% helium, and 2% other elements. The central temperature of the sun is about 15×10^6 °K, and the central density about 130 gm/cm^3. The outer 18% of the solar radius consists of a convective zone in which the variation of temperature with pressure obeys Eq. (6-5). In the inner 82%, the so-called condition of *radiative equilibrium* is found. The temperature gradient is given by Eq. (6-4). Energy is produced near the center of the sun mainly by the proton-proton cycle [1], a series of thermonuclear reactions in which four protons combine to produce a nucleus of helium plus a positron, a neutrino, and gamma radiation.

These studies of the internal structure of the sun require that (1) the mass, (2) the luminosity, and (3) the radius of the sun be known. (1) The *mass* can be derived from the estimates for the earth's mass in solar mass units and in grams, which were discussed in Section 2–3. From Eqs. (2–14) and (2–17) we find the mass of the sun in grams to be $1.987 \pm 0.001 \times 10^{33}$ gm. (2) The total rate of energy generation, *the luminosity*, of the sun can be estimated from the solar constant discussed in Section 2–7. It is 3.79×10^{33} ergs·sec^{-1}. The apparent visual magnitude of the sun is -26.73 [4]. (3) The visible "surface" of the sun is called the *photosphere*. The high opacity of the photospheric material causes the disk of the sun to show a sharp edge. Although the sun extends farther out than the photosphere, its limb is used for determining the *radius*, the result being $6.957 \pm 0.001 \times 10^5$ km.

From these data we can derive the following results:

mean density of the sun: 1.408 gm·cm^{-3},
acceleration due to gravity at the photosphere: 2.738×10^4 cm·sec^{-2},
velocity of escape from the photosphere: 617.2 km·sec^{-1}.

(In each case, the last digit quoted is uncertain by about one unit.)

II. Structure of the Outer Layers

The purpose of theoretical studies of the outer layers of the sun is to account quantitatively for the energy distribution in the solar spectrum, for the details of the spectral lines, and for the darkening toward the limb observed on the solar disk (see Plate VI). The structure of the outer solar layers is obtained from models which satisfy the observations, and whose structural parameters obey known physical principles. Spectroscopic observations, as well as measurements of the drop of intensity at various wavelengths from the center of the solar disk to the limb, are used in the computations. The analysis is very complex, and we can present here only an outline of the method.

In theory, the solar atmosphere can be considered semi-infinite, plane-parallel, and symmetric about the normal direction. If the source function J and the optical depth τ are defined as in Eqs. (2–47) and (2–48), the intensity of radiation emerging at the heliocentric angle θ from the center of the apparent solar disk is given by

$$I\,(\cos\,\theta) = \int_0^\infty J(\tau) e^{-\tau/|\cos\,\theta|}\,\frac{d\tau}{|\cos\,\theta|}\,. \qquad (6\text{–}9)$$

This solution of the equation of transfer [Eq. (2–49)] for the emergent radiant energy must equal the sum of the individual energies emitted in the θ-direction by each element of mass along the line of sight going inward from $\tau = 0$ to $\tau = \infty$, taking into account the absorption losses. According to Eqs. (2–45) and (2–47), each such contribution is given by

$$j \, \frac{\rho ds}{|\cos \theta|} = J \, \frac{d\tau}{|\cos \theta|},$$

and each such parcel of energy is attenuated on the way out by the factor $e^{-\tau/|\cos \theta|}$. In the sun, I ($\cos \theta$) can be observed from $\cos \theta = 1$ at the disk's center to $\cos \theta = 0$ at the limb. Equation (6–9) then may be treated as an integral equation for $J(\tau)$.

The results indicate that, approximately,

$$J(\tau) = \tfrac{1}{2}F(1 + \tfrac{3}{2}\tau), \tag{6–10}$$

F being a constant. This relationship can be explained theoretically if one assumes the solar atmosphere to be in radiative equilibrium in addition to the other conditions listed so far. In radiative equilibrium, the energy arriving from the solar interior is carried through the atmosphere by radiation alone. Since the energy sources are in the deep interior, the net outward flux of radiation is constant with depth. The constant F equals $1/\pi$ times the total flux, or net amount of energy of all wavelengths crossing a normally oriented *square centimeter* per second. According to Stefan's law, $\pi F = \sigma T_e^4$, where T_e is the effective temperature which may be derived from the solar constant (see Section 2–7).

In a state of radiative equilibrium, the energy variation represented by either side of the equation of transfer must appear elsewhere with an opposite sign. Hence, integrating the right-hand side of Eq. (2–49) over an entire sphere and equating to zero, we obtain

$$2\pi \int_{-1}^{+1} I \, d\mu = 4\pi J.$$

The source function is thus related to the intensity of radiation averaged over all directions. According to the blackbody radiation laws, the temperature in the radiation field is related to J through the expression

$$J = \frac{\sigma}{\pi} \, T^4.$$

Inserting these results into Eq. (6–10), we have

$$T^4(\tau) = \tfrac{1}{2}T_e^4[1 + \tfrac{3}{2}\tau]. \tag{6–11}$$

Thus we must penetrate to $\tau = \tfrac{2}{3}$ before encountering the layer whose temperature equals the effective temperature of the sun.

The quantity τ is related to the linear depth in the atmosphere by Eq. (2–48). Hence, in order to derive the temperature as a function of linear depth, the mass absorption coefficient and the density must be known. The mass absorption coefficient (κ_ν) can be derived in principle if the chemical composition and the state of excitation and ionization of the atmosphere are known. This part of the analysis is difficult. Not only is the theory of the absorption coefficient complex, but transition probabilities and atomic cross sections are required which often are not known. The lack of information is most seriously felt in the theory of the detailed solar spectrum. Furthermore, the state of ionization and excitation depends in part on the kinetic temperature, and this is not the same as the radiation temperature, for the atmosphere is not an ideal blackbody. The question of how the computed values of κ_ν should be weighted when they are integrated with respect to frequency presents an additional problem in determining the value of κ required in τ as used in Eq. (6–11).

Because of these difficulties, only simple models of the solar atmosphere have been investigated. For example, the assumption that κ_ν/κ does not vary with depth permits a derivation of this ratio from observations of the center-to-limb variation of brightness, or *limb-darkening*. The results indicate that the opacity of the solar atmosphere, except at the wavelengths of absorption lines, is principally due to the negative hydrogen ion described below. The models, to be realistic, must obey the gas law [Eq. (2–39)] and the equation of hydrostatic equilibrium [Eq. (2–38)]. The reader will find further details in references 5, 6, and 7. We now summarize some of the results of these theoretical considerations.

The atmosphere of the sun consists almost entirely of hydrogen. The number of hydrogen atoms and the number of atoms of other elements are in the ratio 10^4:1. At the temperature of the solar atmosphere, the heavier atoms are ionized, while most of the hydrogen remains neutral. A free electron thus may attach itself to a hydrogen atom to form a negative ion. The negative hydrogen ion is a very effective absorber of radiant energy in the wavelength interval 4000–20,000 A. It causes the high opacity responsible for the illusion of a photospheric surface. The photosphere is actually a deep layer of semiopaque gases.

In describing the structure of the photosphere, it is more convenient to use optical depth than radial distance. According to Eq. (3–25), a beam of radiation traveling straight out from an optical depth of 1.0 will suffer enough absorption to have its intensity reduced by 1.085 magnitudes. The temperature in the photosphere can be derived readily as a function of optical depth from observations of the limb-darkening of the solar disk. At an optical depth of 0.5, the pressure is about 7×10^4 dynes/cm^2, roughly a tenth of an atmosphere, and the density is about 15×10^{-8} gm/cm^3. This layer is some 300 km below the level where the optical depth may be taken as zero, and about 45 km above the layer where the optical depth is 1.0.

III. Phenomena in the Outer Layers

In order to describe the phenomena observed in the outer layers of the sun, it is first necessary to name and define the various zones into which this region is customarily divided. The lines of demarcation between these zones are indefinite, and the zones may overlap each other. They do not constitute geometric entities but represent regions where certain types of phenomena are observed.

The term "atmosphere" as applied to the sun is reserved for all the outer layers beginning with the photosphere. The atmosphere may be divided into the *inner* and the *outer atmosphere*. The inner atmosphere is the photosphere, and the outer atmosphere consists of the *chromosphere* and the *corona*.

The chromosphere is the region responsible for the emission spectrum that is observed during a solar eclipse, after the light from the photosphere has been blocked by the moon. The H- and K-lines of ionized calcium have been detected in emission up to 14,000 km above the photosphere, but most spectral emission lines disappear after the moon's limb covers the first 3000 km. In the chromosphere the temperature increases from about 6000° in the lower levels to the extremely high temperatures found in the corona. The lower part of the chromosphere is sometimes called the *reversing layer*, and it is here that radiation from the deeper layers undergoes most of the atomic and molecular absorption seen in the spectrum of the solar disk. Since, during an eclipse, the chromosphere is not seen projected against this intense background radiation, it shows the emission spectrum characteristic of a hot rarefied gas.

The corona is visible during solar eclipses and may extend out to a distance of several solar radii. Its spectrum shows the presence of highly ionized gases such as Fe XII and Fe XV, indicating kinetic temperatures of the order of 10^6 °K. Emission lines of such highly ionized atoms have been observed down to about 5000 km above the photosphere.

The shape of the corona varies markedly during the well-known sunspot cycle, being roughly circular in outline during minimum activity and elongated along the sun's equator during maximum activity. The spectrum of the corona indicates that its light originates from more than one source. In the notation of Van de Hulst [8], the component of the spectrum showing the emission lines from highly ionized atoms is called the *E-component*. Another component of the spectrum shows a reflection of the solar spectrum and is identical with that of zodiacal light (see Section 6–3) observed at large distances from the sun. This inner-zodiacal light component is designated by F. There is another component, K, which is similar to F, but in which the spectral lines are greatly broadened or washed out. The light that contributes to this component is polarized. The K-component is due to light scattered by the free electrons in the corona. The F- and K-components together are called the *white corona*. They are much more intense than the E-component, which is only observable within one solar radius of the limb during an eclipse, where it constitutes less than 1% of the total light. Within one solar radius from the limb, the K-component dominates the corona, but outside of this range the F-component is the principal one (see Section 6–3).

The phenomena observed in the solar atmosphere can be grouped in two classes according to whether they occur in the *quiet sun* or in the *active sun*. The quiet sun refers to the permanent solar features, the active sun to transient disturbances. *Granulations* are observed in the photosphere of the quiet sun, while the chromosphere shows *spicules*. Phenomena of the quiet sun also include the limb-darkening which has been discussed earlier, the absorption lines of the quiescent solar atmosphere, the undisturbed magnetic field of the sun, and the permanent solar radio emissions. Examples of transient phenomena observed in the active sun are: *sunspots, faculae, flares, filaments,* and *prominences.*

A. *Phenomena in the quiet sun*

(a) The undisturbed photosphere shows a mottled grainlike structure, called *granulation* (see Plate VII), consisting of bright patches of irregular shape, whose diameters average about 700 km. The brightness of the granules compared with that of the surrounding area corresponds to a temperature difference of 100°K to 200°K. They are constantly being re-formed, the lifetime of a granule being a few minutes. The granulation is produced by turbulence in the photosphere.

(b) *Spicules* are spikelike bright projections that may be seen on the limb of the sun (see Plate VIII). The *coronagraph*, a telescope designed to observe the activity on the solar limb by obscuring the solar disk and minimizing scattered light, permits the observation of spicules outside of eclipses. The spicules occur in the chromosphere and extend to heights of 5000 to 10,000 km above the photosphere. Their lifetime is only a few minutes.

(c) The *absorption lines* in the solar spectrum are numerous, being so crowded in the ultraviolet regions that the continuum is observed only in isolated spectral regions. The intensity of light in the center of some absorption lines may be as much as 20% below that of the continuum. Measurement of this intensity as a function of wavelength across a line yields the so-called *line profile*. The analysis of profiles yields information about the abundance of chemical elements in the sun and the physical conditions which produce them. Center-to-limb variations in the profiles play an important role in the theory of the structure of the solar atmosphere.

(d) The splitting of absorption lines by the Zeeman effect supplies data concerning the *magnetic field* of the sun. H. W. Babcock and H. D. Babcock [9] have developed a *solar magnetograph* which utilizes the Zeeman effect to measure and record the component of the photospheric magnetic field in the line of sight with a precision of 0.3 gauss. Observations made from 1952 to 1954, spanning a period of minimum solar activity, revealed the existence in the sun of a general magnetic field with a mean intensity of 1 gauss. The observed field was bipolar with a polarity opposite to that of the magnetic field of the earth, and its axis agreed with that of the solar rotation; furthermore, its observable effects are limited to solar latitudes higher than $\pm 55°$ and show random fluctuations in intensity. That this general field perhaps varies with the well-known solar cycle is indicated by Babcock's results for 1959 [10], which show that the polarity of the general field was opposite to that determined in previous studies.

B. *Phenomena in the active sun*

(a) The earliest known features of the active sun were the *sunspots* (see Plate VI). In general, a sunspot observed near the center of the sun's disk shows a dark central region called the *umbra* surrounded by a gray *penumbra* with radially oriented filaments (see Plate VII). The diameter of the umbra may be 75,000 km in a very large spot, the penumbra being then roughly twice as large. Spots observed close to the limb of the sun show an asymmetry between the umbra and penumbra known as the *Wilson* effect. In such a spot the side of the penumbra facing the disk's center is thinner than the one facing the limb. The effect indicates that the sunspot is a depression in which the inward-sloping sides form the penumbra. Summaries of the numerous observations of sunspots are presented in references 11 and 12.

The period of rotation of the sun can be determined from observations of the positions of long-lived sunspots on the solar disk. The angular rotation per day is given by

$$\omega = 14°.4 - 2°.8 \sin^2 \phi, \qquad (6\text{--}12)$$

where ϕ is the solar latitude. The period of rotation thus depends on the latitude. The rotation may also be measured spectroscopically by means of the Doppler principle. Within the errors of the measurements, the two methods give similar results.

The number of sunspots and their total area varies with time. The *sunspot number*, or *Wolf number*, at a given time can be estimated by means of the formula

$$N = k(10g + f), \qquad (6\text{--}13)$$

where g is the number of sunspot groups and f is the total number of

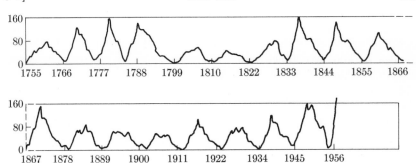

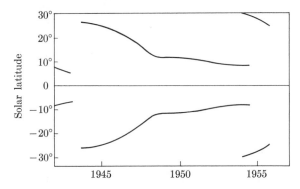

Fig. 6–1. Zürich sunspot numbers.

individual spots. The constant k is chosen for a given observer such that his estimate, which clearly will depend on his equipment, will match that of other observers. Daily observations of Wolf numbers are published by the Zürich Observatory [13], the U.S. Naval Observatory [14], and other agencies. Yearly means of these numbers are plotted in Fig. 6–1.

In addition to the well-known 11-year cycle shown by sunspot numbers, there is a 27-day cycle due to the rotation of the sun. The 27-day variation is made more pronounced by the existence, in the photosphere, of *centers of activity* in which spots tend to recur. The 11-year cycle is somewhat irregular both in amplitude and period. The annual mean number of spots in the years of maximum activity may vary by a factor of 2, while the duration of a single cycle may be 11 ± 1 years. Predictions of sunspot numbers that are based on previous solar activity have failed to agree with subsequent events. The characteristics of a new cycle cannot be predicted from those of past cycles. There is evidence, as shown in Fig. 6–1, that an 80-year cycle is superposed on the 11-year one.

If one plots the average latitude in which spots occur as a function of time, the result shown in Fig. 6–2 is obtained. Clearly, the distribution

Fig. 6–2. Mean latitude of sunspots from 1942 to 1956.

of sunspots in latitude varies as the sunspot cycle progresses, a fact known as *Spörer's law*. Few sunspots are observed either within 5° of the solar equator or above ±30° latitude. The first spots of a new cycle appear near the higher latitude limit when some spots of the previous cycle may still be found near ±5° latitude. At maximum activity the spots are concentrated at about ±15° latitude.

Numerous studies of the physical nature of sunspots have been made. These are summarized in references 6 and 11. Photometric studies indicate that the intensity of light from the umbra is about 25% of that from the photosphere. The umbra has considerable brightness, and its apparent darkness is due to the contrast with the brighter photosphere. The lesser umbral intensity indicates a lower temperature. This is confirmed by spectroscopic analysis. Temperatures of about 4000°K are found from the state of excitation of the atoms as shown by their lines in the sunspot spectrum.

Magnetic fields which are much stronger than the general magnetic field are present in sunspots. The field intensity varies with the size of the spot, the range being from 100 gauss for the smaller spots to 4000 gauss for the larger ones. The lines of force are vertical in the center of the umbra and nearly horizontal at the edge of the sunspot. The field-intensity distribution is quite regular, even though the visible spot may be irregular in shape. Bipolar groups of spots are often formed in which the polarity of the leading spots (in the sense of the solar rotation) is opposite to that of the following ones. The polarities of single spots or of leading spots are opposite in the northern and southern hemispheres. Also, at the beginning of a cycle, the general sense of polarity in a given hemisphere is the reverse of that found during the previous cycle.

Doppler shifts in the spectral lines of the penumbra of spots located near the solar limb indicate that there exists an outward flow of mass from the umbra. This is known as the *Evershed* effect.

De Jager [6] has reviewed the theories developed to explain the sunspots. It is difficult to account for all observational features. Modern theories are based on hydromagnetic considerations.

(b) The areas surrounding the sunspots are lighter than the rest of the photosphere. These light areas are called *faculae;* they are best observed in white light near sunspots approaching the limb. Faculae are strikingly enhanced in photographs made in the light of the ionized calcium K-line or the hydrogen H_α-line (see Plate IX). In these photographs the light areas, sometimes called *plages*, are conspicuous at the center of the disk as well as near the limb. Photographs of the solar image made with light in very narrow wavelength intervals centered in one of the strong absorption lines are called *spectroheliograms*. A spectroheliogram shows how the atoms that produce the spectral line used in the observations are distributed on the solar disk. Excluding the *polar faculae,* which are observed occasionally at about ±70° latitude, faculae are nearly always associated with sunspots.

With the Babcocks' magnetograph, the magnetic fields of faculae may be readily studied. In addition to the general magnetic field of the sun observed at high latitudes, these observations show regions of bipolar magnetic polarity called *BM-regions* [9]. The appearance of a BM-region is usually followed by the appearance of faculae and sunspots, and the BM-region outlasts these. A strong BM-region may last for ten solar rotations.

(c) In the neighborhood of sunspots, sudden flashlike *flares* occur unexpectedly, the total occurrence persisting at most for about two hours (see Plate X). They strongly affect the upper atmosphere and the magnetic field of the earth, and their occurrence has been correlated with intensification of cosmic rays. These solar-terrestrial effects are discussed more extensively in Section 6–2. According to the size of the area which they cover, flares are classified as of *importance 1, 2, or 3.* Finer classifications, i.e., 1^-, 2^+, etc. are also used. A flare of importance class 3 may last for two hours and cover $\frac{1}{1000}$ of the area of the visible hemisphere of the sun. A given sunspot region may show repeated flare activity.

(d) *Filaments* or ribbonlike dark markings are found near sunspot groups in spectroheliograms. They sometimes attain the phenomenal length of 200,000 km. Viewed against the solar limb, filaments are called *prominences* and can be seen to project outward many thousand kilometers from the photosphere. There is a great deal of variety in size, intensity, and mode of development among prominences. Several classification schemes have been presented and a great deal of data have been collected that show their frequency and distribution on the surface of the sun. References 6 and 11 contain summaries of these data.

C. *The nature of solar radiation*

The radiation emitted by the sun is both electromagnetic and corpuscular. The electromagnetic radiation extends in wavelength from x-rays, of wavelength 1 to 150 A, to radio waves several meters long.

In the near-ultraviolet region, the absorption lines are so crowded that the spectral intensity distribution resembles very little that of a blackbody. From about $\lambda = 6000$ A toward the red, the spectrum of the sun is fairly free of absorption lines. Measurements of the energy of the solar continuum as a function of wavelength have been made by many observers [5]. The results can be approximated by two blackbody-radiation curves, one with a temperature of 6000°K suitable for wavelengths shorter than 3700 A, and one with a temperature of 7200°K suitable for wavelengths of visible light. At $\lambda = 3700$ A there is a discontinuity called the *Balmer jump*, which is caused by the continuous absorption at the head of the Balmer series of hydrogen. If I^- is the intensity on the short-wavelength side of λ 3700 A and I^+ that on the long-wavelength side, the Balmer jump is customarily measured by the quantity log (I^+/I^-). In the solar spectrum this quantity is close to 0.13.

At radio wavelengths longer than 0.5 m, the radiation received from the sun originates in the corona. Although the corona is very transparent in the visible spectral region, it is extremely opaque at radio wavelengths due to its high state of ionization. In the quiet sun the blackbody temperature of the corona, indicated by the variation of intensity with wavelength, is of the order of 10^4 to 10^6 °K. It will be recalled that the kinetic temperature indicated by the spectrum of the corona is also of the order of 10^6 °K. The emission in the corona appears to be caused by free-free transitions of electrons in the fields of coronal ions.

The solar radio emission shows complex variations. Sudden intensification or *bursts* lasting from a fraction of a second to several minutes occur. The intensity increase in a burst may be tenfold. The intensities and durations of the bursts vary with the radio frequency. H. Dodson and other investigators [15, 16] have summarized these characteristics for wavelengths in the centimeter, decimeter, and meter ranges. Centimeter radiation is quite uniform with time, whereas decimeter radiation exhibits a variation related to the solar rotation period. Bursts of a few minutes' duration are found. At meter wavelengths very complex variations appear. Disturbances observed at meter wavelengths have been called *noise storms* or *outbursts*. In a noise storm a long series of bursts may be accompanied by an increase in the normal level of radiation found between bursts, and a storm may last from a few hours to several days. The storms appear to originate over large sunspots. Outbursts occur sporadically and may last for a fraction of an hour, with as much as a millionfold increase in intensity. Some outbursts are related to solar flares. After the onset of a flare, observations of outbursts at different frequencies have shown that, as the frequency decreases, a progressive delay occurs in the time of reception of the outbursts. This remarkable effect has been interpreted to be due to the fact that the region which produces the outburst rises from the flare upwards. As time progresses, this rise permits longer and longer wavelengths to filter through the corona. The speed for the upward travel is of the order of 1000 km/sec. Streams of corpuscular radiation shot from the flare region appear to be the cause of this phenomenon.

6–2 Solar-terrestrial relationships. The sun has regulatory effects on the earth by virtue of the night-day cycle. In addition, there are many geophysical phenomena of a transient nature that are related to the active sun. These phenomena, which include geomagnetic storms, auroral displays, ionospheric disturbances, irregularities in the intensity of cosmic rays, intensification of the Van Allen radiation belts, and variations in the airglow, are discussed briefly in this section. All are directly or indirectly related to the earth's magnetic field. The earliest known of these solar-terrestrial effects were variations in the earth's magnetic field.

The general magnetic field of the earth approximates that of a dipole, *the equivalent dipole*, of short length placed near the center of the earth and oriented so that its extended axis pierces the northern hemisphere at latitude 78°.6 and west longitude 70°.1. This axis is called the *geomagnetic axis*, and the points where it cuts the surface are called the *geomagnetic poles*.

The magnetic field that is actually observed appears to be the resultant of this dipole field and a number of other fields. The departures from the dipole field are in part of local origin and are caused by the presence of magnetic material in the earth's crust. Local anomalies may have strengths as great as 3% of the normal field and extend along the surface up to 100 miles. When these local anomalies are smoothed out, there still remain departures from the dipole field. If these deviations are plotted against latitude and longitude, they show a complex pattern which is not correlated with the major geographic entities such as continents and oceans. The residual field originates in part in the earth's interior and in part in the upper atmosphere. Currents of charged particles can account for the departures. In the upper atmosphere such currents have been determined and studied (see below). Attempts to explain the components of the earth's field originating in the interior have not been completely successful.

The characteristics of the magnetic field at a given point are determined by measurement of its *components*, or *elements*, which determine the direction and magnitude of the magnetic field vector **H**. In practice one measures the magnetic intensity in the horizontal plane (H-component) and the one in the vertical direction (V-component). The direction of the field is determined by observing the angular deviation from true north of a compass needle balanced on its center of gravity. This angle, D, is the magnetic *declination* or *deviation*. In addition, the compass needle indicates the *dip*, the angle between the needle's axis and the horizontal, which depends on the relative magnitudes of H and V. At the geomagnetic poles the value of V is 0.63 gauss. At the geomagnetic equator the value of H is 0.31 gauss. One of the objectives of many polar explorations has been the determination of the exact location of the *magnetic poles*. These are defined as the points where the dip is 90° or the horizontal component is zero. The magnetic and geomagnetic poles do not agree in geographic position. The north magnetic pole drifts northwestward about 4 miles per year; in 1950 it was located near latitude 73°N, longitude 100°W. The position of the south magnetic pole is not diametrically opposite to that of the north magnetic pole; in 1950 it was located at 68°S, 144°E.

A technique developed by Gauss* in the first half of the nineteenth century permits the measurement of **H** in absolute units. Gauss's technique led to the establishment of magnetic observatories, some of which have been in continuous operation for well over a hundred years. With the advent of photography in the second half of the last century, continuous recording of the field elements became possible. These records† have revealed a number of variations in the magnetic field.

Secular variations in field direction and intensity have been detected. For example, from 1840 to 1940, the strength of the equivalent dipole decreased slowly. In 1940 the rate of decrease became smaller, whereas at present the field strength appears to be increasing.

Diurnal variations related to the atmospheric tides produced by the sun and the moon have also been detected. A diffuse ionospheric current, or wind, of charged particles which circles the earth is affected by these tides, causing the diurnal variations. Such an ionospheric wind is, in effect, an electric current which contributes to the earth's magnetic field. Observations [17] made in 1949 at a chain of magnetic observatories along the western coast of South America showed that the amplitude of the daily variation of the horizontal component decreases markedly with distance from the equator. These data indicate the presence of a concentrated electric current in the upper atmosphere which is called the *electrojet* and which is superposed on the widespread currents associated with the daily variation.

Besides the above variations, sudden violent disturbances, called *magnetic storms*, are also observed. These cover the entire earth and begin almost simultaneously, within one or two minutes, regardless of location. During a magnetic storm, currents induced in cables cause interruption of telegraphic communications. Accompanying ionospheric disturbances create anomalous reception of radio waves. Chapman [18] has analyzed the variations in the earth's field during 40 magnetic storms recorded simultaneously at magnetic observatories whose average geomagnetic latitude was 37°. Although the variations appear to be quite·irregular during any one storm, the average variations of the *H*- and *V*-components for the 40 storms show a definite trend when plotted against time. Chapman's results for the *H*-component, which exhibits the greatest departure

---

* The physical principles are presented in various textbooks. See, for example, L. Page, *Introduction to Theoretical Physics*, New York: Van Nostrand, 1932, p. 362.

† The publication *Results of the Magnetic Observations of the Royal Greenwich Observatory*, issued yearly by Her Majesty's Stationery Office, London, is an example of the collections of data made by magnetic observatories.

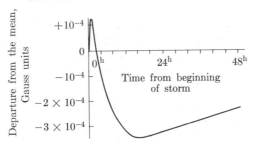

FIG. 6–3. Intensity variations of horizontal field from mean, in gauss. (After Chapman)

from normal, are shown in Fig. 6–3. The average variation of the V-component during a storm is opposite in sign to that of the H-component, and its amplitude is less than half the variation of the H-component. The averaged deviations in declination show an irregular pattern when plotted against time.

An intimate correlation, known since 1741, exists between magnetic storms and auroral displays. Auroras appear most frequently in two regions, called *auroral zones*, which are 23° from the geomagnetic poles. A classification of the various types of auroras is given in reference 19. The heights of auroras above sea level have been measured by triangulation. Maximum occurrence is found at 105 km, but auroras at heights up to 1100 km have been reported.

Auroras are caused by the bombardment of the earth's atmosphere by electrically charged particles. The magnetic field of the earth deflects these particles and guides them to the auroral zones. Many difficult problems are involved in the explanation of the auroral characteristics, but the basic mechanism is well understood. From observations of the aurora, conclusions can be drawn about the structure of the upper atmosphere. Their occurrence at heights of 1000 km indicates that the atmospheric density is not negligible at that elevation.

The spectrum of the aurora and its analysis are described in references 8 and 20. In 1950, Meinel [21] observed the H_α emission line of hydrogen in an aurora by means of a spectrograph directed along the magnetic lines of force. A marked Doppler shift was observed in this direction, while at 90° from it no displacement was found. The shift indicated that a stream of protons had entered the upper atmosphere with speeds ≥ 330 km/sec. This observation confirmed the relationship of the aurora with extra-terrestrial corpuscular streams. The theory of the interaction of charged particles with the earth's magnetic field is summarized in reference 22. That these streams originate in the sun is indicated by the correlation between solar activity on the one hand and auroras and magnetic storms on the other.

In 1943, H. W. Newton [23] found that magnetic storms tend to occur about one day after the observation of an intense flare in the inner seven tenths of the sun's disk. Since then, numerous other studies have failed to isolate any one feature of solar activity that can explain all features of magnetic storms. Although there is a definite tendency for an intense flare to be followed by magnetic storms, there are frequent exceptions. Attempts are being made to classify the storms and to correlate these classes with various phenomena observable in the active sun.

The theory of the production of corpuscular streams in the active solar areas is imperfect. However, it has been observed [24] that the streams leave these areas.

Another type of magnetic field disturbance is the *crochet* [25], which consists of small displacements in amplitude of the *H*-, *V*-, and *D*-elements. The onset of crochets is sudden, their total duration is 30 min, and they are confined to the sunlit side of the earth. They have been observed to occur synchronously with solar flares of importance classes 3 and 3^+. The simultaneity of flares and crochets indicates that the latter are caused by electromagnetic radiation from the flare. The disturbance can be explained by the increase in the level of ionization in the upper atmosphere caused by ultraviolet and x-ray radiations from the flare.

In the 5 to 20 mc/sec range, channels of radio communication located in the sunlit hemisphere are also disturbed by flare activity. These waves are usually reflected from the ionospheric *F*-layer (see Section 3–2). An increase in ionization of the *D*-layer by ionizing radiation from the flare causes absorption in the above frequency range and results in a fade-out. On the other hand, long waves of frequencies about 30 kc/sec, which are usually reflected from the *D*-layer, are enhanced on such occasions. These effects provide a very sensitive method for detecting the arrival on earth of solar ionizing radiation.

Low-energy cosmic rays also show a correlation with solar activity [26]. Their intensity has been known to increase after the onset of an intense flare. These solar cosmic rays must leave the sun at speeds close to that of light. The mechanism of acceleration is not fully understood.

Charged particles may be concentrated by the magnetic field of the earth to form the Van Allen belts [27], which are regions of high-intensity corpuscular radiation. Two separate radiation belts have been found whose cross sections are illustrated in Fig. 6–4. The structure of the belts may be more complex than that shown in the figure. Their origin is explained by the "trapping theory" outlined below.

A charge *e* injected with velocity **v** in a uniform magnetic field of intensity *H* will suffer an acceleration

$$\mathbf{a} = \frac{e}{m}\, \mathbf{v} \times \mathbf{H}, \qquad (6\text{–}14)$$

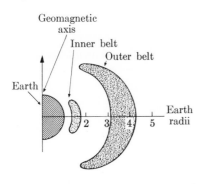

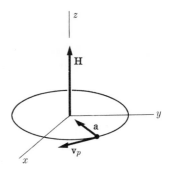

Fig. 6–4. Regions of high intensity in the Van Allen belts. (After Van Allen)

Fig. 6–5. Motion of a positive particle about a line of force.

where **H** is the magnetic field vector of magnitude H, directed along the lines of force. Vector **a** is perpendicular to both **v** and **H** and is directed along the line of advance of a right-handed screw turning from **v** to **H**. If **v** is broken into two components, one along **H** and the other perpendicular to it, the first component suffers no change, and hence the particle advances at a uniform speed in the direction of the lines of force. The perpendicular component v_p, however, produces a constant acceleration with magnitude

$$a = \frac{e}{m} v_p H.$$

Since this is perpendicular to both v_p and **H**, circular motion must result, as shown in Fig. 6–5. We know that in uniform circular motion with speed v_p the centripetal acceleration is v_p^2/r, where r is the radius of the circle. Therefore we have

$$\frac{v_p^2}{r} = \frac{e}{m} v_p H,$$

from which the radius of the circle and the angular velocity ω can be derived. These are

$$r = \frac{mv_p}{eH} \tag{6–15}$$

and

$$\omega = \frac{eH}{m}. \tag{6–16}$$

The resultant motion of the charged particle can be found by combining its constant speed along the line of force with this circular motion. The path is helical, like that followed by the tips of a propeller advancing at a constant speed.

The sense of rotation of the particle about the line of force depends on its charge. In the example considered here we have followed the motion of a *positive* particle. A *negative* particle with an initial velocity component v_p directed in the sense shown in Fig. 6–5 will suffer an outward acceleration. In this case, as the particle approaches a line of force, it is deflected away from it. Eventually the particle will approach a line of force with a v_p-component appropriate to its charge for the production of the helical motion. For example, for a negative particle, the component v_p in Fig. 6–5 would have to be directed in an opposite sense.

The above results are obtained for a uniform field. In the dipole field of the earth, as the charged particle advances toward a geomagnetic pole, the field strength increases. This has the effect of retarding the forward progress of the particles. In a converging field, $\mathbf{H}$ has a component directed inward, in the xy-plane of Fig. 6–5. This component will cause a deceleration of the forward motion of the particle. A *mirror point* is eventually reached where the helical path has flattened completely, and then the particle winds back in a similar path toward a mirror point near the other pole. The particle will oscillate between the two hemispheres with periods of the order of a fraction of a second to seconds, depending on its speed. The theory of the trapping process was worked out by Störmer in a series of investigations starting in 1907, and later by Alfvén. This theory, developed to explain auroral and magnetic disturbances, and the experiments made to test its validity are summarized by Mitra [22].

According to the theory, the particles also drift in geomagnetic longitude. Negative particles drift eastward, while positive particles drift westward. This effect is caused by the increase in field intensity toward the earth, encountered by the particle as it winds about a given line of force. As mentioned previously, the actual field of the earth is not strictly that of a dipole, as assumed in Störmer's theory. This fact, together with the magnetic field variations, will result in very complicated paths for trapped particles.

Eventually, a trapped charged particle will suffer a collision with an air particle. Because the densest air is encountered by the particle at the mirror points, most collisions will occur there; the visible effects of the collisions are the auroras. The Van Allen belts then act as temporary reservoirs in which charged particles are gradually lost in the vicinity of the auroral zones.

Why there should be two radiation belts instead of one is a question of great interest. Intensity measurements [28] made on December 6, 1958, during an especially quiet period of solar activity, and on March 3, 1959, during disturbed solar conditions, showed on the second date a strong enhancement of the intensity of the outer belt but no change in the inner belt. The energy spectrum of the particles trapped in the two zones has been measured by observing the ability of the particles to penetrate shields of different thickness. Van Allen's results [28] indicate that the outer belt is composed exclusively of electrons with relatively low energies (upper limit 100 kev), whereas the inner belt is composed of higher-energy electrons and protons. Van Allen's findings, obtained in 1959, for the intensity values in the hearts of the inner and outer zones located at 3600 km and 16,000 km, respectively, above the geomagnetic equator are as follows. In the inner belt, the maximum number of electrons with energies greater than 20 kev arriving from any one direction is $2 \times 10^9 \mathrm{cm}^{-2} \cdot \mathrm{sec}^{-1}$/steradian. The corresponding rate for electrons with energies greater than 600 kev is 1×10^7. The rate at which protons with energies greater than 40 Mev arrive from all directions is $2 \times 10^4 \mathrm{cm}^{-2} \cdot \mathrm{sec}^{-1}$. In the outer zone, the intensity of electrons arriving from all directions with energies greater than 20 kev is $1 \times 10^{11} \mathrm{cm}^{-2} \cdot \mathrm{sec}^{-1}$. The corresponding rate for electrons with energies greater than 200 kev is equal to or less than $1 \times 10^8 \mathrm{cm}^{-2} \cdot \mathrm{sec}^{-1}$. The upper limit for the rate at which protons with energies greater than 60 Mev arrive from all directions is $100 \ \mathrm{cm}^{-2} \cdot \mathrm{sec}^{-1}$. The electrons in the outer belt are of solar origin. Several authors [28] have suggested that the protons and high-energy electrons in the inner belt result from neutron decay. The neutrons originate in the bombardment of the upper atmosphere by cosmic rays.

On a clear, moonless night, radiation from the sky can be detected from the ground. The radiation is composed of zodiacal light (see Section 6–3), scattered starlight, and light originating in the atmosphere at a height of about 100 km. The latter component is called the *night airglow*. The strongest feature in its spectrum at optical wavelengths is an emission line at $\lambda = 5577$ A which has been identified as a forbidden line of atomic oxygen. The sodium D-line and infrared molecular bands characteristic of hydroxyl (OH) have also been observed.

The night airglow, having at any given time an irregular intensity distribution, is not uniformly bright over the sky. These irregularities permit an estimation of the height of the emitting level by triangulation since they are observed in different sky positions by observers separated by sufficient distances.

The night airglow is widely distributed in latitude, being observable at the geomagnetic equator, and shows no longitude effect. Its geographic distribution and its spectrum indicate that it is not of auroral origin. The energy of the airglow is believed to come from solar radiation absorbed during the day and released at night. During the evening twilight hours, the airglow increases in brightness, resembling at times a faint aurora. Airglow exists also during the daylight hours, but it is difficult to detect.

Striking variations of the night airglow which are closely related to magnetic field disturbances have been observed. The mechanism of the airglow and of these variations is not well understood. The reader will find a more detailed discussion of this phenomenon in reference 22.

6–3 Interplanetary dust and gas. The observations discussed in the last section indicate that sporadic streams of electrically charged particles of solar origin flow from the sun to the earth. In this section we shall consider the more homogeneously distributed matter known or supposed to exist in the space between the planets.

Information about interplanetary matter is derived directly from observations of the sunlight that it scatters, as in the case of *zodiacal dust*. Its existence and properties are also inferred from theory and supported by indirect observation. Our present knowledge of the amount of dust and gas in interplanetary space is still meager.

On a clear, moonless night just after evening twilight or just before dawn, one may observe a pale light oriented along the ecliptic and comparable to the Milky Way in intensity. This is known as *zodiacal light* (see Section 6–1). It is best seen at low latitudes, where the ecliptic is more nearly perpendicular to the horizon. The zodiacal light is more intense and more extensive in the neighborhood of the sun. Under the best observational conditions, the light is observed to circle the sky throughout the night. Observations of the brightness and of the spectrum of zodiacal light have been summarized by Mitra [22] and by Elsässer [29].

The spectrum shows emission features which are identical to those of the airglow, but it also shows a continuum with a strong similarity to the solar spectrum. The airglow component of the spectrum does not appear to be particularly enhanced in the zodiacal light [30]. We may conclude that zodiacal light is scattered sunlight on which the normal airglow in the earth's atmosphere is superimposed.

Observations of zodiacal light close to the sun are difficult because of the foreground illumination of the sky. Blackwell [31] has obtained measurements of brightness in absolute units from a high-flying aircraft during a solar eclipse. As discussed in Section 6–1, the inner zodiacal light blends with the solar corona, forming its F-component, which provides most of the intensity at angular distances of one solar radius or more from the sun. Along the ecliptic, Blackwell's observations of the intensity at an effective wavelength of $\lambda = 6300$ A can be represented by

$$I = 3.9 \times 10^{-8} r^{-2.2}, \tag{6–17}$$

where r is the angular distance from the sun in solar radii and the unit of I is the mean intensity of the solar disk. Equation (6–17) also fits closely the night-time measurements of the zodiacal-light intensity at elongations from 30° to 100° [30]. These findings support the identification of the F-corona with the inner zodiacal light.

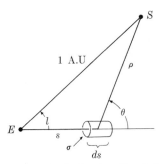

Fig. 6–6. Analysis of zodiacal light.

Behr and Siedentopf [32] have plotted isophotes of the zodiacal light based on photoelectric observations. These authors identified a polarized component of the light and observed its intensity distribution in the sky. About 15% of the light is polarized, with the plane of polarization passing through the sun. The isophotes exhibit a marked symmetry with respect to the ecliptic for polarized light, whereas the light observed without polarizing filters shows an asymmetry. This asymmetry has been confirmed more recently by other research [33]. The observations indicate that the material which produces the zodiacal light has two constituents: (a) free electrons which produce Thomson scattering of sunlight and are responsible for the polarized component, and (b) dust particles with radii $> 10^{-4}$ cm.

Measurements of the intensity of zodiacal light can be used to derive information about the space density of the scattering particles. Let I be the intensity of solar radiation in ergs $\cdot$ sec$^{-1} \cdot$ cm^{-2} at a distance of 1 A.U. from the sun. Consider an elementary (see Fig. 6–6) cylindrically shaped volume of length ds and cross section σ located at distance ρ from the sun (S) and distance s from the earth (E). The radiant energy received per second by the particles in the volume $\sigma \, ds$, where the space density is N particles/cm^3, is

$$\frac{I}{\rho^2} \, N \pi r^2 \sigma \, ds.$$

Here πr^2 is the mean cross-sectional area of the scattering particles, and ρ is expressed in A.U. Let $f(\theta)/s^2$ be the fraction of this energy scattered toward 1 cm^2 facing the scattering particles and located on the earth's surface; here $f(\theta)$ is the *phase function*. The radiation flux received at E from the elementary volume is

$$\frac{I}{\rho^2} \, N \pi r^2 \, \frac{\sigma}{s^2} \, f(\theta) \, ds.$$

Now, σ/s^2 is the solid angle in steradians of σ as seen from E. Dividing the above expression by σ/s^2, we obtain the contribution of the material in the elementary volume to the surface brightness of the zodiacal light, that is, the radiation flux per steradian. The surface brightness of the zodiacal light that is actually observed at elongation angle l is obtained by integration, i.e.,

$$B(l) = I \int_0^\infty \frac{1}{\rho^2} N \pi r^2 f(\theta) \, ds.$$

It is convenient to change the integration variable to θ with the help of

$$\rho = \frac{\sin l}{\sin \theta}, \qquad ds = \frac{\sin l}{\sin^2 \theta} \, d\theta.$$

Thus we find

$$B(l) = \frac{I}{\sin l} \int_l^\pi N \pi r^2 f(\theta) \, d\theta \qquad (6\text{--}18)$$

as the integral equation for N.

The number of scattering particles/cm^3 can be derived as a function of position in space from photometric observations if $\pi r^2 f(\theta)$ is known. For the polarized component of light scattered by free electrons, the Thomson scattering formula yields

$$\pi r^2 f(\theta) = \left(\frac{e^2}{mc^2} \right)^2 \frac{\sin^2 \theta}{2}, \qquad (6\text{--}19)$$

where e, m, c are respectively the charge and mass of the electron and the speed of light. Substituting in Eq. (6–18), one obtains the relationship for computing the space density of electrons, N_e, from the observed brightness due to the free electrons, $B_e(l)$.

Behr and Siedentopf [32] derive from their photometric observations the following results: On the ecliptic at 1 A.U. from the sun, the electron density is 600/cm^3. Toward the sun, at 0.6 A.U., the electron density increases to 1000/cm^3. Away from the earth's orbit, the electron density falls to 120/cm^3 at 1.3 A.U. The data of Behr and Siedentopf have also been analyzed by Elsässer [34], who has made estimates of dust and free-electron densities in directions perpendicular to the plane of the ecliptic. North of the ecliptic plane, the dust density decreases at 0.24 A.U. to half the value found in the ecliptic. South of the ecliptic plane, the corresponding distance is 0.13 A.U. At a distance of 0.17 A.U. north or south of the ecliptic the free electron density decreases to half the value found in the ecliptic plane.

The analysis of the distribution of dust particles in space by means of Eq. (6–18) is more difficult than that of the distribution of free electrons.

Not only is the theory of the phase function very complex, but *a priori* we have only a very rough idea of what we should use for πr^2 in the equation. These problems have been studied by Van de Hulst [35] and Allen [36].

The various models proposed for the scattering medium which gives rise to the unpolarized component are reviewed by Van de Hulst. He finds that scattering caused by solid particles is the only plausible explanation, as had been proposed earlier by Grotrian. Simple reflection of light by the particles, however, cannot explain the observed intensity of the solar corona, a conclusion arrived at independently by Allen and by Van de Hulst. This mode of scattering requires the presence of solid particles close to the sun if the observations are to be satisfied. It is unlikely, however, that the particles can exist in the immediate solar neighborhood without being vaporized. Allen and Van de Hulst find that diffraction of the light by the particles is a way out of the difficulty. As θ approaches l in Fig. 6–6, the intensity of the diffracted light increases, and one does not have to assume large numbers of solid particles close to the sun to explain the intensity increase of the F-corona. In other words, light diffracted by particles located at a distance from the sun will satisfy the observations.

In a study of the spectral intensity distribution of the F-corona, Allen [36] finds a close similarity to the spectral distribution of sunlight. According to diffraction theory, particles with a radius of 10^{-4} cm produce a bluing effect on the scattered light. Particles with $r \geq 10^{-2}$ cm produce a reddening effect, while particles with $r \approx 10^{-3}$ cm cause no change in the intensity distribution in the wavelength range studied by Allen.

As a model, Behr and Siedentopf [32] used spherical dust particles with $r = 10^{-3}$ cm and density of 3 gm·cm^{-3}. They found a space density of 10^{-23} gm·cm^{-3} for the dust particles in the vicinity of the earth's orbit. This density is constant in the ecliptic plane at distances beyond the orbit of Venus and decreases as one moves from Venus toward the sun. On the other hand, if we assume that for each electron in space there is a proton, the mass/cm^3 is of the order of 10^{-21} gm in the vicinity of the earth's orbit.

During the most favorable viewing conditions, one may observe on the ecliptic a faint patch of light located at an angular distance from the sun of about 180°. This light, called the *gegenschein* or *counterglow*, is another manifestation of the zodiacal dust. As mentioned in Section 5–5, the gegenschein may represent an actual concentration of particles near a Lagrangian point. Reported variations in the angular separation of the gegenschein from the sun are apparently related to Jupiter's position [37].

The sources of the dust component of zodiacal light have been studied by Piotrowski [38], Fesenkov [39], and Whipple [40]. The results of

Piotrowski and Fesenkov indicate that collisional fragmentation in the asteroid belt (see Section 6–6) may be an adequate source. Whipple, on the other hand, finds that material ejected from comets may be an important contributor. In these studies, the Poynting-Robertson effect plays an important role.

In 1903, Poynting reasoned that a particle orbiting about the sun and reflecting, or absorbing and reradiating, solar radiation will gradually lose its orbital angular momentum and spiral inward toward the sun so long as the gravitational solar attraction is greater than the solar radiation pressure. The theory of the effect was studied by Robertson [41], who took relativistic effects into account. The time of the fall, according to Robertson, for a spherical particle of radius r and density Δ placed a distance R from the sun, where R is large compared with the solar radius, is

$$t = 7.0 \times 10^6 r R^2 \, \Delta \text{ years.} \tag{6-20}$$

Here R is in A.U., r in cm, Δ in gm·cm^{-3}. A particle with a radius of 10^{-3} cm and a density of 3 gm·cm^{-3}, such as that considered by Behr and Siedentopf, will fall into the sun from $R = 1$ A.U. in 2.1×10^4 years, an interval of time that is very short compared with the age of the solar system.

In order to estimate the importance of radiation pressure relative to gravitational attraction, we assume the particle to be a spherical blackbody. Let the particle's cross section to the radiation be πr^2. Since the solar constant is 1.35×10^6 ergs·cm^{-2}·sec^{-1}, the amount of radiant energy falling per second on the particle is $(1.35 \times 10^6 \pi r^2)/\rho^2$, where ρ is its distance from the sun in astronomical units. This radiation carries with it the momentum $(1.35 \times 10^6 \pi r^2)/\rho^2 c$. The gravitational attraction between the particle and the sun, on the other hand, is $GM_\odot m/\rho^2$. Substituting $G = 6.67 \times 10^{-8}$, $M_\odot = 1.99 \times 10^{33}$ gm, and 1 A.U. $= 1.49 \times 10^{13}$ cm, we obtain $0.6 m/\rho^2$ for the rate of change of the radial momentum due to solar attraction. Equating this rate to that produced by radiation, we have

$$0.6 m = \frac{1.35 \times 10^6 \pi r^2}{c}$$

as the required relationship between m and r if radiation and gravitation are to balance. Now $m = \frac{4}{3}\pi r^3 \, \Delta$ and $c = 3 \times 10^{10}$ cm/sec; hence in cgs units, assuming the geometric cross section to equal the cross section to radiation, we obtain

$$r \, \Delta \approx 5.6 \times 10^{-5}. \tag{6-21}$$

If $r \, \Delta$ is larger than this limit, gravitational attraction will predominate in interplanetary space; otherwise radiation pressure will be dominant. Using $\Delta = 3$ gm·cm^{-3}, which is appropriate for a silicate particle, we find that particles with $r \approx 10^{-3}$ cm will be subject to the Poynting-Robertson effect.

From this analysis it follows that the particles of zodiacal light are being constantly replenished. As mentioned previously, comet disintegration and fragmentation of meteorites in the asteroid belt are possible mechanisms for the replenishment.

Ionized material in comet tails (see Section 6–4) has been observed to leave the nucleus and move radially away from the sun with accelerations that are higher than those expected from the pressure of radiation. Biermann [42] finds that the accelerations can be explained by the impulse mechanism of solar corpuscular radiation. The impulse is found by Biermann to be due to free electrons. The observed effects in cometary tails indicate electron space densities of the same order of magnitude as those found from the work of Behr and Siedentopf [32].

The spatial environment of the earth has also been explored by the study of *whistlers*, or sounds of decreasing pitch, that may be heard in a radio receiver tuned to low frequencies. Radio noise from a lightning stroke is propagated through the ionosphere with a speed that depends on the frequency. The timing of lightning strokes and the beginning of whistlers show that the signal has traveled at least 25,000 km. According to the theory of whistlers, the signal from a lightning stroke is guided by the lines of force of the earth's magnetic field, while the component frequencies are dispersed because of their different speeds. As a result, a lightning stroke in the southern hemisphere may be heard in the northern hemisphere. The higher frequencies arrive first and are followed by lower and lower frequencies. Repeated reflections from the ground may give rise to a family of whistlers following one lightning stroke. Whistler studies indicate that at 11,000 km above the equator there must be electron densities of about 400/cm^3. For further details on this subject see reference 43.

Parker [44] has summarized several observations from which he concludes that the ionized component of the solar corona extends outward beyond the earth. According to Parker's computations, the material in the corona is expanding outward with a speed in excess of 500 km/sec, forming a *solar wind*. The solar wind explains Bierman's findings of the outward acceleration of material in comet tails. The material in the corona is a *plasma* or ionized gas, and hence it carries with it the lines of force of the general magnetic field of the sun. The field strength in the vicinity of the earth is about 2×10^{-5} gauss.

Any cloud of plasma originating in a region of solar activity will carry outward the lines of force from the field in which it originates. Gold [45] has studied the consequences of this effect. According to Gold, charged particles can be trapped in the drawn-out field. They behave like particles trapped in a Van Allen belt, but in this case the field has its source in the region of solar activity.

Some of the constituents of interplanetary material, such as charged particles and dust, produce the various effects described so far from which we have gained some information about their space distribution. However, nonionized gas in interplanetary space is so tenuous that it has proved difficult to detect.

Kupperian, Byram, Chubb, and Friedman [46] have found that the night sky observed from high altitudes is glowing with hydrogen Lyman-α radiation ($\lambda = 1216$ A). Observations obtained from rockets showed that at an elevation of 75 km the entire hemisphere above the rocket was bright with this radiation. At higher altitudes, the Lyman-α radiation was also detected in the atmosphere below the rocket. The phenomena were interpreted to be due in part to scattering of solar Lyman-α radiation in interplanetary space and in part to a further scattering of this light in the earth's atmosphere.

The average flux of the Lyman-α radiation from space was found to be 3.2 erg$\cdot$cm^{-2}/steradian. Neutral hydrogen in space can absorb and re-radiate solar Lyman-α radiation effectively, since it is resonant to the wavelength of this radiation. An analysis of the observations of Kupperian *et al* has been made by Brandt and Chamberlain [47]. They find that the interplanetary neutral hydrogen in the vicinity of the earth has a space density of at least 0.2 atoms/cm^3.

6–4 Meteors. According to Webster's dictionary, the term "meteor" is used to describe the observable effect on the atmosphere produced when an extraterrestrial object or *meteoroid* encounters the earth. However, it has become customary to call the object itself a meteor. A *fireball* is an exceptionally bright meteor capable of casting strong shadows. A *bolide* is a fireball that is accompanied by an explosion. If the object falls to the earth's surface, it is referred to as a *meteorite*.

Most meteors waste away during their flight through the atmosphere. If a meteoroid is sufficiently small, by virtue of its high ratio of surface area to mass it may radiate heat at a sufficiently high rate to stay relatively cool as it traverses the atmosphere. These bodies are called *micrometeorites*. The space density of meteoritic material, the space velocities of meteors, and their nature are topics of special interest in space research.

Meteors have been studied by visual and photographic means as well as by radar. Some data also have been collected from artificial satellites. Visual observations yield rates of fall. Photographic observations which have been made with specially designed wide-angle cameras yield rates of fall as well as velocities and heights above the surface of the earth. To obtain this information, simultaneous observations are made from two locations so that the distance to the meteor can be derived by triangulation. The velocities are determined by interrupting the exposure at a predetermined rate by a sector wheel placed in front of the photographic plate. The photographed trail of the meteor then is a series of dashes whose spacing enables one to derive the velocity, provided two simultaneous photographs are obtained from different locations. The photographic method has been developed and applied by Whipple and his collaborators [48]. Observations for about 500 meteors have been made and reduced.

In the radar method of observation, pulses from the transmitter are reflected by the ionized air left in the path of a meteor. Most of the reflected energy comes from the region immediately behind the meteoroid where the ionization is greatest. Several methods of observation have been devised employing one, two, or three separate radar stations. A description of these methods is beyond the scope of this book. An excellent source of information is the book *Meteor Astronomy* by Lovell [49]. A more recent summary of the methods has been presented by Whipple and Hawkins [48]. Under most weather conditions, the radar technique may be employed during the day as well as at night. Fainter meteors can be detected by radar than is possible if photographic techniques are used.

Many meteors are known to travel through space in swarms or *streams*. When the earth crosses such a stream, the observed tracks appear to diverge from a small area in the sky whose center is known as the *radiant*. The high number of meteors observed on these occasions form a *meteor shower*. These streams have been known for a long time to be associated with comets, and appear to be scattered irregularly along the respective cometary orbits. The orbital motion of the earth causes it to cross these meteor streams approximately on the same date from year to year. Fifty such streams and their orbital elements have been listed by Whipple and Hawkins [48]. Many more streams exist in which the meteors are increasingly diffused in space, and it is hard to decide whether a meteor is a stream meteor or a *sporadic* one.

Sporadic meteors are those not associated with showers. They appear to have originated in the asteroid belt. Most ordinary meteors are of cometary origin, whereas fireballs and meteorites are sporadic.

Meteors have also been considered to originate in interstellar space. Such meteors would approach the earth in hyperbolic orbits, and there would be no upper limit to their velocity. For a meteor originating in the solar system, the maximum velocity with which it will arrive in the earth's atmosphere can be obtained by assuming that its velocity, as it falls toward the earth, equals the sum of the orbital velocity of the earth (30 km/sec)

and the velocity of escape from the gravitational field of the sun at the
earth's distance (42 km/sec). This velocity is somewhat increased by the
earth's attraction. By Eq. (4–76) the initial speed of the meteor v_1 and
the observed speed v_2 are related by

$$v_2^2 = v_1^2 + 2Rg,$$

where R is the radius of the earth and g is the gravitational acceleration
on the earth. When $v_1 = 72$ km/sec, v_2 is about 73 km/sec.

Several thousand meteor velocities have been determined accurately
by the radar method, and only a small fraction appears to exceed
73 km/sec [50]. In those cases, the excess above this speed is a few km/sec.
Of the 500 or so meteor orbits determined by Whipple and his collaborators [48], none has been found to be hyperbolic. The conclusion is that
if there are any meteors which originate in interstellar space, they are
exceedingly rare. The lowest speed with which a meteor may arrive in
the earth's atmosphere is 11.2 km/sec.

The number of meteors observable with the naked eye in a 24-hour
period is about 100 million. More meteors are observed after midnight
than earlier in the night. This is explained by the position of the observer
relative to the direction of orbital motion of the earth. In the morning
hours the observer looks along the direction of the earth's motion, and the
meteors overtaken by the earth increase the observable number.

The number of meteors, dn, of apparent visual magnitude, m, entering
the earth's atmosphere per day has been determined by several investigators. Lovell [49] has summarized and intercompared the results. If dm
is an interval of given magnitude, the observations yield

$$dn = A^m \, dm. \tag{6–22}$$

The value of A is between 2.0 and 2.7, and the observations, which have
been made on sporadic meteors, cover the magnitude range -4.0 to $+9.0$.

The mass distribution of meteors can be determined from the above
results if the relationship between mass and m is known. The brightness
of a meteor, however, is a function of its speed, its mass, its density, and
the density of the air. The function also involves a number of parameters
that are not well known [48, 51]. Fortunately, the mass can also be determined from the drag effects of the atmosphere. If the deceleration is
observed, it is possible in principle to derive the mass, but the equations
still contain some parameters that are only approximately known [48]. A
rough estimate derived by Lovell [49] from the data for sporadic meteors
yields a space density of meteoritic material in the environment of the
earth of 10^{-24} to 10^{-25} gm/cm^3.

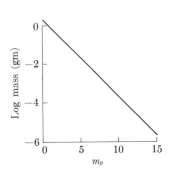

FIG. 6–7. Relationship between brightness (in visual magnitudes) and log mass, in grams. (After Whipple)

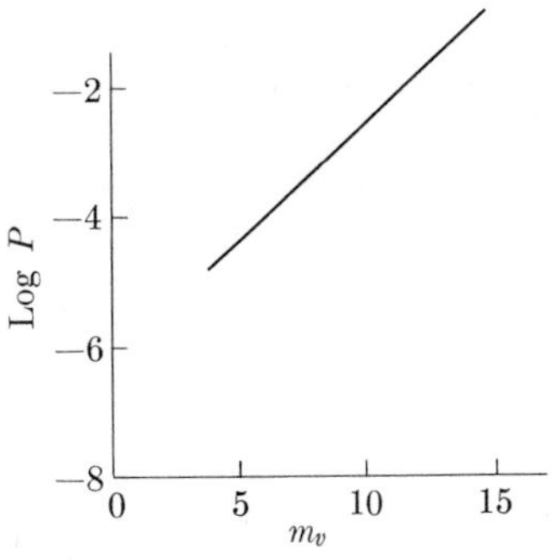

FIG. 6–8. Probability with which meteors of a given brightness class (in visual magnitudes) in the environment of the earth would strike a sphere 3 m in diameter in 24 hr. (After Whipple)

For meteors associated with showers, Lovell arrives at a density inside the stream which is about ten times higher than the above result. For the total mass distributed along the orbit of some typical streams, Lovell obtains 10^{12} kgm.

The frequency of meteoroids of different masses has been summarized by Whipple [52, 53]. His estimates are based on observations of the frequency distribution in brightness and the available theoretical relationships between mass, brightness, and deceleration rates. Whipple's results are summarized in Figs. 6–7 and 6–8, where we have plotted the visual brightness of the meteor against the logarithm of the mass (Fig. 6–7) and against the logarithm of the number of meteors per day that would strike a sphere 3 m in diameter located 1 A.U. from the sun on the ecliptic plane (Fig. 6–8). Masses larger than 1 gm are exceedingly rare. Near the earth's orbit (according to Brown [54]) the density of meteoric bodies with masses of 1 gm or more is about 10^{-12} particles/km^3. Collisions that occur more frequently than once a year will involve only meteoroids of masses less than 10^{-4} gm. Computations beyond the ninth magnitude are based on extrapolations of meteor-brightness counts. For faint meteors, observational data are badly needed.

Direct sampling of interplanetary meteoritic material is possible by means of artificial satellites and space probes. Microphones, wire grids, and crystalline transducers have been used as sensing devices. If microphones are used, the impact of a meteorite is detected by its audible effect. In the wire-grid method, the breaking of one or more tightly wound wires is detected by the change in resistance of the grid. With crystalline sensors, the piezoelectricity due to impact is recorded. The direct calibration of these devices is a difficult problem, since laboratory production of dust particles moving at sufficiently high speeds has not been possible. The results obtained so far by this method are not reliable.

According to their chemical composition, meteorites are classed as *irons* or *stones*. *Irons* are composed almost entirely of pure iron, with some nickel, the percentage of nickel being usually about 7%, but occasionally over 50%; 1% of the total mass is composed of other elements. When a flat surface of an iron meteorite is polished and etched with acid, a characteristic crystalline structure known as *Widmanstatten's* pattern stands out. This pattern is not found in pieces of iron of terrestrial origin, and it has not been produced artificially in the scale shown by meteorites. Diamonds have been found in iron meteorites, which indicates that at some time the meteorites were under high pressure. Irons are relatively rare. Of 350 meteorites which were recovered after being observed to fall, only 10 were irons [55]. However, most of the finds made independently of an observed fall are irons, for stony meteorites are not so conspicuous as irons and also are affected more seriously by weathering.

Stony meteorites are subdivided into *achondrites*, which are quite similar to terrestrial rocks, and *chondrites*, which contain a conglomerate of metallic particles suspended in silicates. *Stony irons* also exist; in these, silicate nodules are suspended in a nickel-iron medium.

The age of meteorites has been determined by several methods making use of radioactive decay. The results indicate an age of 4.5×10^9 years [56]. An explanation of the structure and composition of meteorites makes a fascinating and difficult field of study. It is generally agreed that meteorites are fragments of larger bodies which appear to have undergone a process of heating that has separated the material into layers of different composition.

Meteorites can be regarded as a sample from which the nature of interplanetary particles may be deduced. However, the sample favors the more massive meteoroids. In deducing the composition of the smaller particles in space from meteorite data, we must allow for this selection effect. The selection can work in two ways: (1) The proportion of stones to irons may increase as we turn to the investigation of smaller meteors. If the meteoroids are fragmentation products, the stony particles may become proportionally more numerous than the metallic ones. (2) There may be, besides stones and irons, a type of meteoroid that cannot survive passage through the atmosphere. Whipple and Hawkins [48] have computed meteoroid densities of 0.1 gm/cm^3 from double-station photographic observations. The evidence indicates that a large majority of the observable meteors is in this class. An interpretation of these results in terms of the composition of comets is given in Section 6–5.

The general subject covered in this section has a very extensive bibliography. Excellent brief summaries are given by Whipple and Hawkins [48] and by Whipple [57]. Detailed presentations are given in Lovell's *Meteor Astronomy* [48] and in Öpik's book [51]. Watson's *Between the Planets* [58] includes an excellent descriptive presentation. The interpretation of the composition of meteorites is covered by Urey [54], Iavnel [59], and Jacobs and Allan [60], where additional references may be found. Reference 61 is a collection of papers on this subject. Finally, we should mention Brown's *Bibliography on Meteorites* [62] which covers the world literature published between 1491 and 1950 on meteorites and related subjects.

6–5 Comets. With the advent of more and larger telescopes, the frequency with which comets are found has increased through the years. At present the rate is about seven per year. On the average, only one comet every other year may be seen with the naked eye, and about one or two per century are so bright that they may be seen in daylight hours. About one-third of the comets that are detected turn out to be previously known objects paying a return visit to the neighborhood of the sun.

When a comet is first discovered, it is assigned a letter indicating its order among the comets discovered in a given year, together with the name of the discoverer(s). For example, Comet Arend-Roland 1956h was the eighth comet discovered in 1956. After the orbits of the comets have been computed, it is customary to give them a permanent designation consisting of the year followed by a Roman numeral indicating the order in which they reach perihelia during that year. For example, 1936 I was originally named 1935d. Names of discoverers or of persons who have done extensive studies of a given comet are also used.

Two centers have been organized for the reporting of new comets and other astronomical events requiring urgent attention from observers. The eastern hemisphere of the earth is served by the Copenhagen Observatory and the western hemisphere by the Harvard College Observatory. Both institutions supply a telegram service and an announcement-card service by subscription.

About 75% of the comets whose orbits have been computed have nearly parabolic orbits. The absence of definite hyperbolic orbits indicates that comets are members of the solar system. The rest, that is, the *periodic* comets, have orbits with major axes comparable to those of the planets. Among these there is a definite subgroup with orbital periods of three to nine years. This group, known as *short-period* comets, have aphelia which fall close to Jupiter's orbit. These are also known as *Jupiter's family* of comets.

The distribution of orbital inclinations confirms the reality of the above groups. The parabolic comets have inclinations distributed at random and hence equally probable direct or retrograde orbital motion. The periodic orbits usually have $i < 90°$, Comet Halley being one of the exceptions.

The eccentricities of the short-period comets average about 0.6, and their perihelion distances 1.3 A.U. The orbital motions are all direct, the inclinations being usually less than 20°. The attraction of Jupiter has been shown to be sufficient to change the orbit of any parabolic comet into a short-period one if the comet passes sufficiently close. A number of short-period comets have been discovered a few years after passing close to Jupiter, suggesting that they had been recently captured into Jupiter's family. Other short-period comets, passing close to Jupiter, have been seen to suffer appreciable orbital changes. How the capture takes place will be discussed in a later paragraph. These facts indicate that the inner part of the solar system is not the natural habitat of comets.

Comets at distances greater than 5 A.U. from the sun are exceedingly faint and show little or no tail. The *coma*, or nebulous head, decreases in size as the comet approaches the sun, while the tail lengthens. A starlike nucleus is often seen in the coma. The diameters of nuclei play an important role in the determination of the masses of comets. However, since the nuclei are quite small, they are very difficult to measure directly. The dimensions of the comae cover a large range, with a maximum well over one million kilometers. It is impossible to quote a lower limit for the size of a coma, since small comets are difficult to detect. Tails as long as 1 A.U. have been observed occasionally.

In spite of their large dimensions, comets are not massive. No comet has ever been observed to cause a detectable perturbation on a planet or satellite. In 1886 Brooks' comet passed within Jupiter's satellite system without producing any visible alteration in the satellites' motion. On the other hand, the orbit of this comet was changed from one with a 29-year period to one with a 7-year period. These observations gave an upper limit of 10^{20} gm for the mass of Brooks' comet. A similar passage near Jupiter was made by Comet Lexell (1770 I).

Vorontsov-Velyaminov [63] has derived the mass of Halley's comet. He estimated the volume of the nucleus from its brightness by assuming that the light received from the nucleus is reflected sunlight. This assumption is supported by spectroscopic observations which will be described later. The albedo was assumed to be equal to that of the asteroid Ceres, and the phase-angle effects similar to those for Mercury. To ensure that the brightness was not affected seriously by fluorescence, Vorontsov-Velyaminov used the brightness of the nucleus (not of the coma) when the comet was at a distance of 3.4 A.U. from the sun. As will be explained later, fluorescence affects the brightness of a comet, principally of the coma and tail, when the comet is close to the sun. A diameter of the order of 30 km was obtained for the nucleus. The mass was derived by estimating the rate of loss of gases from the nucleus. This rate in turn was determined from an analysis of the spectral features of the comet. The gas content of meteorites was determined next. When heated in the laboratory in a vacuum, meteorites expel gases of the kind that must exist in the nucleus of the comet in order to produce the observed spectral features. A comparison of the amount of gas expelled by the comet and the amount of gas expelled by a given meteoritic mass yielded an estimate of 3×10^{19} gm for the mass of the nucleus of Halley's comet. This figure may also be taken as the total mass, since the coma and tail are extremely tenuous.

Vorontsov-Velyaminov's computed radius for Comet Halley has been found by Whipple [64] to be consistent with nuclei diameters derived for other comets. He finds nuclei diameters ranging from 0.4 to 164 km.

The mass determination made by Vorontsov-Velyaminov can be criticized because the density of the laboratory samples of meteorites may not be identical with that of the material found in cometary nuclei. An alternate approach uses the density of 0.1 gm/cm^3 quoted by Whipple and Hawkins [48] for the majority of meteors, a density which is much lower than that of meteorites. Assuming that the light meteoroids mentioned by Whipple and Hawkins are of cometary origin, we find the mass of Halley's comet to be about 1.4×10^{18} gm. However, Whipple and Hawkins [48] quote a result by Jacchia according to which, for the Geminid stream, the density of the meteoroids may be higher by 2.4 than the average 0.1 used here. In view of this and of the tentative nature of the density 0.1 quoted by Whipple and Hawkins, we may regard the mass estimate for Halley's comet as a rough approximation. The nuclear diameters quoted by Whipple [64] indicate that the masses of comets have a range of 10^3 or more.

A lower limit for the mass of a comet may be obtained from the total mass of a meteoroid stream which Lovell has estimated to be about 10^{15} gm [49] (see also Section 6–4).

From the inverse-square law one would expect the light intensity of a comet, provided it was due to reflected sunlight, to be given by

$$ I = \frac{I_0}{r^2 \, \Delta^2}, $$

where I_0 is the light intensity of the comet at 1 A.U., r its distance from the sun, and Δ its distance from the earth. Instead, one observes the relationship

$$ I = \frac{I_0}{r^n \, \Delta^2}, \tag{6–23} $$

where n, in general, has a value between 2 and 6 and may vary with time for a given comet. This indicates that the brightness of the comet is due

256 THE SUN AND INTERPLANETARY SPACE [CHAP. 6

to some cause other than reflected sunlight. In different comets I_0 may vary by a factor of 10^6. Oort and Schmidt [65] have found that n is smaller for parabolic comets than for short-period comets. Their findings have been confirmed by Vanýsek [66], who finds, on the average, $n = 4.9$ for short-period comets and 3.1 for near-parabolic ones. These values of n mean that comets distant from the sun are exceedingly faint, which explains why they are usually discovered only near perihelion.

Various attempts have been made to correlate the brightnesses of comets with solar activity. Beyer [67] obtains not only a correlation of brightness changes in a given comet with sunspot numbers, but he also finds that when a comet is at a high heliocentric latitude, its brightness variations are less marked than when it is near the ecliptic. These results are reminiscent of Biermann's findings discussed in the last section. It would be interesting to determine what correlation exists between brightness variations in two comets observed simultaneously. A tendency for the two comets to vary in a similar manner could only be explained by solar effects.

During perihelion passage several comets have been observed to split. The classical example is Biela's comet. This short-period comet, which had been observed on various approaches to the sun, was seen in 1846 to divide into two comets which continued to travel in practically identical orbits while separating from each other. The two comets were observed to return to perihelion in 1852. In the next return, their position with respect to the sun was unfavorable for observation, but in 1865, when the comets should have been very easy to detect, they did not appear, and have never been seen since.

Another similar case involved a group of very bright nearly parabolic comets that had been observed approaching the sun in almost identical orbits. In each of the years 1668, 1843, 1880, 1882, and 1887 one comet of the group passed perihelion within one solar radius of the sun. The comet of 1882 was seen to approach the sun as a single object, but it came away from the sun split into four comets, which gradually separated as they receded from the sun on a common orbit. It is likely that the original group of comets was itself the product of a similar splitting process.

Comet tails are often quite irregular and may be multiple. An interesting case is that of Comet Arend-Roland (1956h), which apparently developed a tail pointed in the direction of the sun, contrary to the well-known trend of orientation of comet tails (see Plate XI). This phenomenon was explained by the peculiar arrangement in space of the comet, the earth, and the sun (Fig. 6–9), on the assumption that the comet was leaving a trail of scattering particles along its orbital path [68].

Comets are often observed to undergo outbursts originating in the nuclei. The outbursts appear to be of an explosive nature. They produce spherical shells on the comae and knotlike condensations in the tails.

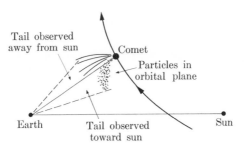

Fig. 6–9. Circumstances in which a comet tail appears directed toward the sun.

Donn and Urey [69] have interpreted the outbursts as the result of explosive chemical reactions involving unstable molecules.

The spectrum of a comet shows a large number of molecular emission bands plus several atomic emission lines superimposed on a continuum similar to that of the solar spectrum. The relative strengths of these components vary from comet to comet, and with time in any given comet. The solar component of the spectrum, which must be due to reflection or scattering of sunlight, is strongest in the nucleus. When the comet's head shows a strong emission spectrum, the tail is usually quite bright. If the tail is absent, the head has a continuous spectrum. In the spectrum of the coma, emission features from neutral and ionized molecules are present; in the tail only ionized molecules are observed. Under standard laboratory conditions the molecular compounds identified in cometary spectra would be chemically unstable, i.e., they would associate immediately with other compounds. Their existence in the comet's coma is a result of a long mean free path. This indicates that the coma and tail are extremely rarefied.

The above-mentioned unstable radicals can be produced by dissociation from stable molecules such as H_2O, NH_3, N_2, CH_4, and CO. Solar radiation is the dissociating agent.

The foregoing spectroscopic results and the data so far presented in this section and in Section 6–4 suggest a physical model for a comet. On the basis of such a model Whipple [70, 64] concludes that the nucleus consists of a conglomerate of ices of the lighter elements in compounds such as H_2O, NH_3, CO_2, CH_4, CO, or their hydrates, mixed with meteoritic material. The initial temperature of the conglomerate is less than 50°K. By the action of solar heat, the ices are sublimated when the comet approaches perihelion. The model accounts for

(a) the observed spectral characteristics,
(b) the low structural strength of comets, indicated by their tendency to split when near perihelion,
(c) the slow disintegration of comets.

In addition, some heretofore unexplained effects in the motions of comets can be accounted for if the nucleus is rotating. For example, Comet Encke has been observed to undergo an acceleration in its mean motion and a decrease in its orbital eccentricity. The jet action of escaping gases from a rotating nucleus will explain these effects, provided there is some time lag between absorption of radiation and the jet action.

As the gas is sublimated from the nucleus, it is dissociated and ionized by solar radiation. In this way, the unstable radicals (for example, C_2, CN, CH, NH, OH, CO^+) identified in the spectrum of the coma are produced. The action of sunlight also causes the excitation required to account for the emission features. The absorbed quanta may be re-emitted in the same wavelength or on longer wavelengths, in which case fluorescence occurs. This mechanism, which provides a good explanation for the observed brightnesses of comets, has been examined in detail by several authors [71].

In an important paper, Oort [72] has summarized the available evidence on the distribution and frequency of comets in the outer regions of the solar system. From well-observed orbits of new comets, he concludes that these usually come from distances of 50,000 to 150,000 A.U. The new comets form a cloud which accompanies the sun as it moves through interstellar space. In this cloud, according to computations performed by Van Woerkom [73], the comets that may originally have had orbits passing near the sun would have been eliminated long ago, mainly by the perturbative action of Jupiter. Hence the new comets arriving now must have been disturbed from their roughly circular orbits in the comet cloud. The disturbing action is found by Oort to be produced by nearby stars. The influence of the stars on the cloud is such that the cloud could not have survived to date if it extended beyond 200,000 A.U. from the sun. This puts an upper limit on the radius of the comet cloud. Furthermore, the stellar perturbations would have changed the shape of the comet cloud from any form it may originally have had to a spherical one, which explains the random inclinations of near-parabolic comets.

For a comet to leave the cloud and approach the sun, where it may be discovered, its motion must be changed so that its new orbit will move it to approximately 1 A.U. from the sun. Oort has estimated the frequency with which comets in the cloud will be directed toward the observable sphere around the sun as a result of stellar encounters for a given total number of comets in the cloud. Combining this estimate with the observed discovery rate, a total comet population of the order of 10^{11} is found for the cloud. By assuming an average mass per comet of 10^{16} gm, Oort deduced that the total mass of the cloud is about $\frac{1}{10}$ to $\frac{1}{100}$ that of the earth. These comets, however, are "new" ones in the sense that they have not been repeatedly exposed to the disrupting action of sunlight.

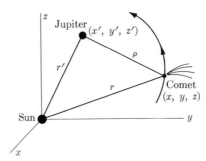

FIG. 6–10. Approach of a comet toward Jupiter.

Therefore their mass may be assumed to be appreciably higher than the minimum mass of 10^{15} gm quoted earlier from Lovell's work on meteor streams. It is thus possible that the total mass of the comet cloud is at least equal to that of the earth.

In the foregoing discussion, reference has been made to perturbations of a cometary orbit by the giant planet Jupiter. Other planets also perturb the motions of these objects if the comets approach sufficiently close. We shall conclude our section on comets by a brief analysis of the "capture" of a comet by Jupiter.

This is essentially a three-body problem for which the theory was developed in Chapter 5. Let (x, y, z) be the heliocentric coordinates of the comet (Fig. 6–10); let (x', y', z') be those of Jupiter. We denote by m' the mass of Jupiter in units of the solar mass. The mass of the comet is negligible compared with that of the sun. Then the disturbing function is

$$R = k^2 m' \left\{ \frac{1}{\rho} - \frac{xx' + yy' + zz'}{r'^3} \right\}, \qquad (6\text{–}24)$$

and the equations of motion are

for Jupiter: *for the comet:*

$$\ddot{x}' + \frac{k^2(1 + m')x'}{r'^3} = 0, \qquad \ddot{x} + \frac{k^2 x}{r^3} = k^2 m' \left[\frac{x' - x}{\rho^3} - \frac{x'}{r'^3} \right],$$

$$\ddot{y}' + \frac{k^2(1 + m')y'}{r'^3} = 0, \qquad \ddot{y} + \frac{k^2 y}{r^3} = k^2 m' \left[\frac{y' - y}{\rho^3} - \frac{y'}{r'^3} \right], \qquad (6\text{–}25)$$

$$\ddot{z}' + \frac{k^2(1 + m')z'}{r'^3} = 0, \qquad \ddot{z} + \frac{k^2 z}{r^3} = k^2 m' \left[\frac{z' - z}{\rho^3} - \frac{z'}{r'^3} \right].$$

These expressions are in effect Eqs. (5–13), (5–14), and (5–15) of Chapter 5.

Let $\xi = x - x'$, $\eta = y - y'$, $\zeta = z - z'$ be the coordinates of the comet relative to Jupiter. Then the equations of motion of the comet relative to Jupiter are

$$\ddot{\xi} + \frac{k^2 m' \xi}{\rho^3} = k^2 \left[\frac{x'}{r'^3} - \frac{x}{r^3} \right], \tag{6-26}$$

$$\ddot{\eta} + \frac{k^2 m' \eta}{\rho^3} = k^2 \left[\frac{y'}{r'^3} - \frac{y}{r^3} \right], \tag{6-27}$$

$$\ddot{\zeta} + \frac{k^2 m' \zeta}{\rho^3} = k^2 \left[\frac{z'}{r'^3} - \frac{z}{r^3} \right]. \tag{6-28}$$

The right-hand sides of these equations represent the force components due to the perturbative action of the sun on the comet during its motion around Jupiter.

For simplicity, let the accelerations of the comet, due both to the sun and to the perturbations by Jupiter, be

$$A_1 = \frac{k^2 x}{r^3}, \qquad P_1 = k^2 m' \left[\frac{x' - x}{\rho^3} - \frac{x'}{r'^3} \right],$$

$$A_2 = \frac{k^2 y}{r^3}, \qquad P_2 = k^2 m' \left[\frac{y' - y}{\rho^3} - \frac{y'}{r'^3} \right], \tag{6-29}$$

$$A_3 = \frac{k^2 z}{r^3}, \qquad P_3 = k^2 m' \left[\frac{z' - z}{\rho^3} - \frac{z'}{r'^3} \right].$$

Similarly, in Eqs. (6–26), (6–27), and (6–28) let the accelerations of the comet in the Jovicentric system be

$$A_1' = \frac{k^2 m' \xi}{\rho^3}, \qquad P_1' = k^2 \left[\frac{x'}{r'^3} - \frac{x}{r^3} \right],$$

$$A_2' = \frac{k^2 m' \eta}{\rho^3}, \qquad P_2' = k^2 \left[\frac{y'}{r'^3} - \frac{y}{r^3} \right], \tag{6-30}$$

$$A_3' = \frac{k^2 m' \zeta}{\rho^3}, \qquad P_3' = k^2 \left[\frac{z'}{r'^3} - \frac{z}{r^3} \right].$$

Then the ratios P_j/A_j and P_j'/A_j' ($j = 1, 2, 3$) serve as a measure of the influence of the disturbing force on the otherwise Keplerian motion. If P_j/A_j is small, the perturbations by Jupiter are small and the comet moves essentially under the influence of the sun. At the same time, P_j'/A_j' would be large. Conversely, if P_j/A_j is large, Jupiter has a great perturbative effect on the comet and P_j'/A_j' is small. We take as a criterion of the transfer of control from the sun to Jupiter the condition

$$\frac{P_j}{A_j} \geq \frac{P'_j}{A'_j}. \tag{6-31}$$

Using the equality sign in Eq. (6–31), together with the definitions (6–29) and (6–30), we could obtain the equation for a surface about Jupiter. This is approximately a sphere [74] centered on Jupiter, of radius

$$R_i = (m')^{2/5} r'. \tag{6-32}$$

Here m' is in units of the solar mass and r' is in astronomical units. This is called the *sphere of influence* or *sphere of activity* of the planet. For the various planets we find:

Planet	R_i, A.U.	Planet	R_i, A.U.
Mercury	6.71×10^{-4}	Jupiter	0.322
Venus	4.11×10^{-3}	Saturn	0.365
Earth	6.19×10^{-3}	Uranus	0.346
Mars	3.87×10^{-3}	Neptune	0.576
		Pluto	0.255

We have assumed for Pluto a mass equal to that of the earth.

The procedure for analyzing the motion of a comet near a large planet can be summarized as follows. As the comet moves at some distance from the planet, its coordinates, together with those of the planet, are calculated at suitable time intervals. The values of ρ are thus obtained. By interpolation one can then determine when $\rho = R_i$. Hence the velocity components of the comet and of the planet can be found at the instant of entrance into the sphere of influence. Then the Jovicentric coordinates and velocities can be computed, together with the elements of the hyperbolic orbit in which the comet swings by the planet. The comet will obviously leave the sphere of influence with the same speed and magnitude of true anomaly with which it entered. The planetocentric coordinates of exit from the sphere of influence and the velocities can be found and, after correction for the motion of the planet, the continued heliocentric motion of the comet is determinable.

When a comet passes sufficiently close to Jupiter and is captured, its orbit will be so altered that its aphelion distance is about 5 A.U. We shall illustrate this by a very simple pictorial example. Consider a comet approaching the sun along a parabolic path (Fig. 6–11). If Jupiter were not present, the comet would pass around the sun along the dashed path.

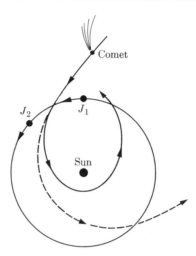

FIG. 6–11. Capture of a comet by Jupiter.

Because of Jupiter's presence (J_1), the comet is deflected into the nearly elliptical path with aphelion near the orbit of the planet. However, during the passage of the comet through perihelion and upon its return to Jupiter's orbit, the planet has moved along to position J_2. The comet continues to move in its new elliptic orbit. If at some future time the comet and Jupiter are close together, another perturbation may again change the cometary orbit completely. At present there are about 30 comets with orbital aphelia near Jupiter's orbit. The suggestion is strong that they have been captured in the manner described.

6–6 Asteroids or minor planets. A great many asteroids have been found since the discovery of the first asteroid (Ceres) at the beginning of the last century. Initially a determined effort was made to keep track of the detected asteroids by computing their orbital elements. The rapid succession of discoveries, however, outpaced the computation of orbits until, at present, the number of asteroids known to exist far exceeds that for which orbital elements have been computed.

A list of the asteroids with known orbital elements is published yearly by the Institute of Theoretical Astronomy at Leningrad [75]. These asteroids are designated by numbers which indicate approximately the order of discovery. A name given by the discoverer is also used. Examples of the usual designation are: Achilles(588), Thule (279), Betulia (1580). Before an asteroid is included in the above list, it is designated by the year of discovery followed by two capital letters giving the order in which the discovery is announced, for example, 1960 AA, 1960 AB, etc. Some of these temporary designations have been adopted instead of proper names in the list published by the Institute of Theoretical Astronomy.

This yearly publication is titled *Ephemerides of Minor Planets*. In addition to the list of orbital elements, the publication includes the right ascensions and declinations at 10-day intervals for the asteroids that will pass through opposition during the year. These ephemerides cover 60-day periods centered about the epochs of opposition.

The *Ephemerides of Minor Planets* for 1961 contains 1633 asteroids with assigned numbers and three asteroids without numbers (Apollo, Adonis, Hermes). At present the practice of assigning a permanent number is limited to asteroids that have been observed in two oppositions and for which satisfactory orbits have been obtained. When a new minor planet approaches the earth within Mars' orbit, a number may be assigned after a single opposition, provided a satisfactory orbit has been calculated.

As a clearinghouse for the rapid communication of news about minor planets, the International Astronomical Union established a Minor Planet Center at the University of Cincinnati Observatory. This Center issues *Minor Planet Circulars* [76] containing observations of position, corrections for published ephemerides, new orbital elements, etc.

A study of the completeness of the list of numbered asteroids has been made by Kuiper in collaboration with a group of investigators [77]. They find that all asteroids with photographic magnitude at opposition equal to 14.0 or less are probably known and numbered. There are 437 such asteroids in the 1961 *Ephemerides of Minor Planets*.

Kuiper and his collaborators have established a formula for the number of asteroids of photographic magnitude p_0 at opposition in the range $p_0 - \frac{1}{2}$ to $p_0 + \frac{1}{2}$:

$$\log N(p_0) = -2.38 + 0.35 p_0. \tag{6–33}$$

The computed number of asteroids brighter than $p_0 = 19.5$ is 33,600. Equation (6–33) is based on a systematic survey up to and including magnitude 16.5 and on an estimate previously made by Baade [78] of the number of asteroids brighter than magnitude 19.0.

A convenient summary of the distribution of the orbital elements of asteroids has been given by Watson [58]. Most asteroids have mean distances between 2.3 and 3.3 A.U. The asteroid with the greatest mean distance from the sun, 5.79 A.U., is Hidalgo (944). The smallest mean distance, 1.077 A.U., is that of Icarus (1566). The eccentricity of Icarus' orbit (0.827) is the largest among the asteroids. Its perihelion distance is 0.186 A.U., well within Mercury's orbit. Icarus can approach the earth within 0.04 A.U.

The orbital periods of the asteroids usually fall in the range 3.5 to 6 years. Icarus and Hidalgo, having the least and the greatest mean distances, have the minimum and maximum periods, 1.12 years and 13.7 years, respectively. A frequency distribution of the periods shows conspicuous voids at 3.97, 4.76, and 5.95 years, which are called *Kirkwood's gaps* after their discoverer. These periods are exactly $\frac{1}{3}$, $\frac{2}{5}$, and $\frac{1}{2}$ of Jupiter's period (11.9 years). The gaps are supposed to be caused by a resonance effect similar in nature to the effect which causes the gaps in Saturn's rings (see Section 2–10). An asteroid with one of the above periods will cross the line sun-Jupiter at regular intervals at the same heliocentric longitude. As a result, the effects of the perturbations will accumulate and eventually change the orbit to one with a period that is not in resonance with Jupiter. This simple explanation, however, fails for minor planets with semimajor axes greater than 3.5 A.U. Instead of gaps, one then finds concentrations at the periods that are commensurable with that of Jupiter. For example, although only one asteroid is known with mean distance, a, between 3.65 and 3.88 A.U., there are 20 known asteroids forming the so-called Hilda group with $3.88 \leq a \leq 3.99$. The average orbital period of these asteroids is close to two-thirds of Jupiter's period. Curiously, one of these asteroids is Kirkwood (1578). The reasons for these concentrations are not understood at present.

One interesting group is that formed by asteroids Achilles (588), Patroclus (617), Hector (624), Nestor (659), Priamus (884), Agamemnon (911), Odysseus (1143), Aeneas (1172), Anchises (1173), Troilus (1208), Ajax (1404), Diomedes (1437), and Antilochus (1583). These are the Trojan asteroids which are concentrated near the equilateral Lagrangian points (Section 5–5) of Jupiter's orbit.

The orbital inclinations of the asteroids as a whole average about 10°, the maximum inclination so far known being 52° for asteroid Betulia (1580). There is a correlation between inclination and eccentricity, the more inclined orbits tending to be the more eccentric ones. The eccentricities average about 0.15.

The longitudes of perihelion show a tendency to agree with Jupiter's longitude of perihelion. There are about twice as many asteroids having ω-values within 90° of that of Jupiter as there are asteroids having ω-values within 90° of the longitude of Jupiter's aphelion direction.

Several investigators have searched for mathematical functions of the orbital elements that are not affected by perturbations. Brouwer [79] has listed these functions. When one of these is plotted against another, several conspicuous groups of asteroids appear. Since the semimajor axis is the most stable orbital element, the members of these groups have values of a falling into well-defined narrow intervals. These groups are called *Hirayama families* after the Japanese astronomer who first detected them. According to Brouwer, there are about 29 such groups. The members of a given Hirayama family may have originated from the fragmentation of a parent asteroid. Kuiper [80] has reviewed the various ways in which the parent asteroids may have been disrupted. Collisions between asteroids appear to be the most likely explanation. According to Kuiper, five to ten large asteroids were formed at a distance of between 2 and $3\frac{1}{2}$ A.U. from the sun when the solar system originated. The collision probability was low at first, but once a collision occurred, other collisions became more and more likely. The fragmentation products of these collisions can account for the sporadic meteors and the zodiacal light. Comets may also have originated in the asteroid belt [72].

Conclusions about the physical properties of the asteroids can be derived in most cases only from measurements of their brightness. The observed brightness of an asteroid is a function of (a) the asteroid's distance from the sun, (b) the asteroid's distance from the earth, (c) the phase function or percentage of light reflected in the earth's direction, (d) the surface area exposed to the sun, and (e) the albedo.

For the purpose of intercomparing asteroids, we define a mean absolute magnitude at opposition, g, that is independent of the first three factors listed above. By definition,

$$g = p_0 - 5 \log a(a - 1), \qquad (6\text{–}34)$$

where p_0 is the mean magnitude at opposition and a is the semimajor axis in astronomical units. The values of g and of p_0 for each asteroid are listed in the *Ephemerides of Minor Planets*.

The frequency distribution of g-values has been investigated by Kuiper and his collaborators [77]. In the interval $2.0 < a < 3.5$ A.U., completeness in the available data can only be assumed for magnitudes $g < 10.0$. The frequency distribution of g-values in this interval is shown in Fig. 6–12. Kuiper and his co-workers find that if the interval $2.0 < a < 3.5$ A.U. is subdivided, the characteristics of the g-distributions vary with distance from the sun.

Albedos have been determined only for the largest asteroids, whose sizes can be measured from the angular diameters they exhibit telescopically. The diameters of Ceres (1), Pallas (2), Juno (3), and Vesta (4) have been observed by Barnard [81]. Russell [82] has used Barnard's observations to derive albedos by means of Eq. (2–25). Since the phase functions of the asteroids can be observed only for phase angles as great as 20° to 30°, considerable extrapolation is necessary, and the results are uncertain. For the asteroids, Russell uses the value of 0.55 for the quantity

$$2\int_0^\pi \phi(\alpha) \sin \alpha \, d\alpha$$

appearing in Eq. (2–25). The diameters and albedos for the larger asteroids are presented in Table 6–1. These albedos have been computed from the

TABLE 6–1

VISUAL ABSOLUTE MAGNITUDES g_v,
DIAMETERS, AND ALBEDOS OF THE LARGEST ASTEROIDS

Asteroid	g_v	Diameter, km	Albedo
Ceres (1)	3.40	770	0.07
Pallas (2)	4.38	490	0.07
Juno (3)	5.63	200	0.14
Vesta (4)	3.54	390	0.25

formulas presented by Russell, but with the newer photometric data published jointly by Groeneveld and Kuiper [83], [84]. The pertinent absolute visual magnitudes (g_v) are also included in the table.

It should be remarked that for the largest asteroid, Ceres, the angular diameter observed from a distance of 1 A.U. is only about 1″.0. Thus the measurements of the diameters of asteroids are quite difficult. The albedos, which according to Eq. (2–25) vary as the square of the radius, suffer therefore from the uncertainties in the determination of the radii.

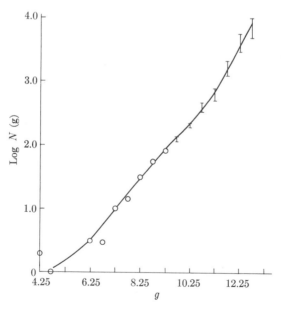

FIG. 6–12. Frequency distribution of asteroids according to absolute photographic magnitude.

For Juno, a not unlikely error of 10% in the measured diameter will change the computed albedo by 12%.

A relationship between the absolute magnitude, g, of an asteroid and its radius may be derived from Eq. (2–25) if we substitute the value of 0.55 given by Russell [82] for twice the value of the integral appearing in that equation. In this case, we obtain from Eq. (2–25) the relationship

$$\log \frac{A}{0.55} = -2 \log \bar{r} + \log \Delta^2 a^2 + \log \frac{P(0)}{S}.$$

We recall that the distance of the asteroid from the earth, Δ, and the mean radius of the asteroid, $\bar{r}$, must be measured in the same units. We choose kilometers. The quantity a is the distance in astronomical units from sun to asteroid. $P(0)$ and S are, respectively, the brightness of the asteroid at opposition and the brightness of the sun. The absolute visual magnitude of an asteroid, g_v, is defined for $\Delta = a = 1$ A.U. Furthermore, according to Eq. (2–35), the term $\log P(0)/S$ is given by

$$m_s - g_v = 2.5 \log \frac{P(0)}{S},$$

where m_s is the sun's apparent visual magnitude, -26.73 [4]. Combining these relationships, we have

$$\log \bar{r} = 2.82 + \frac{1}{2} \log \frac{0.55}{A} - 0.2 g_v.$$

Using the albedo of Ceres or Pallas, we finally obtain

$$\log \bar{r} = 3.27 - 0.2 g_v, \tag{6–35}$$

where $\bar{r}$ is given in kilometers. An error by a factor of 2 in the quantity $0.55/A$ will change the constant 3.27 by ± 0.15.

One can make estimates of asteroid masses by combining volumes derived from photometric radii and some plausible density. A density of 3.0 gm/cm^3, corresponding to an average for the moon, Jupiter's satellites, and also for stony meteorites, does not appear unlikely. Assuming this density, we may derive from Eq. (6–35) an expression for the mass in kilograms:

$$\log M = 22.91 - 0.6 g_v. \tag{6–36}$$

According to this formula, the masses of Ceres, Eros, and Adonis are about 7.4×10^{20}, 10^{16}, and 10^9 kgm, respectively. Adonis has the faintest absolute magnitude listed in the *Ephemerides of Minor Planets*.

In order to estimate the total mass of the asteroid belt, the distribution of absolute magnitudes must be known accurately. Kuiper and his co-workers [77] find that the number of faint asteroids is not sufficiently well known for this purpose.

In general, an asteroid shows regular variations in brightness. On the assumption that the asteroid rotates, these variations can be ascribed to irregularities in the shape or to differences in the albedo on the surface of a given asteroid. Disentangling these effects by means of photometric observations is not possible. Another difficulty in interpreting the photometric data is the present imperfect knowledge of the phase function (see Section 2–7). A list of papers presenting photoelectric observations of asteroids is given in reference 77.

Some asteroids which at times pass close to the earth deserve special mention. The importance of asteroid Eros (433) for the calibration of the astronomical unit and the determination of the masses of the planets was discussed in Chapter 2. Other asteroids may pass closer to the earth than Eros, whose closest distance is about 0.15 A.U. For example, Icarus(1566) can approach the earth to a distance of 0.04 A.U., as noted earlier. Another asteroid of interest in this respect is Amor (1221), whose perihelion distance from the sun is 1.083 A.U. This asteroid may approach the earth to a distance of 0.11 A.U. Even closer passages have been observed for asteroids Apollo, Adonis, and Hermes, all of which are extremely faint objects, and all of which were discovered during their close approach to the earth. Hermes passed the earth at a distance of 800,000 km, or a little more than 0.005 A.U.

REFERENCES

1. M. Schwarzschild, *Structure and Evolution of Stars.* Princeton: Princeton University Press, 1958.

2. Marshal H. Wrubel, "Stellar Interiors," *Handbuch der Physik,* vol. 51, p. 1. Berlin: Springer, 1958.

3. B. Strömgren, "The Sun as a Star," *The Sun,* Chapter 2, G. P. Kuiper, editor. Chicago: Chicago University Press, 1953.

4. Joel Stebbins and Gerald E. Kron, "Six-Color Photometry of Stars. X. The Stellar Magnitude and Color Index of the Sun," *Astrophys. J.,* **126,** 266, 1957.

5. Leo Goldberg and A. Keith Pierce, "The Photosphere of the Sun," *Handbuch der Physik,* vol. 52, p. 1. Berlin: Springer, 1959.

6. C. de Jager, "Structure and Dynamics of the Solar Atmosphere," *Handbuch der Physik,* vol. 52, p. 80. Berlin: Springer, 1959.

7. M. Minnaert, "The Photosphere," Chapter 3 of reference 3.

8. H. C. van de Hulst, "The Chromosphere and the Corona," Chapter 5 of reference 3.

9. Horace W. Babcock and Harold D. Babcock, "The Sun's Magnetic Field, 1952–1954," *Astrophys. J.,* **121,** 349, 1955.

10. HAROLD D. BABCOCK, "The Sun's Polar Magnetic Field," *Astrophys. J.*, **130**, 364, 1959.

11. K. O. KIEPENHEUER, "Solar Activity," Chapter 6 of reference 3.

12. GIORGIO ABETTI, *The Sun*. New York: MacMillan, 1957.

13. *Quarterly Bulletins on Solar Activity*. Zürich: Eidgenössische Sternwarte.

14. *U. S. Naval Observatory Circulars*. Washington: U. S. Naval Observatory.

15. H. DODSON, "Studies at the McMath-Hulbert Observatory of Radio Frequency Radiation at the Time of Solar Flares," *Proc. IRE*, **46**, 149, 1958.

16. J. L. PAWSEY and S. F. SMERD, "Solar Radio Emission," Chapter 7 of reference 3.

17. National Academy of Sciences, *IGY Bulletins*, No. 17, p. 2, 1958.

18. SYDNEY CHAPMAN, "The Morphology of Geomagnetic Storms and Bays: a General Review," *Vistas in Astronomy*, vol. 2, p. 912, A. Beer, editor. London: Pergamon, 1956.

19. No. 12, p. 7, 1958, of reference 17.

20. EDISON PETTIT, "The Sun and Stellar Radiation," *Astrophysics*, Chapter 6, J. A. Hynek, editor. New York: McGraw-Hill, 1951.

21. A. B. MEINEL, "Doppler-Shifted Auroral Hydrogen Emission," *Astrophys. J.*, **113**, 50, 1951.

22. S. K. MITRA, *The Upper Atmosphere*. Calcutta: The Asiatic Society, 1952.

23. H. W. NEWTON, "Solar Flares and Magnetic Storms," *Monthly Notices Roy. Astron. Soc.*, **103**, 244, 1943.

24. HELEN W. DODSON, RUTH E. HEDEMAN, and JOSEPH CHAMBERLAIN, "Ejection of Hydrogen and Ionized Calcium Atoms with High Velocity at the Time of Solar Flares," *Astrophys. J.*, **117**, 66, 1953.

25. M. A. ELLISON, "Solar Flares," p. 799 of reference 18.

26. P. M. S. BLACKETT, "Cosmic Rays and the Sun," p. 820 of reference 18.

27. JAMES A. VAN ALLEN and LOUIS A. FRANK, "Radiation Around the Earth to a Radial Distance of 107,400 km," *Nature*, **183**, 430, 1959.

28. JAMES A. VAN ALLEN, "The Geomagnetically Trapped Corpuscular Radiation," *Journal Geophys. Research*, **64**, 1683, 1959. See also no. 30, p. 1, 1959, of reference 17.

29. H. ELSÄSSER, "Interplanetare Materie," *Mitt. Astronom. Ges. 1957 II*, p. 61, 1958.

30. F. E. ROACH, HELEN B. PETTIT, E. TANDBERG-HANSSEN, and DOROTHY N. DAVIS, "Observations of the Zodiacal Light," *Astrophys. J.*, **119**, 253, 1954.

31. D. E. BLACKWELL, "A Study of the Outer Corona from a High Altitude Aircraft at the Eclipse of 1954 June 30," *Monthly Notices Roy. Astron. Soc.*, **115**, 629, 1955.

32. A. BEHR and H. SIEDENTOPF, "Untersuchungen über Zodiakallicht und Gegenschein nach lichtelektrischen Messungen auf dem Jungfraujoch," *Z. Astrophys.*, **32**, 19, 1953.

33. N. B. DIVARI and A. C. ASAAD, *Astronomicheskiĭ Zhurnal, Akademiia Nauk SSSR*, **36**, 356, 1959.

34. Hans Elsässer, "Die räumliche Verteilung der Zodiakallichtmaterie," *Z. Astrophys.*, **33**, 274, 1954.

35. H. C. van de Hulst, "Zodiacal Light in the Solar Corona," *Astrophys. J.*, **105**, 471, 1947.

36. C. W. Allen, "The Spectrum of the Corona at the Eclipse of 1940 October 1," *Monthly Notices Roy. Astronom. Soc.*, **106**, 137, 1946.

37. C. Hoffmeister, "Untersuchungen über das Zodiakallicht," *Veröffentlichungen der Universitätssternwarte zu Berlin-Babelsberg*, **10**, 56, 1935.

38. S. Piotrowski, "The Collisions of Asteroids," *Acta Astron. Series a*, **5**, 115, 1953.

39. V. G. Fesenkov, "Zodiacal Light as the Product of Disintegration of Asteroids," *Soviet Astron.–AJ*, (English translation of *Astronomicheskii Zhurnal*), **2**, 303, 1959.

40. Fred L. Whipple, "A Comet Model. III. The Zodiacal Light," *Astrophys. J.*, **121**, 750, 1955.

41. H. P. Robertson, "Dynamical Effects of Radiation in the Solar System," *Monthly Notices Roy. Astron. Soc.*, **97**, 423, 1937.

42. L. Biermann, "Kometenschweife und solare Korpuskularstrahlung," *Z. Astrophys.*, **29**, 274, 1951.

43. F. J. Kerr, "Radio Echoes from Sun, Moon and Planets," *Handbuch der Physik*, vol. 52, p. 462. Berlin: Springer, 1959.

44. Eugene Parker, "Extension of the Solar Corona into Interplanetary Space," *J. Geophys. Research*, **64**, 1675, 1959.

45. Thomas Gold, "Plasma and Magnetic Fields in the Solar System," *J. Geophys. Research*, **64**, 1665, 1959.

46. J. E. Kupperian, Jr., E. Byram, T. A. Chubb, and H. Friedman, "Far Ultraviolet Radiation in the Night Sky," *Planetary and Space Science*, **1**, 3, 1959.

47. John C. Brandt and Joseph W. Chamberlain, "Interplanetary Gas. I. Hydrogen Radiation in the Night Sky," *Astrophys. J.*, **130**, 670, 1959.

48. See p. 519 of reference 43.

49. A. C. B. Lovell, *Meteor Astronomy*. New York: Oxford University Press, 1954.

50. D. W. R. McKinley, "Meteor Velocities Determined by Radio Observations," *Astrophys. J.*, **113**, 225, 1951.

51. E. J. Öpik, *Physics of Meteor Flight in the Atmosphere*. New York: Interscience, 1958.

52. Fred L. Whipple, "The Meteoritic Risk to Space Vehicles," *Proc. VIIIth Intern. Astronautical Congress, Barcelona 1957*, p. 418. Vienna: Springer, 1958.

53. *Harvard College Observatory Reprint*, no. 495.

54. H. Brown, "The Density and Mass Distribution of Meteoretic Bodies in the Neighborhood of the Earth's Orbit," *J. Geophys. Research*, **65**, 1661, 1960.

55. H. N. Russell, R. S. Dugan, and J. Q. Stewart, *Astronomy*, rev. edition, vol. 1, p. 454. Boston: Ginn & Co., 1945.

56. Johannes Geiss and David C. Hess, "Argon-Potassium Ages and the Isotopic Composition of Argon from Meteorites," *Astrophys. J.*, **127**, 224, 1958.

57. Fred L. Whipple, *Advances in Geophysics*, vol. 1, p. 119, H. E. Landsberg, editor. New York: Academic Press, 1952.

58. F. G. Watson, *Between the Planets*, rev. ed. Cambridge: Harvard University Press, 1956.

59. A. A. Iavnel, "The Characteristics of Chemical Composition of Meteoritic Matter and the Origin of Meteorites," *Soviet Astron.–AJ*, (English translation of *Astronomicheskii Zhurnal*), **1**, 435, 1957.

60. J. A. Jacobs and D. W. Allan, "Thermal Aspects of the Origin of Meteorites," *J. Roy. Astron. Soc. Canada*, **50**, 122, 1956.

61. T. R. Kaiser, editor, *Meteors*. London: Pergamon, 1955.

62. H. Brown, editor, *A Bibliography on Meteorites*. Chicago: Chicago University Press, 1953.

63. B. Vorontsov-Velyaminov, "Structure and Mass of Cometary Nuclei," *Astrophys. J.*, **104**, 226, 1946.

64. Fred L. Whipple, "A Comet Model. II. Physical Relations for Comets and Meteors," *Astrophys. J.*, **113**, 464, 1951.

65. J. H. Oort and M. Schmidt, "Differences Between New and Old Comets," *Bull. Astron. Inst. Netherlands*, **11**, 259, 1951.

66. V. Vanýsek, "The Photometric Parameters of Comets," *Collection de Mémoires in 8°, Institut d'Astrophysique, Université de Liége (Belgique)*, vol. 15, no. 352, p. 30, 1955.

67. M. Beyer, "Brightness of Comets and Solar Activity," *Collection de Mémoires in 8°, Institut d'Astrophysique, Université de Liége (Belgique)*, vol. 15, no. 352, p. 237, 1955.

68. B. L. Meek, "On the Shape of Cometary Tails," *Monthly Notices Roy. Astron. Soc.*, **119**, 3, 1959.

69. Bertram Donn and Harold C. Urey, "On the Mechanism of Comet Outbursts and the Chemical Composition of Comets," *Astrophys. J.*, **123**, 339, 1956.

70. Fred L. Whipple, "A Comet Model. I. The Acceleration of Comet Encke," *Astrophys. J.*, **111**, 375, 1950.

71. K. Wurm, "Die Kometen," p. 465 of reference 43.

72. J. H. Oort, "The Structure of the Cloud of Comets Surrounding the Solar System, and a Hypothesis Concerning its Origin," *Bull. Astron. Inst. Netherlands*, **11**, 91, 1950.

73. A. F. F. van Woerkom, "On the Origin of Comets," *Bull. Astron. Inst. Netherlands*, **10**, 445, 1948.

74. F. Tisserand, *Traité de Mécanique Céleste*, vol. 4, p. 200. Paris: Gauthier-Villars, 1896.

75. *Éfemeridy Malykh Planet*, Institut Teoreticheskoi Astronomii. Leningrad: Akademii Nauk SSSR.

76. Cincinnati Observatory, *Minor Planet Circulars*.

77. G. P. Kuiper, Y. Fujita, T. Gehrels, I. Groeneveld, J. Kent, G. van Biesbroeck, and C. J. van Houten, "Survey of Asteroids," *Astrophys. J. Suppl. Series*, **3**, 289–428, 1958.

78. W. Baade, "On the Number of Asteroids Brighter than Photographic Magnitude 19.0," *Publ. Astron. Soc. Pacific*, **46**, 54, 1934.

79. Dirk Brouwer, "Secular Variations of the Orbital Elements of Minor Planets," *Astron. J.*, **56**, 9, 1951.

80. G. P. KUIPER, "On the Origin of Asteroids," *Astron. J.*, **55,** 164, 1950.

81. E. E. BARNARD, "On the Dimensions of the Planets and Satellites," *Astron. Nachr.*, **157,** 261, 1902.

82. HENRY NORRIS RUSSELL, "On the Albedo of the Planets and their Satellites," *Astrophys. J.*, **43,** 173, 1916.

83. INGRID GROENEVELD and GERARD P. KUIPER, "Photometric Studies of Asteroids. I," *Astrophys. J.*, **120,** 200, 1954.

84. INGRID GROENEVELD and GERARD P. KUIPER, "Photometric Studies of Asteroids. II," *Astrophys. J.*, **120,** 529, 1954.

APPENDIX

Tables I through IV reproduced by permission of the U. S. Naval Observatory.

TABLE I(a)

CONVERSION OF ARC TO TIME

°	h m	°	h m	°	h m	′	m s	″	s	″	s	″	s
0	0 00	60	4 00	120	8 00	0	0 00	0	0.000	0.00	0.000	0.50	0.033
1	0 04	61	4 04	121	8 04	1	0 04	1	0.067	.01	.001	.51	.034
2	0 08	62	4 08	122	8 08	2	0 08	2	0.133	.02	.001	.52	.035
3	0 12	63	4 12	123	8 12	3	0 12	3	0.200	.03	.002	.53	.035
4	0 16	64	4 16	124	8 16	4	0 16	4	0.267	.04	.003	.54	.036
5	0 20	65	4 20	125	8 20	5	0 20	5	0.333	0.05	0.003	0.55	0.037
6	0 24	66	4 24	126	8 24	6	0 24	6	0.400	.06	.004	.56	.037
7	0 28	67	4 28	127	8 28	7	0 28	7	0.467	.07	.005	.57	.038
8	0 32	68	4 32	128	8 32	8	0 32	8	0.533	.08	.005	.58	.039
9	0 36	69	4 36	129	8 36	9	0 36	9	0.600	.09	.006	.59	.039
10	0 40	70	4 40	130	8 40	10	0 40	10	0.667	0.10	0.007	0.60	0.040
11	0 44	71	4 44	131	8 44	11	0 44	11	0.733	.11	.007	.61	.041
12	0 48	72	4 48	132	8 48	12	0 48	12	0.800	.12	.008	.62	.041
13	0 52	73	4 52	133	8 52	13	0 52	13	0.867	.13	.009	.63	.042
14	0 56	74	4 56	134	8 56	14	0 56	14	0.933	.14	.009	.64	.043
15	1 00	75	5 00	135	9 00	15	1 00	15	1.000	0.15	0.010	0.65	0.043
16	1 04	76	5 04	136	9 04	16	1 04	16	1.067	.16	.011	.66	.044
17	1 08	77	5 08	137	9 08	17	1 08	17	1.133	.17	.011	.67	.045
18	1 12	78	5 12	138	9 12	18	1 12	18	1.200	.18	.012	.68	.045
19	1 16	79	5 16	139	9 16	19	1 16	19	1.267	.19	.013	.69	.046
20	1 20	80	5 20	140	9 20	20	1 20	20	1.333	0.20	0.013	0.70	0.047
21	1 24	81	5 24	141	9 24	21	1 24	21	1.400	.21	.014	.71	.047
22	1 28	82	5 28	142	9 28	22	1 28	22	1.467	.22	.015	.72	.048
23	1 32	83	5 32	143	9 32	23	1 32	23	1.533	.23	.015	.73	.049
24	1 36	84	5 36	144	9 36	24	1 36	24	1.600	.24	.016	.74	.049
25	1 40	85	5 40	145	9 40	25	1 40	25	1.667	0.25	0.017	0.75	0.050
26	1 44	86	5 44	146	9 44	26	1 44	26	1.733	.26	.017	.76	.051
27	1 48	87	5 48	147	9 48	27	1 48	27	1.800	.27	.018	.77	.051
28	1 52	88	5 52	148	9 52	28	1 52	28	1.867	.28	.019	.78	.052
29	1 56	89	5 56	149	9 56	29	1 56	29	1.933	.29	.019	.79	.053
30	2 00	90	6 00	150	10 00	30	2 00	30	2.000	0.30	0.020	0.80	0.053
31	2 04	91	6 04	151	10 04	31	2 04	31	2.067	.31	.021	.81	.054
32	2 08	92	6 08	152	10 08	32	2 08	32	2.133	.32	.021	.82	.055
33	2 12	93	6 12	153	10 12	33	2 12	33	2.200	.33	.022	.83	.055
34	2 16	94	6 16	154	10 16	34	2 16	34	2.267	.34	.023	.84	.056
35	2 20	95	6 20	155	10 20	35	2 20	35	2.333	0.35	0.023	0.85	0.057
36	2 24	96	6 24	156	10 24	36	2 24	36	2.400	.36	.024	.86	.057
37	2 28	97	6 28	157	10 28	37	2 28	37	2.467	.37	.025	.87	.058
38	2 32	98	6 32	158	10 32	38	2 32	38	2.533	.38	.025	.88	.059
39	2 36	99	6 36	159	10 36	39	2 36	39	2.600	.39	.026	.89	.059
40	2 40	100	6 40	160	10 40	40	2 40	40	2.667	0.40	0.027	0.90	0.060
41	2 44	101	6 44	161	10 44	41	2 44	41	2.733	.41	.027	.91	.061
42	2 48	102	6 48	162	10 48	42	2 48	42	2.800	.42	.028	.92	.061
43	2 52	103	6 52	163	10 52	43	2 52	43	2.867	.43	.029	.93	.062
44	2 56	104	6 56	164	10 56	44	2 56	44	2.933	.44	.029	.94	.063
45	3 00	105	7 00	165	11 00	45	3 00	45	3.000	0.45	0.030	0.95	0.063
46	3 04	106	7 04	166	11 04	46	3 04	46	3.067	.46	.031	.96	.064
47	3 08	107	7 08	167	11 08	47	3 08	47	3.133	.47	.031	.97	.065
48	3 12	108	7 12	168	11 12	48	3 12	48	3.200	.48	.032	.98	.065
49	3 16	109	7 16	169	11 16	49	3 16	49	3.267	.49	.033	.99	.066
50	3 20	110	7 20	170	11 20	50	3 20	50	3.333	0.50	0.033	1.00	0.067
51	3 24	111	7 24	171	11 24	51	3 24	51	3.400				
52	3 28	112	7 28	172	11 28	52	3 28	52	3.467				
53	3 32	113	7 32	173	11 32	53	3 32	53	3.533				
54	3 36	114	7 36	174	11 36	54	3 36	54	3.600		90° =	6ʰ	
55	3 40	115	7 40	175	11 40	55	3 40	55	3.667				
56	3 44	116	7 44	176	11 44	56	3 44	56	3.733		180° =	12ʰ	
57	3 48	117	7 48	177	11 48	57	3 48	57	3.800				
58	3 52	118	7 52	178	11 52	58	3 52	58	3.867		270° =	18ʰ	
59	3 56	119	7 56	179	11 56	59	3 56	59	3.933				

Table I(b)

CONVERSION OF TIME TO ARC

m	0ʰ (° ′)	1ʰ (° ′)	2ʰ (° ′)	3ʰ (° ′)	4ʰ (° ′)	5ʰ (° ′)	s	(′ ″)
0	0 00	15 00	30 00	45 00	60 00	75 00	0	0 00
1	0 15	15 15	30 15	45 15	60 15	75 15	1	0 15
2	0 30	15 30	30 30	45 30	60 30	75 30	2	0 30
3	0 45	15 45	30 45	45 45	60 45	75 45	3	0 45
4	1 00	16 00	31 00	46 00	61 00	76 00	4	1 00
5	1 15	16 15	31 15	46 15	61 15	76 15	5	1 15
6	1 30	16 30	31 30	46 30	61 30	76 30	6	1 30
7	1 45	16 45	31 45	46 45	61 45	76 45	7	1 45
8	2 00	17 00	32 00	47 00	62 00	77 00	8	2 00
9	2 15	17 15	32 15	47 15	62 15	77 15	9	2 15
10	2 30	17 30	32 30	47 30	62 30	77 30	10	2 30
11	2 45	17 45	32 45	47 45	62 45	77 45	11	2 45
12	3 00	18 00	33 00	48 00	63 00	78 00	12	3 00
13	3 15	18 15	33 15	48 15	63 15	78 15	13	3 15
14	3 30	18 30	33 30	48 30	63 30	78 30	14	3 30
15	3 45	18 45	33 45	48 45	63 45	78 45	15	3 45
16	4 00	19 00	34 00	49 00	64 00	79 00	16	4 00
17	4 15	19 15	34 15	49 15	64 15	79 15	17	4 15
18	4 30	19 30	34 30	49 30	64 30	79 30	18	4 30
19	4 45	19 45	34 45	49 45	64 45	79 45	19	4 45
20	5 00	20 00	35 00	50 00	65 00	80 00	20	5 00
21	5 15	20 15	35 15	50 15	65 15	80 15	21	5 15
22	5 30	20 30	35 30	50 30	65 30	80 30	22	5 30
23	5 45	20 45	35 45	50 45	65 45	80 45	23	5 45
24	6 00	21 00	36 00	51 00	66 00	81 00	24	6 00
25	6 15	21 15	36 15	51 15	66 15	81 15	25	6 15
26	6 30	21 30	36 30	51 30	66 30	81 30	26	6 30
27	6 45	21 45	36 45	51 45	66 45	81 45	27	6 45
28	7 00	22 00	37 00	52 00	67 00	82 00	28	7 00
29	7 15	22 15	37 15	52 15	67 15	82 15	29	7 15
30	7 30	22 30	37 30	52 30	67 30	82 30	30	7 30
31	7 45	22 45	37 45	52 45	67 45	82 45	31	7 45
32	8 00	23 00	38 00	53 00	68 00	83 00	32	8 00
33	8 15	23 15	38 15	53 15	68 15	83 15	33	8 15
34	8 30	23 30	38 30	53 30	68 30	83 30	34	8 30
35	8 45	23 45	38 45	53 45	68 45	83 45	35	8 45
36	9 00	24 00	39 00	54 00	69 00	84 00	36	9 00
37	9 15	24 15	39 15	54 15	69 15	84 15	37	9 15
38	9 30	24 30	39 30	54 30	69 30	84 30	38	9 30
39	9 45	24 45	39 45	54 45	69 45	84 45	39	9 45
40	10 00	25 00	40 00	55 00	70 00	85 00	40	10 00
41	10 15	25 15	40 15	55 15	70 15	85 15	41	10 15
42	10 30	25 30	40 30	55 30	70 30	85 30	42	10 30
43	10 45	25 45	40 45	55 45	70 45	85 45	43	10 45
44	11 00	26 00	41 00	56 00	71 00	86 00	44	11 00
45	11 15	26 15	41 15	56 15	71 15	86 15	45	11 15
46	11 30	26 30	41 30	56 30	71 30	86 30	46	11 30
47	11 45	26 45	41 45	56 45	71 45	86 45	47	11 45
48	12 00	27 00	42 00	57 00	72 00	87 00	48	12 00
49	12 15	27 15	42 15	57 15	72 15	87 15	49	12 15
50	12 30	27 30	42 30	57 30	72 30	87 30	50	12 30
51	12 45	27 45	42 45	57 45	72 45	87 45	51	12 45
52	13 00	28 00	43 00	58 00	73 00	88 00	52	13 00
53	13 15	28 15	43 15	58 15	73 15	88 15	53	13 15
54	13 30	28 30	43 30	58 30	73 30	88 30	54	13 30
55	13 45	28 45	43 45	58 45	73 45	88 45	55	13 45
56	14 00	29 00	44 00	59 00	74 00	89 00	56	14 00
57	14 15	29 15	44 15	59 15	74 15	89 15	57	14 15
58	14 30	29 30	44 30	59 30	74 30	89 30	58	14 30
59	14 45	29 45	44 45	59 45	74 45	89 45	59	14 45

SECONDS

s	″	s	″
0.00	0.00	0.50	7.50
.01	0.15	.51	7.65
.02	0.30	.52	7.80
.03	0.45	.53	7.95
.04	0.60	.54	8.10
0.05	0.75	0.55	8.25
.06	0.90	.56	8.40
.07	1.05	.57	8.55
.08	1.20	.58	8.70
.09	1.35	.59	8.85
0.10	1.50	0.60	9.00
.11	1.65	.61	9.15
.12	1.80	.62	9.30
.13	1.95	.63	9.45
.14	2.10	.64	9.60
0.15	2.25	0.65	9.75
.16	2.40	.66	9.90
.17	2.55	.67	10.05
.18	2.70	.68	10.20
.19	2.85	.69	10.35
0.20	3.00	0.70	10.50
.21	3.15	.71	10.65
.22	3.30	.72	10.80
.23	3.45	.73	10.95
.24	3.60	.74	11.10
0.25	3.75	0.75	11.25
.26	3.90	.76	11.40
.27	4.05	.77	11.55
.28	4.20	.78	11.70
.29	4.35	.79	11.85
0.30	4.50	0.80	12.00
.31	4.65	.81	12.15
.32	4.80	.82	12.30
.33	4.95	.83	12.45
.34	5.10	.84	12.60
0.35	5.25	0.85	12.75
.36	5.40	.86	12.90
.37	5.55	.87	13.05
.38	5.70	.88	13.20
.39	5.85	.89	13.35
0.40	6.00	0.90	13.50
.41	6.15	.91	13.65
.42	6.30	.92	13.80
.43	6.45	.93	13.95
.44	6.60	.94	14.10
0.45	6.75	0.95	14.25
.46	6.90	.96	14.40
.47	7.05	.97	14.55
.48	7.20	.98	14.70
.49	7.35	.99	14.85
0.50	7.50	1.00	15.00

6ʰ = 90°

12ʰ = 180°

18ʰ = 270°

TABLE II

CONVERSION OF MEAN SIDEREAL INTO MEAN SOLAR TIME

m	0ʰ	1ʰ	2ʰ	3ʰ	4ʰ	5ʰ	6ʰ	7ʰ	SECONDS	
	m s	m s	m s	m s	m s	m s	m s	m s	s	s
0	0 00.000	0 09.830	0 19.659	0 29.489	0 39.318	0 49.148	0 58.977	1 08.807	0	0.000
1	0 00.164	0 09.993	0 19.823	0 29.653	0 39.482	0 49.312	0 59.141	1 08.971	1	.003
2	0 00.328	0 10.157	0 19.987	0 29.816	0 39.646	0 49.475	0 59.305	1 09.135	2	.005
3	0 00.491	0 10.321	0 20.151	0 29.980	0 39.810	0 49.639	0 59.469	1 09.298	3	.008
4	0 00.655	0 10.485	0 20.314	0 30.144	0 39.974	0 49.803	0 59.633	1 09.462	4	.011
5	0 00.819	0 10.649	0 20.478	0 30.308	0 40.137	0 49.967	0 59.796	1 09.626	5	0.014
6	0 00.983	0 10.813	0 20.642	0 30.472	0 40.301	0 50.131	0 59.960	1 09.790	6	.016
7	0 01.147	0 10.976	0 20.806	0 30.635	0 40.465	0 50.295	1 00.124	1 09.954	7	.019
8	0 01.311	0 11.140	0 20.970	0 30.799	0 40.629	0 50.458	1 00.288	1 10.118	8	.022
9	0 01.474	0 11.304	0 21.134	0 30.963	0 40.793	0 50.622	1 00.452	1 10.281	9	.025
10	0 01.638	0 11.468	0 21.297	0 31.127	0 40.956	0 50.786	1 00.616	1 10.445	10	0.027
11	0 01.802	0 11.632	0 21.461	0 31.291	0 41.120	0 50.950	1 00.779	1 10.609	11	.030
12	0 01.966	0 11.795	0 21.625	0 31.455	0 41.284	0 51.114	1 00.943	1 10.773	12	.033
13	0 02.130	0 11.959	0 21.789	0 31.618	0 41.448	0 51.278	1 01.107	1 10.937	13	.035
14	0 02.294	0 12.123	0 21.953	0 31.782	0 41.612	0 51.441	1 01.271	1 11.100	14	.038
15	0 02.457	0 12.287	0 22.117	0 31.946	0 41.776	0 51.605	1 01.435	1 11.264	15	0.041
16	0 02.621	0 12.451	0 22.280	0 32.110	0 41.939	0 51.769	1 01.599	1 11.428	16	.044
17	0 02.785	0 12.615	0 22.444	0 32.274	0 42.103	0 51.933	1 01.762	1 11.592	17	.046
18	0 02.949	0 12.778	0 22.608	0 32.438	0 42.267	0 52.097	1 01.926	1 11.756	18	.049
19	0 03.113	0 12.942	0 22.772	0 32.601	0 42.431	0 52.260	1 02.090	1 11.920	19	.052
20	0 03.277	0 13.106	0 22.936	0 32.765	0 42.595	0 52.424	1 02.254	1 12.083	20	0.055
21	0 03.440	0 13.270	0 23.099	0 32.929	0 42.759	0 52.588	1 02.418	1 12.247	21	.057
22	0 03.604	0 13.434	0 23.263	0 33.093	0 42.922	0 52.752	1 02.582	1 12.411	22	.060
23	0 03.768	0 13.598	0 23.427	0 33.257	0 43.086	0 52.916	1 02.745	1 12.575	23	.063
24	0 03.932	0 13.761	0 23.591	0 33.420	0 43.250	0 53.080	1 02.909	1 12.739	24	.066
25	0 04.096	0 13.925	0 23.755	0 33.584	0 43.414	0 53.243	1 03.073	1 12.903	25	0.068
26	0 04.259	0 14.089	0 23.919	0 33.748	0 43.578	0 53.407	1 03.237	1 13.066	26	.071
27	0 04.423	0 14.253	0 24.082	0 33.912	0 43.742	0 53.571	1 03.401	1 13.230	27	.074
28	0 04.587	0 14.417	0 24.246	0 34.076	0 43.905	0 53.735	1 03.564	1 13.394	28	.076
29	0 04.751	0 14.581	0 24.410	0 34.240	0 44.069	0 53.899	1 03.728	1 13.558	29	.079
30	0 04.915	0 14.744	0 24.574	0 34.403	0 44.233	0 54.063	1 03.892	1 13.722	30	0.082
31	0 05.079	0 14.908	0 24.738	0 34.567	0 44.397	0 54.226	1 04.056	1 13.886	31	.085
32	0 05.242	0 15.072	0 24.902	0 34.731	0 44.561	0 54.390	1 04.220	1 14.049	32	.087
33	0 05.406	0 15.236	0 25.065	0 34.895	0 44.724	0 54.554	1 04.384	1 14.213	33	.090
34	0 05.570	0 15.400	0 25.229	0 35.059	0 44.888	0 54.718	1 04.547	1 14.377	34	.093
35	0 05.734	0 15.563	0 25.393	0 35.223	0 45.052	0 54.882	1 04.711	1 14.541	35	0.096
36	0 05.898	0 15.727	0 25.557	0 35.386	0 45.216	0 55.046	1 04.875	1 14.705	36	.098
37	0 06.062	0 15.891	0 25.721	0 35.550	0 45.380	0 55.209	1 05.039	1 14.868	37	.101
38	0 06.225	0 16.055	0 25.885	0 35.714	0 45.544	0 55.373	1 05.203	1 15.032	38	.104
39	0 06.389	0 16.219	0 26.048	0 35.878	0 45.707	0 55.537	1 05.367	1 15.196	39	.106
40	0 06.553	0 16.383	0 26.212	0 36.042	0 45.871	0 55.701	1 05.530	1 15.360	40	0.109
41	0 06.717	0 16.546	0 26.376	0 36.206	0 46.035	0 55.865	1 05.694	1 15.524	41	.112
42	0 06.881	0 16.710	0 26.540	0 36.369	0 46.199	0 56.028	1 05.858	1 15.688	42	.115
43	0 07.045	0 16.874	0 26.704	0 36.533	0 46.363	0 56.192	1 06.022	1 15.851	43	.117
44	0 07.208	0 17.038	0 26.867	0 36.697	0 46.527	0 56.356	1 06.186	1 16.015	44	.120
45	0 07.372	0 17.202	0 27.031	0 36.861	0 46.690	0 56.520	1 06.350	1 16.179	45	0.123
46	0 07.536	0 17.366	0 27.195	0 37.025	0 46.854	0 56.684	1 06.513	1 16.343	46	.126
47	0 07.700	0 17.529	0 27.359	0 37.188	0 47.018	0 56.848	1 06.677	1 16.507	47	.128
48	0 07.864	0 17.693	0 27.523	0 37.352	0 47.182	0 57.011	1 06.841	1 16.671	48	.131
49	0 08.027	0 17.857	0 27.687	0 37.516	0 47.346	0 57.175	1 07.005	1 16.834	49	.134
50	0 08.191	0 18.021	0 27.850	0 37.680	0 47.510	0 57.339	1 07.169	1 16.998	50	0.137
51	0 08.355	0 18.185	0 28.014	0 37.844	0 47.673	0 57.503	1 07.332	1 17.162	51	.139
52	0 08.519	0 18.349	0 28.178	0 38.008	0 47.837	0 57.667	1 07.496	1 17.326	52	.142
53	0 08.683	0 18.512	0 28.342	0 38.171	0 48.001	0 57.831	1 07.660	1 17.490	53	.145
54	0 08.847	0 18.676	0 28.506	0 38.335	0 48.165	0 57.994	1 07.824	1 17.654	54	.147
55	0 09.010	0 18.840	0 28.670	0 38.499	0 48.329	0 58.158	1 07.988	1 17.817	55	0.150
56	0 09.174	0 19.004	0 28.833	0 38.663	0 48.492	0 58.322	1 08.152	1 17.981	56	.153
57	0 09.338	0 19.168	0 28.997	0 38.827	0 48.656	0 58.486	1 08.315	1 18.145	57	.156
58	0 09.502	0 19.331	0 29.161	0 38.991	0 48.820	0 58.650	1 08.479	1 18.309	58	.158
59	0 09.666	0 19.495	0 29.325	0 39.154	0 48.984	0 58.814	1 08.643	1 18.473	59	.161

Subtract tabular amount from mean sidereal time interval to obtain equivalent mean solar time interval.

(cont.)

Table II (cont.)

CONVERSION OF MEAN SIDEREAL INTO MEAN SOLAR TIME

m	8h	9h	10h	11h	12h	13h	14h	15h	SECONDS	
	m s	m s	m s	m s	m s	m s	m s	m s	s	s
0	1 18.636	1 28.466	1 38.296	1 48.125	1 57.955	2 07.784	2 17.614	2 27.443	0	0.000
1	1 18.800	1 28.630	1 38.459	1 48.289	1 58.119	2 07.948	2 17.778	2 27.607	1	.003
2	1 18.964	1 28.794	1 38.623	1 48.453	1 58.282	2 08.112	2 17.941	2 27.771	2	.005
3	1 19.128	1 28.958	1 38.787	1 48.617	1 58.446	2 08.276	2 18.105	2 27.935	3	.008
4	1 19.292	1 29.121	1 38.951	1 48.780	1 58.610	2 08.440	2 18.269	2 28.099	4	.011
5	1 19.456	1 29.285	1 39.115	1 48.944	1 58.774	2 08.603	2 18.433	2 28.263	5	0.014
6	1 19.619	1 29.449	1 39.279	1 49.108	1 58.938	2 08.767	2 18.597	2 28.426	6	.016
7	1 19.783	1 29.613	1 39.442	1 49.272	1 59.101	2 08.931	2 18.761	2 28.590	7	.019
8	1 19.947	1 29.777	1 39.606	1 49.436	1 59.265	2 09.095	2 18.924	2 28.754	8	.022
9	1 20.111	1 29.940	1 39.770	1 49.600	1 59.429	2 09.259	2 19.088	2 28.918	9	.025
10	1 20.275	1 30.104	1 39.934	1 49.763	1 59.593	2 09.423	2 19.252	2 29.082	10	0.027
11	1 20.439	1 30.268	1 40.098	1 49.927	1 59.757	2 09.586	2 19.416	2 29.245	11	.030
12	1 20.602	1 30.432	1 40.261	1 50.091	1 59.921	2 09.750	2 19.580	2 29.409	12	.033
13	1 20.766	1 30.596	1 40.425	1 50.255	2 00.084	2 09.914	2 19.744	2 29.573	13	.035
14	1 20.930	1 30.760	1 40.589	1 50.419	2 00.248	2 10.078	2 19.907	2 29.737	14	.038
15	1 21.094	1 30.923	1 40.753	1 50.583	2 00.412	2 10.242	2 20.071	2 29.901	15	0.041
16	1 21.258	1 31.087	1 40.917	1 50.746	2 00.576	2 10.405	2 20.235	2 30.065	16	.044
17	1 21.422	1 31.251	1 41.081	1 50.910	2 00.740	2 10.569	2 20.399	2 30.228	17	.046
18	1 21.585	1 31.415	1 41.244	1 51.074	2 00.904	2 10.733	2 20.563	2 30.392	18	.049
19	1 21.749	1 31.579	1 41.408	1 51.238	2 01.067	2 10.897	2 20.727	2 30.556	19	.052
20	1 21.913	1 31.743	1 41.572	1 51.402	2 01.231	2 11.061	2 20.890	2 30.720	20	0.055
21	1 22.077	1 31.906	1 41.736	1 51.565	2 01.395	2 11.225	2 21.054	2 30.884	21	.057
22	1 22.241	1 32.070	1 41.900	1 51.729	2 01.559	2 11.388	2 21.218	2 31.048	22	.060
23	1 22.404	1 32.234	1 42.064	1 51.893	2 01.723	2 11.552	2 21.382	2 31.211	23	.063
24	1 22.568	1 32.398	1 42.227	1 52.057	2 01.887	2 11.716	2 21.546	2 31.375	24	.066
25	1 22.732	1 32.562	1 42.391	1 52.221	2 02.050	2 11.880	2 21.709	2 31.539	25	0.068
26	1 22.896	1 32.726	1 42.555	1 52.385	2 02.214	2 12.044	2 21.873	2 31.703	26	.071
27	1 23.060	1 32.889	1 42.719	1 52.548	2 02.378	2 12.208	2 22.037	2 31.867	27	.074
28	1 23.224	1 33.053	1 42.883	1 52.712	2 02.542	2 12.371	2 22.201	2 32.031	28	.076
29	1 23.387	1 33.217	1 43.047	1 52.876	2 02.706	2 12.535	2 22.365	2 32.194	29	.079
30	1 23.551	1 33.381	1 43.210	1 53.040	2 02.869	2 12.699	2 22.529	2 32.358	30	0.082
31	1 23.715	1 33.545	1 43.374	1 53.204	2 03.033	2 12.863	2 22.692	2 32.522	31	.085
32	1 23.879	1 33.708	1 43.538	1 53.368	2 03.197	2 13.027	2 22.856	2 32.686	32	.087
33	1 24.043	1 33.872	1 43.702	1 53.531	2 03.361	2 13.191	2 23.020	2 32.850	33	.090
34	1 24.207	1 34.036	1 43.866	1 53.695	2 03.525	2 13.354	2 23.184	2 33.013	34	.093
35	1 24.370	1 34.200	1 44.029	1 53.859	2 03.689	2 13.518	2 23.348	2 33.177	35	0.096
36	1 24.534	1 34.364	1 44.193	1 54.023	2 03.852	2 13.682	2 23.512	2 33.341	36	.098
37	1 24.698	1 34.528	1 44.357	1 54.187	2 04.016	2 13.846	2 23.675	2 33.505	37	.101
38	1 24.862	1 34.691	1 44.521	1 54.351	2 04.180	2 14.010	2 23.839	2 33.669	38	.104
39	1 25.026	1 34.855	1 44.685	1 54.514	2 04.344	2 14.173	2 24.003	2 33.833	39	.106
40	1 25.190	1 35.019	1 44.849	1 54.678	2 04.508	2 14.337	2 24.167	2 33.996	40	0.109
41	1 25.353	1 35.183	1 45.012	1 54.842	2 04.672	2 14.501	2 24.331	2 34.160	41	.112
42	1 25.517	1 35.347	1 45.176	1 55.006	2 04.835	2 14.665	2 24.495	2 34.324	42	.115
43	1 25.681	1 35.511	1 45.340	1 55.170	2 04.999	2 14.829	2 24.658	2 34.488	43	.117
44	1 25.845	1 35.674	1 45.504	1 55.333	2 05.163	2 14.993	2 24.822	2 34.652	44	.120
45	1 26.009	1 35.838	1 45.668	1 55.497	2 05.327	2 15.156	2 24.986	2 34.816	45	0.123
46	1 26.172	1 36.002	1 45.832	1 55.661	2 05.491	2 15.320	2 25.150	2 34.979	46	.126
47	1 26.336	1 36.166	1 45.995	1 55.825	2 05.655	2 15.484	2 25.314	2 35.143	47	.128
48	1 26.500	1 36.330	1 46.159	1 55.989	2 05.818	2 15.648	2 25.477	2 35.307	48	.131
49	1 26.664	1 36.493	1 46.323	1 56.153	2 05.982	2 15.812	2 25.641	2 35.471	49	.134
50	1 26.828	1 36.657	1 46.487	1 56.316	2 06.146	2 15.976	2 25.805	2 35.635	50	0.137
51	1 26.992	1 36.821	1 46.651	1 56.480	2 06.310	2 16.139	2 25.969	2 35.798	51	.139
52	1 27.155	1 36.985	1 46.815	1 56.644	2 06.474	2 16.303	2 26.133	2 35.962	52	.142
53	1 27.319	1 37.149	1 46.978	1 56.808	2 06.637	2 16.467	2 26.297	2 36.126	53	.145
54	1 27.483	1 37.313	1 47.142	1 56.972	2 06.801	2 16.631	2 26.460	2 36.290	54	.147
55	1 27.647	1 37.476	1 47.306	1 57.136	2 06.965	2 16.795	2 26.624	2 36.454	55	0.150
56	1 27.811	1 37.640	1 47.470	1 57.299	2 07.129	2 16.959	2 26.788	2 36.618	56	.153
57	1 27.975	1 37.804	1 47.634	1 57.463	2 07.293	2 17.122	2 26.952	2 36.781	57	.156
58	1 28.138	1 37.968	1 47.797	1 57.627	2 07.457	2 17.286	2 27.116	2 36.945	58	.158
59	1 28.302	1 38.132	1 47.961	1 57.791	2 07.620	2 17.450	2 27.280	2 37.109	59	0.161

Subtract tabular amount from mean sidereal time interval to obtain equivalent mean solar time interval.

(cont.)

TABLE II (*cont.*)

CONVERSION OF MEAN SIDEREAL INTO MEAN SOLAR TIME

m	16^h	17^h	18^h	19^h	20^h	21^h	22^h	23^h	SECONDS	
	m s	m s	m s	m s	m s	m s	m s	m s	s	s
0	2 37.273	2 47.102	2 56.932	3 06.762	3 16.591	3 26.421	3 36.250	3 46.080	0	0.000
1	2 37.437	2 47.266	2 57.096	3 06.925	3 16.755	3 26.585	3 36.414	3 46.244	1	.003
2	2 37.601	2 47.430	2 57.260	3 07.089	3 16.919	3 26.748	3 36.578	3 46.407	2	.005
3	2 37.764	2 47.594	2 57.424	3 07.253	3 17.083	3 26.912	3 36.742	3 46.571	3	.008
4	2 37.928	2 47.758	2 57.587	3 07.417	3 17.246	3 27.076	3 36.906	3 46.735	4	.011
5	2 38.092	2 47.922	2 57.751	3 07.581	3 17.410	3 27.240	3 37.069	3 46.899	5	0.014
6	2 38.256	2 48.085	2 57.915	3 07.745	3 17.574	3 27.404	3 37.233	3 47.063	6	.016
7	2 38.420	2 48.249	2 58.079	3 07.908	3 17.738	3 27.568	3 37.397	3 47.227	7	.019
8	2 38.584	2 48.413	2 58.243	3 08.072	3 17.902	3 27.731	3 37.561	3 47.390	8	.022
9	2 38.747	2 48.577	2 58.406	3 08.236	3 18.066	3 27.895	3 37.725	3 47.554	9	.025
10	2 38.911	2 48.741	2 58.570	3 08.400	3 18.229	3 28.059	3 37.889	3 47.718	10	0.027
11	2 39.075	2 48.905	2 58.734	3 08.564	3 18.393	3 28.223	3 38.052	3 47.882	11	.030
12	2 39.239	2 49.068	2 58.898	3 08.728	3 18.557	3 28.387	3 38.216	3 48.046	12	.033
13	2 39.403	2 49.232	2 59.062	3 08.891	3 18.721	3 28.550	3 38.380	3 48.210	13	.035
14	2 39.566	2 49.396	2 59.226	3 09.055	3 18.885	3 28.714	3 38.544	3 48.373	14	.038
15	2 39.730	2 49.560	2 59.389	3 09.219	3 19.049	3 28.878	3 38.708	3 48.537	15	0.041
16	2 39.894	2 49.724	2 59.553	3 09.383	3 19.212	3 29.042	3 38.871	3 48.701	16	.044
17	2 40.058	2 49.888	2 59.717	3 09.547	3 19.376	3 29.206	3 39.035	3 48.865	17	.046
18	2 40.222	2 50.051	2 59.881	3 09.710	3 19.540	3 29.370	3 39.199	3 49.029	18	.049
19	2 40.386	2 50.215	3 00.045	3 09.874	3 19.704	3 29.533	3 39.363	3 49.193	19	.052
20	2 40.549	2 50.379	3 00.209	3 10.038	3 19.868	3 29.697	3 39.527	3 49.356	20	0.055
21	2 40.713	2 50.543	3 00.372	3 10.202	3 20.032	3 29.861	3 39.691	3 49.520	21	.057
22	2 40.877	2 50.707	3 00.536	3 10.366	3 20.195	3 30.025	3 39.854	3 49.684	22	.060
23	2 41.041	2 50.870	3 00.700	3 10.530	3 20.359	3 30.189	3 40.018	3 49.848	23	.063
24	2 41.205	2 51.034	3 00.864	3 10.693	3 20.523	3 30.353	3 40.182	3 50.012	24	.066
25	2 41.369	2 51.198	3 01.028	3 10.857	3 20.687	3 30.516	3 40.346	3 50.175	25	0.068
26	2 41.532	2 51.362	3 01.192	3 11.021	3 20.851	3 30.680	3 40.510	3 50.339	26	.071
27	2 41.696	2 51.526	3 01.355	3 11.185	3 21.014	3 30.844	3 40.674	3 50.503	27	.074
28	2 41.860	2 51.690	3 01.519	3 11.349	3 21.178	3 31.008	3 40.837	3 50.667	28	.076
29	2 42.024	2 51.853	3 01.683	3 11.513	3 21.342	3 31.172	3 41.001	3 50.831	29	.079
30	2 42.188	2 52.017	3 01.847	3 11.676	3 21.506	3 31.336	3 41.165	3 50.995	30	0.082
31	2 42.352	2 52.181	3 02.011	3 11.840	3 21.670	3 31.499	3 41.329	3 51.158	31	.085
32	2 42.515	2 52.345	3 02.174	3 12.004	3 21.834	3 31.663	3 41.493	3 51.322	32	.087
33	2 42.679	2 52.509	3 02.338	3 12.168	3 21.997	3 31.827	3 41.657	3 51.486	33	.090
34	2 42.843	2 52.673	3 02.502	3 12.332	3 22.161	3 31.991	3 41.820	3 51.650	34	.093
35	2 43.007	2 52.836	3 02.666	3 12.496	3 22.325	3 32.155	3 41.984	3 51.814	35	0.096
36	2 43.171	2 53.000	3 02.830	3 12.659	3 22.489	3 32.318	3 42.148	3 51.978	36	.098
37	2 43.334	2 53.164	3 02.994	3 12.823	3 22.653	3 32.482	3 42.312	3 52.141	37	.101
38	2 43.498	2 53.328	3 03.157	3 12.987	3 22.817	3 32.646	3 42.476	3 52.305	38	.104
39	2 43.662	2 53.492	3 03.321	3 13.151	3 22.980	3 32.810	3 42.639	3 52.469	39	.106
40	2 43.826	2 53.656	3 03.485	3 13.315	3 23.144	3 32.974	3 42.803	3 52.633	40	0.109
41	2 43.990	2 53.819	3 03.649	3 13.478	3 23.308	3 33.138	3 42.967	3 52.797	41	.112
42	2 44.154	2 53.983	3 03.813	3 13.642	3 23.472	3 33.301	3 43.131	3 52.961	42	.115
43	2 44.317	2 54.147	3 03.977	3 13.806	3 23.636	3 33.465	3 43.295	3 53.124	43	.117
44	2 44.481	2 54.311	3 04.140	3 13.970	3 23.800	3 33.629	3 43.459	3 53.288	44	.120
45	2 44.645	2 54.475	3 04.304	3 14.134	3 23.963	3 33.793	3 43.622	3 53.452	45	0.123
46	2 44.809	2 54.638	3 04.468	3 14.298	3 24.127	3 33.957	3 43.786	3 53.616	46	.126
47	2 44.973	2 54.802	3 04.632	3 14.461	3 24.291	3 34.121	3 43.950	3 53.780	47	.128
48	2 45.137	2 54.966	3 04.796	3 14.625	3 24.455	3 34.284	3 44.114	3 53.943	48	.131
49	2 45.300	2 55.130	3 04.960	3 14.789	3 24.619	3 34.448	3 44.278	3 54.107	49	.134
50	2 45.464	2 55.294	3 05.123	3 14.953	3 24.782	3 34.612	3 44.442	3 54.271	50	0.137
51	2 45.628	2 55.458	3 05.287	3 15.117	3 24.946	3 34.776	3 44.605	3 54.435	51	.139
52	2 45.792	2 55.621	3 05.451	3 15.281	3 25.110	3 34.940	3 44.769	3 54.599	52	.142
53	2 45.956	2 55.785	3 05.615	3 15.444	3 25.274	3 35.104	3 44.933	3 54.763	53	.145
54	2 46.120	2 55.949	3 05.779	3 15.608	3 25.438	3 35.267	3 45.097	3 54.926	54	.147
55	2 46.283	2 56.113	3 05.942	3 15.772	3 25.602	3 35.431	3 45.261	3 55.090	55	0.150
56	2 46.447	2 56.277	3 06.106	3 15.936	3 25.765	3 35.595	3 45.425	3 55.254	56	.153
57	2 46.611	2 56.441	3 06.270	3 16.100	3 25.929	3 35.759	3 45.588	3 55.418	57	.156
58	2 46.775	2 56.604	3 06.434	3 16.264	3 26.093	3 35.923	3 45.752	3 55.582	58	.158
59	2 46.939	2 56.768	3 06.598	3 16.427	3 26.257	3 36.086	3 45.916	3 55.746	59	0.161

Subtract tabular amount from mean sidereal time interval to obtain equivalent mean solar time interval.

TABLE III

CONVERSION OF MEAN SOLAR INTO MEAN SIDEREAL TIME

m	0ʰ	1ʰ	2ʰ	3ʰ	4ʰ	5ʰ	6ʰ	7ʰ	SECONDS	
	m s	m s	m s	m s	m s	m s	m s	m s	s	s
0	0 00.000	0 09.856	0 19.713	0 29.569	0 39.426	0 49.282	0 59.139	1 08.995	0	0.000
1	0 00.164	0 10.021	0 19.877	0 29.734	0 39.590	0 49.447	0 59.303	1 09.160	1	.003
2	0 00.329	0 10.185	0 20.041	0 29.898	0 39.754	0 49.611	0 59.467	1 09.324	2	.005
3	0 00.493	0 10.349	0 20.206	0 30.062	0 39.919	0 49.775	0 59.632	1 09.488	3	.008
4	0 00.657	0 10.514	0 20.370	0 30.227	0 40.083	0 49.939	0 59.796	1 09.652	4	.011
5	0 00.821	0 10.678	0 20.534	0 30.391	0 40.247	0 50.104	0 59.960	1 09.817	5	0.014
6	0 00.986	0 10.842	0 20.699	0 30.555	0 40.412	0 50.268	1 00.124	1 09.981	6	.016
7	0 01.150	0 11.006	0 20.863	0 30.719	0 40.576	0 50.432	1 00.289	1 10.145	7	.019
8	0 01.314	0 11.171	0 21.027	0 30.884	0 40.740	0 50.597	1 00.453	1 10.310	8	.022
9	0 01.478	0 11.335	0 21.191	0 31.048	0 40.904	0 50.761	1 00.617	1 10.474	9	.025
10	0 01.643	0 11.499	0 21.356	0 31.212	0 41.069	0 50.925	1 00.782	1 10.638	10	0.027
11	0 01.807	0 11.663	0 21.520	0 31.376	0 41.233	0 51.089	1 00.946	1 10.802	11	.030
12	0 01.971	0 11.828	0 21.684	0 31.541	0 41.397	0 51.254	1 01.110	1 10.967	12	.033
13	0 02.136	0 11.992	0 21.849	0 31.705	0 41.561	0 51.418	1 01.274	1 11.131	13	.036
14	0 02.300	0 12.156	0 22.013	0 31.869	0 41.726	0 51.582	1 01.439	1 11.295	14	.038
15	0 02.464	0 12.321	0 22.177	0 32.034	0 41.890	0 51.746	1 01.603	1 11.459	15	0.041
16	0 02.628	0 12.485	0 22.341	0 32.198	0 42.054	0 51.911	1 01.767	1 11.624	16	.044
17	0 02.793	0 12.649	0 22.506	0 32.362	0 42.219	0 52.075	1 01.932	1 11.788	17	.047
18	0 02.957	0 12.813	0 22.670	0 32.526	0 42.383	0 52.239	1 02.096	1 11.952	18	.049
19	0 03.121	0 12.978	0 22.834	0 32.691	0 42.547	0 52.404	1 02.260	1 12.117	19	.052
20	0 03.285	0 13.142	0 22.998	0 32.855	0 42.711	0 52.568	1 02.424	1 12.281	20	0.055
21	0 03.450	0 13.306	0 23.163	0 33.019	0 42.876	0 52.732	1 02.589	1 12.445	21	.057
22	0 03.614	0 13.471	0 23.327	0 33.183	0 43.040	0 52.896	1 02.753	1 12.609	22	.060
23	0 03.778	0 13.635	0 23.491	0 33.348	0 43.204	0 53.061	1 02.917	1 12.774	23	.063
24	0 03.943	0 13.799	0 23.656	0 33.512	0 43.368	0 53.225	1 03.081	1 12.938	24	.066
25	0 04.107	0 13.963	0 23.820	0 33.676	0 43.533	0 53.389	1 03.246	1 13.102	25	0.068
26	0 04.271	0 14.128	0 23.984	0 33.841	0 43.697	0 53.554	1 03.410	1 13.266	26	.071
27	0 04.435	0 14.292	0 24.148	0 34.005	0 43.861	0 53.718	1 03.574	1 13.431	27	.074
28	0 04.600	0 14.456	0 24.313	0 34.169	0 44.026	0 53.882	1 03.739	1 13.595	28	.077
29	0 04.764	0 14.620	0 24.477	0 34.333	0 44.190	0 54.046	1 03.903	1 13.759	29	.079
30	0 04.928	0 14.785	0 24.641	0 34.498	0 44.354	0 54.211	1 04.067	1 13.924	30	0.082
31	0 05.093	0 14.949	0 24.805	0 34.662	0 44.518	0 54.375	1 04.231	1 14.088	31	.085
32	0 05.257	0 15.113	0 24.970	0 34.826	0 44.683	0 54.539	1 04.396	1 14.252	32	.088
33	0 05.421	0 15.278	0 25.134	0 34.990	0 44.847	0 54.703	1 04.560	1 14.416	33	.090
34	0 05.585	0 15.442	0 25.298	0 35.155	0 45.011	0 54.868	1 04.724	1 14.581	34	.093
35	0 05.750	0 15.606	0 25.463	0 35.319	0 45.176	0 55.032	1 04.888	1 14.745	35	0.096
36	0 05.914	0 15.770	0 25.627	0 35.483	0 45.340	0 55.196	1 05.053	1 14.909	36	.099
37	0 06.078	0 15.935	0 25.791	0 35.648	0 45.504	0 55.361	1 05.217	1 15.073	37	.101
38	0 06.242	0 16.099	0 25.955	0 35.812	0 45.668	0 55.525	1 05.381	1 15.238	38	.104
39	0 06.407	0 16.263	0 26.120	0 35.976	0 45.833	0 55.689	1 05.546	1 15.402	39	.107
40	0 06.571	0 16.427	0 26.284	0 36.140	0 45.997	0 55.853	1 05.710	1 15.566	40	0.110
41	0 06.735	0 16.592	0 26.448	0 36.305	0 46.161	0 56.018	1 05.874	1 15.731	41	.112
42	0 06.900	0 16.756	0 26.612	0 36.469	0 46.325	0 56.182	1 06.038	1 15.895	42	.115
43	0 07.064	0 16.920	0 26.777	0 36.633	0 46.490	0 56.346	1 06.203	1 16.059	43	.118
44	0 07.228	0 17.085	0 26.941	0 36.798	0 46.654	0 56.510	1 06.367	1 16.223	44	.120
45	0 07.392	0 17.249	0 27.105	0 36.962	0 46.818	0 56.675	1 06.531	1 16.388	45	0.123
46	0 07.557	0 17.413	0 27.270	0 37.126	0 46.983	0 56.839	1 06.695	1 16.552	46	.126
47	0 07.721	0 17.577	0 27.434	0 37.290	0 47.147	0 57.003	1 06.860	1 16.716	47	.129
48	0 07.885	0 17.742	0 27.598	0 37.455	0 47.311	0 57.168	1 07.024	1 16.881	48	.131
49	0 08.049	0 17.906	0 27.762	0 37.619	0 47.475	0 57.332	1 07.188	1 17.045	49	.134
50	0 08.214	0 18.070	0 27.927	0 37.783	0 47.640	0 57.496	1 07.353	1 17.209	50	0.137
51	0 08.378	0 18.234	0 28.091	0 37.947	0 47.804	0 57.660	1 07.517	1 17.373	51	.140
52	0 08.542	0 18.399	0 28.255	0 38.112	0 47.968	0 57.825	1 07.681	1 17.538	52	.142
53	0 08.707	0 18.563	0 28.420	0 38.276	0 48.132	0 57.989	1 07.845	1 17.702	53	.145
54	0 08.871	0 18.727	0 28.584	0 38.440	0 48.297	0 58.153	1 08.010	1 17.866	54	.148
55	0 09.035	0 18.892	0 28.748	0 38.605	0 48.461	0 58.317	1 08.174	1 18.030	55	0.151
56	0 09.199	0 19.056	0 28.912	0 38.769	0 48.625	0 58.482	1 08.338	1 18.195	56	.153
57	0 09.364	0 19.220	0 29.077	0 38.933	0 48.790	0 58.646	1 08.502	1 18.359	57	.156
58	0 09.528	0 19.384	0 29.241	0 39.097	0 48.954	0 58.810	1 08.667	1 18.523	58	.159
59	0 09.692	0 19.549	0 29.405	0 39.262	0 49.118	0 58.975	1 08.831	1 18.688	59	0.162

Add tabular amount to mean solar time interval to obtain equivalent mean sidereal time interval.

(cont.)

TABLE III (*cont.*)

CONVERSION OF MEAN SOLAR INTO MEAN SIDEREAL TIME

m	8ʰ	9ʰ	10ʰ	11ʰ	12ʰ	13ʰ	14ʰ	15ʰ	SECONDS	
	m s	m s	m s	m s	m s	m s	m s	m s	s	s
0	1 18.852	1 28.708	1 38.565	1 48.421	1 58.278	2 08.134	2 17.991	2 27.847	0	0.000
1	1 19.016	1 28.873	1 38.729	1 48.585	1 58.442	2 08.298	2 18.155	2 28.011	1	.003
2	1 19.180	1 29.037	1 38.893	1 48.750	1 58.606	2 08.463	2 18.319	2 28.176	2	.005
3	1 19.345	1 29.201	1 39.058	1 48.914	1 58.771	2 08.627	2 18.483	2 28.340	3	.008
4	1 19.509	1 29.365	1 39.222	1 49.078	1 58.935	2 08.791	2 18.648	2 28.504	4	.011
5	1 19.673	1 29.530	1 39.386	1 49.243	1 59.099	2 08.956	2 18.812	2 28.668	5	0.014
6	1 19.837	1 29.694	1 39.550	1 49.407	1 59.263	2 09.120	2 18.976	2 28.833	6	.016
7	1 20.002	1 29.858	1 39.715	1 49.571	1 59.428	2 09.284	2 19.141	2 28.997	7	.019
8	1 20.166	1 30.022	1 39.879	1 49.735	1 59.592	2 09.448	2 19.305	2 29.161	8	.022
9	1 20.330	1 30.187	1 40.043	1 49.900	1 59.756	2 09.613	2 19.469	2 29.326	9	.025
10	1 20.495	1 30.351	1 40.207	1 50.064	1 59.920	2 09.777	2 19.633	2 29.490	10	0.027
11	1 20.659	1 30.515	1 40.372	1 50.228	2 00.085	2 09.941	2 19.798	2 29.654	11	.030
12	1 20.823	1 30.680	1 40.536	1 50.393	2 00.249	2 10.105	2 19.962	2 29.818	12	.033
13	1 20.987	1 30.844	1 40.700	1 50.557	2 00.413	2 10.270	2 20.126	2 29.983	13	.036
14	1 21.152	1 31.008	1 40.865	1 50.721	2 00.578	2 10.434	2 20.290	2 30.147	14	.038
15	1 21.316	1 31.172	1 41.029	1 50.885	2 00.742	2 10.598	2 20.455	2 30.311	15	0.041
16	1 21.480	1 31.337	1 41.193	1 51.050	2 00.906	2 10.763	2 20.619	2 30.476	16	.044
17	1 21.644	1 31.501	1 41.357	1 51.214	2 01.070	2 10.927	2 20.783	2 30.640	17	.047
18	1 21.809	1 31.665	1 41.522	1 51.378	2 01.235	2 11.091	2 20.948	2 30.804	18	.049
19	1 21.973	1 31.829	1 41.686	1 51.542	2 01.399	2 11.255	2 21.112	2 30.968	19	.052
20	1 22.137	1 31.994	1 41.850	1 51.707	2 01.563	2 11.420	2 21.276	2 31.133	20	0.055
21	1 22.302	1 32.158	1 42.015	1 51.871	2 01.727	2 11.584	2 21.440	2 31.297	21	.057
22	1 22.466	1 32.322	1 42.179	1 52.035	2 01.892	2 11.748	2 21.605	2 31.461	22	.060
23	1 22.630	1 32.487	1 42.343	1 52.200	2 02.056	2 11.912	2 21.769	2 31.625	23	.063
24	1 22.794	1 32.651	1 42.507	1 52.364	2 02.220	2 12.077	2 21.933	2 31.790	24	.066
25	1 22.959	1 32.815	1 42.672	1 52.528	2 02.385	2 12.241	2 22.098	2 31.954	25	0.068
26	1 23.123	1 32.979	1 42.836	1 52.692	2 02.549	2 12.405	2 22.262	2 32.118	26	.071
27	1 23.287	1 33.144	1 43.000	1 52.857	2 02.713	2 12.570	2 22.426	2 32.283	27	.074
28	1 23.451	1 33.308	1 43.164	1 53.021	2 02.877	2 12.734	2 22.590	2 32.447	28	.077
29	1 23.616	1 33.472	1 43.329	1 53.185	2 03.042	2 12.898	2 22.755	2 32.611	29	.079
30	1 23.780	1 33.637	1 43.493	1 53.349	2 03.206	2 13.062	2 22.919	2 32.775	30	0.082
31	1 23.944	1 33.801	1 43.657	1 53.514	2 03.370	2 13.227	2 23.083	2 32.940	31	.085
32	1 24.109	1 33.965	1 43.822	1 53.678	2 03.534	2 13.391	2 23.247	2 33.104	32	.088
33	1 24.273	1 34.129	1 43.986	1 53.842	2 03.699	2 13.555	2 23.412	2 33.268	33	.090
34	1 24.437	1 34.294	1 44.150	1 54.007	2 03.863	2 13.720	2 23.576	2 33.432	34	.093
35	1 24.601	1 34.458	1 44.314	1 54.171	2 04.027	2 13.884	2 23.740	2 33.597	35	0.096
36	1 24.766	1 34.622	1 44.479	1 54.335	2 04.192	2 14.048	2 23.905	2 33.761	36	.099
37	1 24.930	1 34.786	1 44.643	1 54.499	2 04.356	2 14.212	2 24.069	2 33.925	37	.101
38	1 25.094	1 34.951	1 44.807	1 54.664	2 04.520	2 14.377	2 24.233	2 34.090	38	.104
39	1 25.259	1 35.115	1 44.971	1 54.828	2 04.684	2 14.541	2 24.397	2 34.254	39	.107
40	1 25.423	1 35.279	1 45.136	1 54.992	2 04.849	2 14.705	2 24.562	2 34.418	40	0.110
41	1 25.587	1 35.444	1 45.300	1 55.156	2 05.013	2 14.869	2 24.726	2 34.582	41	.112
42	1 25.751	1 35.608	1 45.464	1 55.321	2 05.177	2 15.034	2 24.890	2 34.747	42	.115
43	1 25.916	1 35.772	1 45.629	1 55.485	2 05.342	2 15.198	2 25.054	2 34.911	43	.118
44	1 26.080	1 35.936	1 45.793	1 55.649	2 05.506	2 15.362	2 25.219	2 35.075	44	.120
45	1 26.244	1 36.101	1 45.957	1 55.814	2 05.670	2 15.527	2 25.383	2 35.239	45	0.123
46	1 26.408	1 36.265	1 46.121	1 55.978	2 05.834	2 15.691	2 25.547	2 35.404	46	.126
47	1 26.573	1 36.429	1 46.286	1 56.142	2 05.999	2 15.855	2 25.712	2 35.568	47	.129
48	1 26.737	1 36.593	1 46.450	1 56.306	2 06.163	2 16.019	2 25.876	2 35.732	48	.131
49	1 26.901	1 36.758	1 46.614	1 56.471	2 06.327	2 16.184	2 26.040	2 35.897	49	.134
50	1 27.066	1 36.922	1 46.778	1 56.635	2 06.491	2 16.348	2 26.204	2 36.061	50	0.137
51	1 27.230	1 37.086	1 46.943	1 56.799	2 06.656	2 16.512	2 26.369	2 36.225	51	.140
52	1 27.394	1 37.251	1 47.107	1 56.964	2 06.820	2 16.676	2 26.533	2 36.389	52	.142
53	1 27.558	1 37.415	1 47.271	1 57.128	2 06.984	2 16.841	2 26.697	2 36.554	53	.145
54	1 27.723	1 37.579	1 47.436	1 57.292	2 07.149	2 17.005	2 26.861	2 36.718	54	.148
55	1 27.887	1 37.743	1 47.600	1 57.456	2 07.313	2 17.169	2 27.026	2 36.882	55	0.151
56	1 28.051	1 37.908	1 47.764	1 57.621	2 07.477	2 17.334	2 27.190	2 37.047	56	.153
57	1 28.215	1 38.072	1 47.928	1 57.785	2 07.641	2 17.498	2 27.354	2 37.211	57	.156
58	1 28.380	1 38.236	1 48.093	1 57.949	2 07.806	2 17.662	2 27.519	2 37.375	58	.159
59	1 28.544	1 38.400	1 48.257	1 58.113	2 07.970	2 17.826	2 27.683	2 37.539	59	0.162

Add tabular amount to mean solar time interval to obtain equivalent mean sidereal time interval.

(*cont.*)

TABLE III (cont.)

CONVERSION OF MEAN SOLAR INTO MEAN SIDEREAL TIME

m	16ʰ	17ʰ	18ʰ	19ʰ	20ʰ	21ʰ	22ʰ	23ʰ	s	s
	m s	m s	m s	m s	m s	m s	m s	m s		
0	2 37.704	2 47.560	2 57.417	3 07.273	3 17.129	3 26.986	3 36.842	3 46.699	0	0.000
1	2 37.868	2 47.724	2 57.581	3 07.437	3 17.294	3 27.150	3 37.007	3 46.863	1	.003
2	2 38.032	2 47.889	2 57.745	3 07.602	3 17.458	3 27.315	3 37.171	3 47.027	2	.005
3	2 38.196	2 48.053	2 57.909	3 07.766	3 17.622	3 27.479	3 37.335	3 47.192	3	.008
4	2 38.361	2 48.217	2 58.074	3 07.930	3 17.787	3 27.643	3 37.500	3 47.356	4	.011
5	2 38.525	2 48.381	2 58.238	3 08.094	3 17.951	3 27.807	3 37.664	3 47.520	5	0.014
6	2 38.689	2 48.546	2 58.402	3 08.259	3 18.115	3 27.972	3 37.828	3 47.685	6	.016
7	2 38.854	2 48.710	2 58.566	3 08.423	3 18.279	3 28.136	3 37.992	3 47.849	7	.019
8	2 39.018	2 48.874	2 58.731	3 08.587	3 18.444	3 28.300	3 38.157	3 48.013	8	.022
9	2 39.182	2 49.039	2 58.895	3 08.751	3 18.608	3 28.464	3 38.321	3 48.177	9	.025
10	2 39.346	2 49.203	2 59.059	3 08.916	3 18.772	3 28.629	3 38.485	3 48.342	10	0.027
11	2 39.511	2 49.367	2 59.224	3 09.080	3 18.937	3 28.793	3 38.649	3 48.506	11	.030
12	2 39.675	2 49.531	2 59.388	3 09.244	3 19.101	3 28.957	3 38.814	3 48.670	12	.033
13	2 39.839	2 49.696	2 59.552	3 09.409	3 19.265	3 29.122	3 38.978	3 48.834	13	.036
14	2 40.003	2 49.860	2 59.716	3 09.573	3 19.429	3 29.286	3 39.142	3 48.999	14	.038
15	2 40.168	2 50.024	2 59.881	3 09.737	3 19.594	3 29.450	3 39.307	3 49.163	15	0.041
16	2 40.332	2 50.188	3 00.045	3 09.901	3 19.758	3 29.614	3 39.471	3 49.327	16	.044
17	2 40.496	2 50.353	3 00.209	3 10.066	3 19.922	3 29.779	3 39.635	3 49.492	17	.047
18	2 40.661	2 50.517	3 00.373	3 10.230	3 20.086	3 29.943	3 39.799	3 49.656	18	.049
19	2 40.825	2 50.681	3 00.538	3 10.394	3 20.251	3 30.107	3 39.964	3 49.820	19	.052
20	2 40.989	2 50.846	3 00.702	3 10.559	3 20.415	3 30.271	3 40.128	3 49.984	20	0.055
21	2 41.153	2 51.010	3 00.866	3 10.723	3 20.579	3 30.436	3 40.292	3 50.149	21	.057
22	2 41.318	2 51.174	3 01.031	3 10.887	3 20.744	3 30.600	3 40.456	3 50.313	22	.060
23	2 41.482	2 51.338	3 01.195	3 11.051	3 20.908	3 30.764	3 40.621	3 50.477	23	.063
24	2 41.646	2 51.503	3 01.359	3 11.216	3 21.072	3 30.929	3 40.785	3 50.642	24	.066
25	2 41.810	2 51.667	3 01.523	3 11.380	3 21.236	3 31.093	3 40.949	3 50.806	25	0.068
26	2 41.975	2 51.831	3 01.688	3 11.544	3 21.401	3 31.257	3 41.114	3 50.970	26	.071
27	2 42.139	2 51.995	3 01.852	3 11.708	3 21.565	3 31.421	3 41.278	3 51.134	27	.074
28	2 42.303	2 52.160	3 02.016	3 11.873	3 21.729	3 31.586	3 41.442	3 51.299	28	.077
29	2 42.468	2 52.324	3 02.181	3 12.037	3 21.893	3 31.750	3 41.606	3 51.463	29	.079
30	2 42.632	2 52.488	3 02.345	3 12.201	3 22.058	3 31.914	3 41.771	3 51.627	30	0.082
31	2 42.796	2 52.653	3 02.509	3 12.366	3 22.222	3 32.078	3 41.935	3 51.791	31	.085
32	2 42.960	2 52.817	3 02.673	3 12.530	3 22.386	3 32.243	3 42.099	3 51.956	32	.088
33	2 43.125	2 52.981	3 02.838	3 12.694	3 22.551	3 32.407	3 42.264	3 52.120	33	.090
34	2 43.289	2 53.145	3 03.002	3 12.858	3 22.715	3 32.571	3 42.428	3 52.284	34	.093
35	2 43.453	2 53.310	3 03.166	3 13.023	3 22.879	3 32.736	3 42.592	3 52.449	35	0.096
36	2 43.617	2 53.474	3 03.330	3 13.187	3 23.043	3 32.900	3 42.756	3 52.613	36	.099
37	2 43.782	2 53.638	3 03.495	3 13.351	3 23.208	3 33.064	3 42.921	3 52.777	37	.101
38	2 43.946	2 53.803	3 03.659	3 13.515	3 23.372	3 33.228	3 43.085	3 52.941	38	.104
39	2 44.110	2 53.967	3 03.823	3 13.680	3 23.536	3 33.393	3 43.249	3 53.106	39	.107
40	2 44.275	2 54.131	3 03.988	3 13.844	3 23.700	3 33.557	3 43.413	3 53.270	40	0.110
41	2 44.439	2 54.295	3 04.152	3 14.008	3 23.865	3 33.721	3 43.578	3 53.434	41	.112
42	2 44.603	2 54.460	3 04.316	3 14.173	3 24.029	3 33.886	3 43.742	3 53.598	42	.115
43	2 44.767	2 54.624	3 04.480	3 14.337	3 24.193	3 34.050	3 43.906	3 53.763	43	.118
44	2 44.932	2 54.788	3 04.645	3 14.501	3 24.358	3 34.214	3 44.071	3 53.927	44	.120
45	2 45.096	2 54.952	3 04.809	3 14.665	3 24.522	3 34.378	3 44.235	3 54.091	45	0.123
46	2 45.260	2 55.117	3 04.973	3 14.830	3 24.686	3 34.543	3 44.399	3 54.256	46	.126
47	2 45.425	2 55.281	3 05.137	3 14.994	3 24.850	3 34.707	3 44.563	3 54.420	47	.129
48	2 45.589	2 55.445	3 05.302	3 15.158	3 25.015	3 34.871	3 44.728	3 54.584	48	.131
49	2 45.753	2 55.610	3 05.466	3 15.322	3 25.179	3 35.035	3 44.892	3 54.748	49	.134
50	2 45.917	2 55.774	3 05.630	3 15.487	3 25.343	3 35.200	3 45.056	3 54.913	50	0.137
51	2 46.082	2 55.938	3 05.795	3 15.651	3 25.508	3 35.364	3 45.220	3 55.077	51	.140
52	2 46.246	2 56.102	3 05.959	3 15.815	3 25.672	3 35.528	3 45.385	3 55.241	52	.142
53	2 46.410	2 56.267	3 06.123	3 15.980	3 25.836	3 35.693	3 45.549	3 55.405	53	.145
54	2 46.574	2 56.431	3 06.287	3 16.144	3 26.000	3 35.857	3 45.713	3 55.570	54	.148
55	2 46.739	2 56.595	3 06.452	3 16.308	3 26.165	3 36.021	3 45.878	3 55.734	55	0.151
56	2 46.903	2 56.759	3 06.616	3 16.472	3 26.329	3 36.185	3 46.042	3 55.898	56	.153
57	2 47.067	2 56.924	3 06.780	3 16.637	3 26.493	3 36.350	3 46.206	3 56.063	57	.156
58	2 47.232	2 57.088	3 06.944	3 16.801	3 26.657	3 36.514	3 46.370	3 56.227	58	.159
59	2 47.396	2 57.252	3 07.109	3 16.965	3 26.822	3 36.678	3 46.535	3 56.391	59	0.162

Add tabular amount to mean solar time interval to obtain equivalent mean sidereal time interval.

TABLE IV(a)

COORDINATES OF THE SUN, 1960

FOR 0ʰ EPHEMERIS TIME

Date	Apparent Right Ascension		Apparent Declination		Radius Vector		Semi-diameter	Equation of Time Apparent − Mean	
	h m s	s	° ′ ″	″			′ ″	m s	s
Nov. 16	15 24 59·82	247·85	−18 41 00·3	− 893·7	0·988 7908	−2115	16 12·08	+15 15·78	−11·29
17	15 29 07·67	248·69	18 55 54·0	873·7	·988 5793	2095	16 12·28	15 04·49	12·14
18	15 33 16·36	249·53	19 10 27·7	853·0	·988 3698	2074	16 12·49	14 52·35	12·97
19	15 37 25·89	250·35	19 24 40·7	832·0	·988 1624	2054	16 12·69	14 39·38	13·79
20	15 41 36·24	251·17	19 38 32·7	810·7	·987 9570	2033	16 12·90	14 25·59	14·61
21	15 45 47·41	251·97	−19 52 03·4	− 789·0	0·987 7537	−2009	16 13·10	+14 10·98	−15·40
22	15 49 59·38	252·74	20 05 12·4	766·9	·987 5528	1984	16 13·29	13 55·58	16·18
23	15 54 12·12	253·50	20 17 59·3	744·4	·987 3544	1955	16 13·49	13 39·40	16·95
24	15 58 25·62	254·26	20 30 23·7	721·6	·987 1589	1921	16 13·68	13 22·45	17·70
25	16 02 39·88	254·98	20 42 25·3	698·3	·986 9668	1885	16 13·87	13 04·75	18·43
26	16 06 54·86	255·71	−20 54 03·6	− 674·9	0·986 7783	−1843	16 14·06	+12 46·32	−19·15
27	16 11 10·57	256·41	21 05 18·5	651·0	·986 5940	1799	16 14·24	12 27·17	19·86
28	16 15 26·98	257·09	21 16 09·5	626·8	·986 4141	1751	16 14·42	12 07·31	20·54
29	16 19 44·07	257·78	21 26 36·3	602·4	·986 2390	1700	16 14·59	11 46·77	21·23
30	16 24 01·85	258·44	21 36 38·7	577·7	·986 0690	1647	16 14·76	11 25·54	21·88
Dec. 1	16 28 20·29	259·08	−21 46 16·4	− 552·6	0·985 9043	−1593	16 14·92	+11 03·66	−22·52
2	16 32 39·37	259·70	21 55 29·0	527·4	·985 7450	1537	16 15·08	10 41·14	23·15
3	16 36 59·07	260·32	22 04 16·4	501·8	·985 5913	1480	16 15·23	10 17·99	23·75
4	16 41 19·39	260·90	22 12 38·2	476·0	·985 4433	1425	16 15·38	9 54·24	24·34
5	16 45 40·29	261·46	22 20 34·2	450·0	·985 3008	1367	16 15·52	9 29·90	24·90
6	16 50 01·75	262·00	−22 28 04·2	− 423·8	0·985 1641	−1313	16 15·65	+ 9 05·00	−25·45
7	16 54 23·75	262·52	22 35 08·0	397·2	·985 0328	1258	16 15·78	8 39·55	25·96
8	16 58 46·27	263·01	22 41 45·2	370·5	·984 9070	1205	16 15·91	8 13·59	26·45
9	17 03 09·28	263·47	22 47 55·7	343·7	·984 7865	1153	16 16·03	7 47·14	26·92
10	17 07 32·75	263·91	22 53 39·4	316·5	·984 6712	1102	16 16·14	7 20·22	27·36
11	17 11 56·66	264·32	−22 58 55·9	− 289·3	0·984 5610	−1055	16 16·25	+ 6 52·86	−27·76
12	17 16 20·98	264·70	23 03 45·2	261·9	·984 4555	1008	16 16·36	6 25·10	28·15
13	17 20 45·68	265·05	23 08 07·1	234·3	·984 3547	965	16 16·46	5 56·95	28·50
14	17 25 10·73	265·37	23 12 01·4	206·5	·984 2582	924	16 16·55	5 28·45	28·81
15	17 29 36·10	265·65	23 15 27·9	178·8	·984 1658	885	16 16·64	4 59·64	29·09
16	17 34 01·75	265·90	−23 18 26·7	− 150·8	0·984 0773	− 849	16 16·73	+ 4 30·55	−29·34
17	17 38 27·65	266·11	23 20 57·5	122·8	·983 9924	815	16 16·82	4 01·21	29·55
18	17 42 53·76	266·27	23 23 00·3	94·7	·983 9109	782	16 16·90	3 31·66	29·71
19	17 47 20·03	266·41	23 24 35·0	66·6	·983 8327	747	16 16·98	3 01·95	29·84
20	17 51 46·44	266·49	23 25 41·6	38·4	·983 7580	712	16 17·05	2 32·11	29·93
21	17 56 12·93	266·54	−23 26 20·0	− 10·1	0·983 6868	− 675	16 17·12	+ 2 02·18	−29·98
22	18 00 39·47	266·54	23 26 30·1	+ 18·2	·983 6193	635	16 17·19	1 32·20	29·98
23	18 05 06·01	266·51	23 26 11·9	46·4	·983 5558	591	16 17·25	1 02·22	29·96
24	18 09 32·52	266·45	23 25 25·5	74·7	·983 4967	545	16 17·31	0 32·26	29·90
25	18 13 58·97	266·35	23 24 10·8	103·0	·983 4422	494	16 17·36	+ 0 02·36	29·80
26	18 18 25·32	266·23	−23 22 27·8	+ 131·2	0·983 3928	− 440	16 17·41	− 0 27·44	−29·67
27	18 22 51·55	266·06	23 20 16·6	159·3	·983 3488	384	16 17·46	0 57·11	29·51
28	18 27 17·61	265·87	23 17 37·3	187·3	·983 3104	325	16 17·49	1 26·62	29·31
29	18 31 43·48	265·65	23 14 30·0	215·4	·983 2779	265	16 17·53	1 55·93	29·09
30	18 36 09·13	265·40	23 10 54·6	243·1	·983 2514	203	16 17·55	2 25·02	28·84
31	18 40 34·53	265·11	−23 06 51·5	+ 270·9	0·983 2311	− 139	16 17·57	− 2 53·86	−28·56
32	18 44 59·64		−23 02 20·6		0·983 2172		16 17·59	− 3 22·42	

Table IV(b)

UNIVERSAL AND SIDEREAL TIMES, 1960

Date 0ʰ U.T.	Julian Date	Sidereal Time H.A. of First Point of Aries Apparent	Mean	Equation of Equinoxes	G.S.D. 0ʰ S.T.	Universal Time Transit of First Point of Aries Apparent	Mean
	2437	h m s	s	s	2443	d h m s	s
Nov. 16	254·5	3 40 15·606	16·033	−0·428	929·0	Nov. 16 20 16 24·572	24·143
17	255·5	3 44 12·158	12·589	·431	930·0	17 20 12 28·665	28·233
18	256·5	3 48 08·711	09·144	·433	931·0	18 20 08 32·754	32·324
19	257·5	3 52 05·268	05·699	·431	932·0	19 20 04 36·841	36·414
20	258·5	3 56 01·828	02·255	·427	933·0	20 20 00 40·925	40·505
21	259·5	3 59 58·390	58·810	−0·421	934·0	21 19 56 45·009	44·595
22	260·5	4 03 54·952	55·366	·414	935·0	22 19 52 49·093	48·686
23	261·5	4 07 51·514	51·921	·407	936·0	23 19 48 53·179	52·776
24	262·5	4 11 48·073	48·476	·403	937·0	24 19 44 57·268	56·867
25	263·5	4 15 44·630	45·032	·402	938·0	25 19 41 01·359	00·958
26	264·5	4 19 41·184	41·587	−0·403	939·0	26 19 37 05·453	05·048
27	265·5	4 23 37·736	38·142	·406	940·0	27 19 33 09·547	09·139
28	266·5	4 27 34·288	34·698	·410	941·0	28 19 29 13·641	13·229
29	267·5	4 31 30·839	31·253	·414	942·0	29 19 25 17·734	17·320
30	268·5	4 35 27·393	27·808	·416	943·0	30 19 21 21·826	21·410
Dec. 1	269·5	4 39 23·947	24·364	−0·416	944·0	Dec. 1 19 17 25·915	25·501
2	270·5	4 43 20·504	20·919	·415	945·0	2 19 13 30·003	29·591
3	271·5	4 47 17·063	17·475	·412	946·0	3 19 09 34·089	33·682
4	272·5	4 51 13·622	14·030	·408	947·0	4 19 05 38·175	37·772
5	273·5	4 55 10·183	10·585	·403	948·0	5 19 01 42·261	41·863
6	274·5	4 59 06·743	07·141	−0·398	949·0	6 18 57 46·347	45·953
7	275·5	5 03 03·302	03·696	·394	950·0	7 18 53 50·434	50·044
8	276·5	5 06 59·861	60·251	·391	951·0	8 18 49 54·523	54·134
9	277·5	5 10 56·417	56·807	·389	952·0	9 18 45 58·613	58·225
10	278·5	5 14 52·972	53·362	·390	953·0	10 18 42 02·705	02·316
11	279·5	5 18 49·526	49·917	−0·391	954·0	11 18 38 06·798	06·406
12	280·5	5 22 46·079	46·473	·394	955·0	12 18 34 10·892	10·497
13	281·5	5 26 42·631	43·028	·397	956·0	13 18 30 14·985	14·587
14	282·5	5 30 39·184	39·584	·399	957·0	14 18 26 19·077	18·678
15	283·5	5 34 35·739	36·139	·400	958·0	15 18 22 23·166	22·768
16	284·5	5 38 32·296	32·694	−0·398	959·0	16 18 18 27·253	26·859
17	285·5	5 42 28·856	29·250	·394	960·0	17 18 14 31·336	30·949
18	286·5	5 46 25·419	25·805	·386	961·0	18 18 10 35·418	35·040
19	287·5	5 50 21·983	22·360	·377	962·0	19 18 06 39·500	39·130
20	288·5	5 54 18·547	18·916	·369	963·0	20 18 02 43·583	43·221
21	289·5	5 58 15·109	15·471	−0·362	964·0	21 17 58 47·669	47·311
22	290·5	6 02 11·669	12·026	·358	965·0	22 17 54 51·758	51·402
23	291·5	6 06 08·225	08·582	·357	966·0	23 17 50 55·849	55·493
24	292·5	6 10 04·779	05·137	·358	967·0	24 17 46 59·942	59·583
25	293·5	6 14 01·331	01·693	·361	968·0	25 17 43 04·036	03·674
26	294·5	6 17 57·884	58·248	−0·364	969·0	26 17 39 08·128	07·764
27	295·5	6 21 54·438	54·803	·366	970·0	27 17 35 12·220	11·855
28	296·5	6 25 50·993	51·359	·366	971·0	28 17 31 16·309	15·945
29	297·5	6 29 47·550	47·914	·364	972·0	29 17 27 20·397	20·036
30	298·5	6 33 44·108	44·469	·361	973·0	30 17 23 24·483	24·126
31	299·5	6 37 40·668	41·025	−0·357	974·0	31 17 19 28·569	28·217
32	300·5	6 41 37·229	37·580	−0·351	975·0	32 17 15 32·654	32·307

TABLE V

ORBITAL ELEMENTS OF THE PLANETS FOR EPOCH SEPT. 23.0 1960, U.T. (Variations for 100-day intervals are indicated)

	Mercury	Venus	Earth	Mars	Jupiter	Saturn	Uranus	Neptune	Pluto
Semimajor axis (a), A.U.	0.387099 0	0.723332 0	1.000000	1.523691 0	5.203705 +114	9.580337 +40	19.14103 −769	30.19825 −1874	39.43871 −3455
Mean daily motion (n)	4°.092339 0	1°.602130	0°.9856091	0°.524033	0°.0830696 −27	0°.0332426 −2	0°.0117697 +71	0°.0059394 +55	0°.0039794 +52
Sidereal period (P) of revolution, tropical years	0.24085	0.61521	1.000039	1.88089	11.86532	29.6501	83.7445	165.951	247.687
Synodic period (S), tropical years	0.31726	1.59872	—	2.13539	1.09211	1.03495	1.01209	1.0061	1.0041
Eccentricity (e)	0.205627	0.006792	0.01672592 −11	0.093369	0.0486288 −323	0.0509895 +1252	0.0457866 −4258	0.0045616 +6510	0.2502358 −5150
Orbital inclination to the ecliptic (i)	7°.00400 +1	3°.39424	—	1°.84993 0	1°.30631 −5	2°.48718 0	0°.77225 −2	1°.77327 −1	17°.16987 +65
Longitude of the node (Ω)	47°.86575 +325	76°.32625 +247	—	49°.25464 +211	100°.0623 +24	113°.3236 +31	73°.7034 −56	131°.3549 +134	109°.8856 +63
Longitude of perihelion (ϖ)	76°.84441 +426	131°.01853 +385	102°.26498 +471	335°.33609 +504	13°.2649 −234	91°.2345 −4088	172°.8640 +274	22°.3129 +73	224°.1602 +1180
Mean anomaly Sept. 23.0	152°.303	108°.652	259°.5825	62°.572	268°.7994	197°.9259	331°.6846	196°.4852	317°.4861

TABLE VI — PHYSICAL CHARACTERISTICS OF THE PLANETS*

	Mercury	Venus	Earth
Mass:			
Reciprocal solar units	6,120,000 ± 43,000	408,600 ± 200	332,480 ± 40
Earth units	0.0543	0.8137	unity
Grams	3.24×10^{26}	4.863×10^{27}	5.976×10^{27}
Size:			
Apparent angular diameter (mean or equatorial)	6″.45 at 1 A.U.	16″.82 at 1 A.U.	Eq. 17″.5967 at 1 A.U.
Equatorial radius, km	—	—	6,378.15
Polar radius, km	—	—	6,356.76
Mean radius, km	2,340	6,100	6,371.02
Mean radius (earth = 1)	0.37	0.957	unity
Oblateness	0	0	1/298.2
Period of rotation, mean solar units	88.0 days	—	$23^h56^m4^s.09$
Axial inclination	—	—	23°27′8″.0
Density, gm · cm⁻³	6.03	5.11	5.517
Velocity of escape at equator, (km/sec)	4.30	10.3	11.2
Gravitational acceleration at the equator,			
cm · sec⁻²	395	872	978.039
(earth = 1)	0.40	0.89	unity
Albedo	0.058	0.76	0.39
Theoretical temperatures, °K:			
Spherical blackbody	441°	230°	246°
Hemispheric blackbody	525°	273°	293°
Subsolar blackbody	624°	325°	348°
Apparent magnitude	−0.2 at elongation	−4.08 at elongation	—

*See Chapter 2 for sources of data.

TABLE VI (*cont.*)

Mars	Jupiter	Saturn	Uranus	Neptune	Pluto
3,103,000 ± 7,000	1047.36 ± 0.03	3,499.7 ± 0.3	22,900 ± 200	18,890 ± 60	—
0.1071	317.45	95.00	14.5	17.60	—
6.400×10^{26}	1.8971×10^{30}	5.677×10^{29}	8.67×10^{28}	1.052×10^{29}	—
9″.18 at 1 A.U.	Eq. 37″.84 at 5.2 A.U.	Eq. 17″.44 at 9.54 A.U.	3″.42 at 19 A.U.	2″.04 at 30.1 A.U.	0″.23 at 35.6 A.U.
3,330	71,320	60,300	24,150	22,700	—
3,310	66,610	53,920	22,950	21,900	—
3,324	69,750	58,170	23,750	22,400	3,000
0.522	10.95	9.13	3.73	3.52	0.47
1/191	1/15.4	1/9.5	1/20:	1/30:	0
$4^h37^m22^s.668$	9^h50^m to 9^h55^m	10^h14^m to 10^h38^m	$10^h.7$	12^h43^m	6.39 days
25°12′	3°6′.9	26°44′.7	98°.0	29°	—
4.16	1.34	0.68	1.55	2.23	—
5.06	57.5	35.4	21.9	24.4	—
384	2316	877	946	1366	—
0.39	2.37	0.90	0.97	1.40	—
0.15	0.51	0.50	0.66	0.62	0.16
218°	102°	76°	49°	40°	42°
259°	122°	90°	58°	47°	50°
308°	145°	107°	69°	56°	60°
−1.94 at opposition	−2.4 at opposition	+0.8 at opposition	+5.8 at opposition	+7.6 at opposition	+14.7 at opposition

TABLE VII — ORBITAL AND PHYSICAL CHARACTERISTICS OF THE SATELLITES

| | Mean distance | | Siderial period, days | Orbit inclination to: planet's equator† proper plane* |
	(Equatorial radius of planet = 1)	km		
Moon	60.268	384,405	27.321661	28°35′ to 18°19′†
Mars I Phobos	2.82	9,380	0.318910	1°8′
Mars II Deimos	7.06	23,500	1.262441	1°46′
Jupiter V	2.531	180,500	0.498179	0°24′
Jupiter I Io	5.911	421,600	1.769138	0°2′ *
Jupiter II Europa	9.405	670,800	3.551181	0°28′
Jupiter III Ganymede	15.003	1,070,000	7.154553	0°11′
Jupiter IV Callisto	26.388	1,882,000	16.689018	0°15′
Jupiter VI	160.8	11,470,000	250.57	27°38′
Jupiter X	166	11,850,000	263.55	29°01′
Jupiter VII	165	11,800,000	259.65	24°46′
Jupiter XII	297	21,200,000	631.1	147° †
Jupiter XI	316	22,600,000	692.5	164°
Jupiter VIII	329	23,500,000	738.9	145°
Jupiter IX	332	23,700,000	758.	153°
Saturn I Mimas	3.07	185,400	0.942422	1°31′
Saturn II Enceladus	3.95	237,900	1.370218	0°01′
Saturn III Tethys	4.88	294,500	1.887802	1°06′
Saturn IV Dione	6.25	377,200	2.736916	0°01′
Saturn V Rhea	8.73	526,700	4.517503	0°21′ †
Saturn VI Titan	20.2	1,221,000	15.945452	0°20′
Saturn VII Hyperion	24.53	1,479,300	21.276665	0°26′
Saturn VIII Iapetus	59.01	3,558,400	79.33082	14°43′
Saturn IX Phoebe	215	12,945,500	550.45	150°
Uranus V Miranda	5.10	123,000	1.414	0°
Uranus I Ariel	7.938	191,700	2.520383	0°
Uranus II Umbriel	11.06	267,000	4.144183	0° †
Uranus III Titania	18.14	438,000	8.705876	0°
Uranus IV Oberon	24.26	585,960	13.463262	0°
Neptune I Triton	15.5	353,400	5.876833	159°57′†
Neptune II Nereid	245	5,560,000	359.881	27°48′†

†References are listed on p. 290, other sources of data are given in Section 2–10.

TABLE VII (*cont.*)

Orbit eccentricity	Stellar magnitude at opposition	Diameter, km	Mass (planet = 1)	Mass, gm	Density	Ref.‡
0.05490	−12.5	3476	0.01229	7.344×10^{25}	3.34	[1, 2, 3]
0.0210	10–12	—	—	—	—	[4]
0.0028	11–12	—	—	—	—	[4]
0.003	13.	—	—	—	—	
0.0000	5.3–5.8	3240	0.0000381 ± 30	7.23×10^{25}	4.06	[5]
0.0003	5.7–6.4	2830	0.0000248 ± 5	4.70×10^{25}	3.95	[5]
0.0015	4.9–5.3	4900	0.0000817 ± 10	1.55×10^{26}	2.52	[5]
0.0075	6.1–6.4	4570	0.0000509 ± 40	9.66×10^{25}	1.94	[5]
0.15798	14.7	—	—	—	—	[6]
0.13029	19.	—	—	—	—	
0.20719	17.5–18	—	—	—	—	[7]
0.16870	18.	—	—	—	—	
0.20678	19.	—	—	—	—	
0.378	17.0	—	—	—	—	[8]
0.275	18.6	—	—	—	—	[9]
0.0201	12.1	—	6.68×10^{-8}	—	—	[10]
0.00444	11.7	<940	1.51×10^{-7}	—	—	[10]
0.0000	10.6	—	1.14×10^{-6}	—	—	[10]
0.00221	10.7	—	1.85×10^{-6}	—	—	[10]
0.00098	10.0	1350	4×10^{-6}	2.27×10^{24}	1.76	[11]
0.02890	8.3	4950	2.48×10^{-4}	1.41×10^{26}	2.21	[12]
0.1042	15.	—	2×10^{-7}	—	—	[13]
0.02813	10.8	—	2.7×10^{-6}	—	—	[12]
0.16326	14.	—	—	—	—	
0	16.8	—	—	—	—	
0.0028	13.7	—	—	—	—	
0.0035	14.5	—	—	—	—	
0.0024	13.7	—	—	—	—	
0.0007	13.8	—	—	—	—	
0.000	13.6	—	0.00128 ± 23	1.3×10^{26}	—	[14]
0.749	19.5	—	—	—	—	[15]

REFERENCES FOR TABLE VII

1. U. S. NAUTICAL ALMANAC OFFICE, *The American Ephemeris and Nautical Almanac for the Year 1960*. Washington: U. S. Government Printing Office, 1958.

2. H. N. RUSSELL, R. S. DUGAN, and J. Q. STEWART, *Astronomy*, Vol. 1, rev. ed. Boston: Ginn and Company, 1945.

3. E. RABE, "Derivation of Fundamental Astronomical Constants from the Observations of Eros During 1926–1945," *Astron. J.*, **55**, 112, 1951.

4. E. W. WOOLARD, "The Secular Perturbations of the Satellites of Mars," *Astron. J.*, **51**, 33, 1944.

5. W. DE SITTER, "Jupiter's Galilean Satellites," *Monthly Notices Roy. Astron. Soc.*, **91**, 706, 1931.

6. J. BOBONE, "Tablas del VI (sexto) satélite de Júpiter," *Astron. Nachr.*, **262**, 321, 1937.

7. J. BOBONE, "Tablas del VII (séptimo) satélite de Júpiter, *Astron. Nachr.*, **263**, 401, 1937.

8. SETH B. NICHOLSON, "The Orbit of the Ninth Satellite of Jupiter," *Pub. Astron. Soc. Pacific*, **39**, 242, 1927.

9. *Ibid.*

10. G. STRUVE, "Neue Untersuchungen im Saturnsystem. IV. Die Systeme Mimas-Tethys und Enceladus-Dione, "*Veröffentlichungen der Universitätssternwarte zu Berlin-Babelsberg*, **6**, part 4, 1930.

11. G. STRUVE, "Neue Untersuchungen im Saturnsystem. I. Die Bahn von Rhea," *Veröffentlichungen der Universitätssternwarte zu Berlin-Babelsberg*, **6**, part 1, 1924.

12. G. STRUVE, "Neue Untersuchungen im Saturnsystem. V. Die Beobachtungen der Äusseren Trabanten und die Bahnen von Titan und Japetus," *Veröffentlichungen der Universitätssternwarte zu Berlin-Babelsberg*, **6**, part 5, 1933.

13. J. WOLTJER, JR., "The Motion of Hyperion," *Annalen van de Sterrewacht te Leiden*, **16**, part 3, 1928.

14. H. L. ALDEN, "Observations of the Satellite of Neptune," *Astron. J.*, **50**, 110, 1943.

15. G. VAN BIESBROECK, "The Mass of Neptune from a New Orbit of its Second Satellite Nereid," *Astron. J.*, **62**, 272, 1957.

NAME INDEX

SUBJECT INDEX